EMERGING PERSPECTIVES AND TRENDS IN INNOVATIVE
TECHNOLOGY FOR QUALITY EDUCATION 4.0

T0199671

PROCEEDINGS OF THE 1ST INTERNATIONAL CONFERENCE ON INNOVATION IN EDUCATION AND PEDAGOGY, 5 OCTOBER 2019, JAKARTA, INDONESIA

Emerging Perspectives and Trends in Innovative Technology for Quality Education 4.0

Editors

Udan Kusmawan, Siti Aisyah, Isti Rokhiyah, Andayani, Della Raymena Jovanka & Dodi Sukmayadi

Universitas Terbuka, Indonesia

Routledge
Taylor & Francis Group

LONDON AND NEW YORK

Routledge is an imprint of the Taylor & Francis Group, an informa business

© 2020 Taylor & Francis Group, London, UK

Typeset by Integra Software Services Pvt. Ltd., Pondicherry, India

Library of Congress Cataloging-in-Publication Data

Applied for

Published by: CRC Press/Balkema
        Schipholweg 107C, 2316XC Leiden, The Netherlands
        e-mail: Pub.NL@taylorandfrancis.com
        www.crcpress.com – www.taylorandfrancis.com

ISBN: 978-0-367-25803-0 (Hbk)
ISBN: 978-0-367-54582-6 (pbk)
ISBN: 978-0-429-28998-9 (eBook)
DOI: 10.1201/9780429289989
https://doi.org/10.1201/9780429289989

Emerging Perspectives and Trends in Innovative Technology for Quality Education 4.0 – Kusmawan et al (eds)
© 2020 Taylor & Francis Group, London, ISBN 978-0-367-25803-0

Table of contents

Foreword

Faculty of Teacher Training and Education (read: FKIP) of Universitas Terbuka in cooperation with Research Synergy Foundation held the first International Conference on Innovation in Education and Pedagogy (ICIEP). Four invited keynotes speakers from the four nations were welcomed to given speech in excess of a hundred researchers, academics, and professionals. Prof. Ojat Darojat, M.Bus., Ph.D. (Rector of Universitas Terbuka, Indonesia), Prof. Ruth Reynold (Expert in Education and leader of the GERT, Global Education Research and Teaching Team, Australia), Prof. Dr. Suminto A. Sayuti (Expert in Culture and Language from Universitas Negeri Yogyakarta, Indonesia), Prof. Patricia S. Moyer-Packenham (Director of Mathematics Education and Leadership Program at Utah State University, United State of America), and Prof. Ahmad Rozelan Yunus (Faculty of Management Technology and Technopreneurship of Universiti Teknikal Malaysia Melaka, Malaysia). This first ICIEP was in scope of content and participants strengthening the Annual Seminar of the FKIP named Temu Ilmiah National Guru (TING). The TING that was conjoint with the first ICIEP was for its eleventh TING served by the FKIP-UT. Themed "Innovative Digital Technology for Quality Education 4.0", the 1st ICIEP had a total of 22 sub-themes, which were mostly in every aspect of education, included pedagogy, management, technology, linguistics, sciences, and higher education.

Aside from Universitas Terbuka, selected papers were written by various scholars and nationality, such as Universitas Teknologi Mara (Malaysia), Rajamangala University of Technology Tanyaburi (Thailand), De La Salle University (Philippine), Islanders Education (Maldives), UIN Syarif Hidayatullah Jakarta (Indonesia), Universitas Muhammadiyah Surabaya (Indonesia), IKIP PGRI Pontianak (Indonesia), Universitas Pendidikan Indonesia (Bandung, Indonesia), Universitas Al-Azhar (Indonesia), UIN Ar-Raniry Aceh (Indonesia), Universitas Serambi Mekkah (Indonesia), Universitas Pelita Harapan (Indonesia), Institut Teknologi Adhitama (Surabaya, Indonesia), Universitas Negeri Surabaya (Indonesia), STKIP Kusumanegara Jakarta (Indonesia), Universitas Muhammadiyah Jakarta (Indonesia), Institut Pembina Rohani Islam IPRIJA Jakarta (Indonesia), Universitas Negeri Jakarta (Indonesia), Universitas Palangkaraya (Indonesia), Universitas Multimedia Nusantara Jakarta (Indonesia), Politeknik Negeri Balikpapan (Indonesia), Universitas Muhammadiyah Prof Hamka UHAMKA Jakarta (Indonesia), Universitas Lambung Mangkurat Banjarmasin (Indonesia), Universitas Sebelas Maret Surakarta (Indonesia), SD Negeri Sukamahi 03 (Indonesia), and Yayasan Anak Montessori Cahaya Dunia (Indonesia).

We hope that knowledge and best practices discussed in the seminar and published in its proceeding will enrich the reference for all experts in the field of education and accordingly stimulate further researches in wider areas.

We thank all authors and participants for their contributions.

Prof. Udan Kusmawan, M.A., Ph.D.

Conference Chair

Emerging Perspectives and Trends in Innovative Technology
for Quality Education 4.0 – Kusmawan et al (eds)
© 2020 Taylor & Francis Group, London, ISBN 978-0-367-25803-0

Organizing committee

Conference Chair
Prof. Udan Kusmawan

Co-Conference Chair
Dr. Hendrarti Dwi Mulyaningsih

Committee
Prof. Karnedi, PhD
Isti Rokhiyah, PhD
Dr. Siti Aisyah
Jamaludin, M.Si

Conference Coordinator
Santi Rahmawati
Ani Wahyu Rachmawati

Conference Support
Diah Kusumastuti
Astri Amanda

Information and Technology Support by Scholarvein Team

Scientific review committee

Arief Budiman
Universitas Lambung Mangkurat, Indonesia

Dr. Ismi Rajiani
STIAMAK Barunawati, Indonesia

Prof. Ahmad Rozelan Yunus
Universiti Teknikal Malaysia Melaka

Associate Prof. Masloch Piotr
War Studies University Warsaw, Poland

Daniel Marco-Stefan Kleber
University of Applied Science Kaiserslautern, Germany

Dr. Rukchanok Chumnanmak
Khon Kaen University, Thailand

Dr. Nur Syafiqah A. Rahim
Universiti Kuala Lumpur, Malaysia

Associate Prof. OmKumar Krishnan
Indian Institute of Management Kozhikode, India

Dr. Thanh Hyunh
Bournemouth University, UK

Dr. Wanichcha Narongchai
Khon Kaen University, Thailand

Prof. Li-jinn Chen
Feng Chia University, Chinese Taipei

Dr. Piyanard Ungkawanichakul
Srinakharinwirot University, Thailand

Dr. Ma Huang
Guang Zhou University, China

Prof. Suciati
Universitas Terbuka, Indonesia

Prof. M. Gorky Sembiring
Universitas Terbuka, Indonesia

Dr. Amalia Sapriati
Universitas Terbuka, Indonesia

Dr. Siti Julaeha
Universitas Terbuka, Indonesia

Dr. Sugilar
Universitas Terbuka, Indonesia

Dewi Artati Padmo, PhD
Universitas Terbuka, Indonesia

Assoc Prof Ruth Raynolds
Global Education Research and Teaching Team, Australia

Dr. Dody Sukmayadi
Universitas Terbuka, Indonesia

Dr. Suratinah
Universitas Terbuka, Indonesia

Dr. Ucu Rahayu
Universitas Terbuka, Indonesia

Dr. Adi Suryanto
Universitas Terbuka, Indonesia

Dr. Sandra Sukmaning Adji
Universitas Terbuka, Indonesia

Dr. Mukti Amini
Universitas Terbuka, Indonesia

Emerging Perspectives and Trends in Innovative Technology
for Quality Education 4.0 – Kusmawan et al (eds)
© 2020 Taylor & Francis Group, London, ISBN 978-0-367-25803-0

Ideology in the Indonesian President's speech: Study in the dimension of ontology, epistemology, and axiology

Syarifa Rafiqa
Universitas Negeri Jakarta, Indonesia

ABSTRACT: The study aims at providing an understanding of the Ideology in the state speaker of the president of the Republic of Indonesia, particularly in the education field. The analysis is done from the perspectives of the following disciplines; Ontology, Epistemology, and Axiology. Ontology is a subfield of metaphysics that deals with the study of beings, the method used in this study is referred to as the Critical Discourse Analysis (CDA) model of Norman Fairclough. Epistemology, on the other hand, is concerned with the arrival of scientific truth, through the use of scientific methods. Here data was collected using the following techniques; documentation, literature review, note-taking, and interviews. Data analysis involved the connection of micro, mediate, and macro elements in three dimensions, namely, text, discourse, and socio-cultural practice. Finally, the last area of analysis is axiology with deals with questions pertaining to human values. The questions it seeks to address are whether one is moral or immoral, or whether their values are considered as good or bad. The main aim of this study is to act as a reference point to future studies, besides understanding how the Indonesian president instills values while delivering the state of the nation address.

1 INTRODUCTION

As a political tool, state speeches are designed to inform and convince the nation or society about the implementation of existing socio-economic policies, plans, and actions. These speeches have developed into a regular pattern and become an example for every element of society, particularly for speech learners. The general understanding is that the acceptability of a speech by the masses is reliant on its ability to ascribe to the Pancasila principle. In Indonesia, like any other state, these speeches are an avenue for delivery and implementation of government programs, borrowing heavily on the democratic desires of the masses, especially in the sector of education. All the Indonesian president presented unique characters from one another, and these uniqueness is visible in their attempts to address critical challenges within the state, such as security, issues of integrity as well as reforms on the electoral body and press freedom.

The main sources of research in this paper include the following works; President Soekarno's State Speech at the Non-Aligned Summit in Cairo by Livia Soenarto (2015: 69-78), Humaidi research on the rhetoric of Barack Obama and SBY in dealing with ISIS issues (2016: 115-127). Other relevant documents used in this study include Asmara (2016: 77-95) research about Jokowi's linguistic strategy in instilling Ideology and the Government Manifesto. It is important to note that speech analysis based on methodology, techniques employed, visible variations, and rhetorical devices have not been previously used in the context of Indonesia. It is for this reason that this study employs the fore mentioned scientific disciplines, axiology, epistemology, and ontology, to analyze the stat speech.

Scientific research bears uniqueness from another knowledge form; this is in the sense that for information to be considered scientific, it has to be factual, tested, and proven to be true. The branches of a scientific inquiry, epistemology, axiology, and ontology offer us the ability to understand and analyze the presidential speech using an appraisal system that sheds light on the interpersonal meanings present in these speeches (Fairclogh, 1995: 9). Appraisal theory is hailed for its ability to uncover the hidden agenda in a person's speech that is aimed at influencing their listeners. Norman Fairclough's model of critical discourse offers a better understanding of this theory through the use of a three dimensions' framework, namely: text, discourse practice, and socio-cultural practices.

The originality of this study is based on the fact that it employs the use of appraisal theory, besides using the technique to analyze data from the education speeches of Indonesian heads of state. It is expected that the findings of this study will contribute immensely in the field of critical discourse analysis in particular and also boost the linguistics discipline as a whole.

2   DISCUSSION

Critical discourse analysis is an approach in the analysis of communication aimed at understanding its root meaning. This type of analysis uses a scientific approach to examine the meaning of the language used by society in both written and oral communication.

This type of analysis views language as a fundamental phenomenon in initiating power imbalances that witnessed in society. The aim, therefore, is to establish social change during the communication process among power figures such as the head of state.

2.1   *Ontology*

It deals with questions on existence, having its roots in two Greek words onto and logos, which means being and logic, respectively. (Ekawati, 2013: 75) Concludes that ontology is the study of human existence or in other words, a theory of being. Other scholars such as Suriasumantri (2005: 234) explains that ontology can be interpreted as a study of the nature of reality of an object examined aimed at generating knowledge. In this respect, therefore, scholars are tasked with asking the following fundamental questions; what is the object under study? What is the nature of the object? What is the relationship between the object and the human ability to produce knowledge?

Consequently, the author, while applying the ontological criteria, will begin by analyzing the title of the research paper. Keeping in mind that ontological studies can be categorized as a sub-field of Applied Linguistics, the mode of analysis used will by the Critical Discourse Analysis. Two material form engaged under this study will be the object material and formal object, the former signifying the presidential speech. Formal objects lead to the understanding of speech as material objects through the CDA. Language becomes the key factor of analysis because of its prevalence in highlighting power imbalances that occur within society. The ontological analysis is, therefore, engaged for its ability to uncover the practice of communication aimed at preserving the social world, including social relations that involve unequal power relations. In addition to these, the analysis reveals the relationship between science and power, Darma (2009: 53).

Van Dijk, Fairclough and Wodak in Eriyanto (2005: 8-13) presents the following five characteristics of a critical discourse;

Measures; a dialogue is understood as an action aimed at influencing, debating, persuading, supporting, or reacting.

Context; the focus is given to the context within which communication occurred. These involve factors such as its background, the situation at hand, events that led to it, and conditions under which such talk was held.

Historically, dialogue arises in certain contexts within which it can exclusively be understood. The provision for this is that understanding of discourse is only possible when one pegs it to a particular historical context.

Power; caution has to be taken to identify the influence of communication. For instance, conversation appearing in natural or neutral, it exhibits some form of interference by power.

Ideology; this is a key concern during the analysis of dialogue. The assumption is that any piece of information on a dialogue, written or recorded, is a collection of certain ideologies that must be unearthed.

2.2 *Epistemology*

The word epistemology comes from two Greek episteme referring to knowledge or science and logos meaning discipline or theory. Webster's dictionary defines epistemology as a theory of science that investigates the origin, basis, method, and limits of science (Kristiawan, 2016: 149). According to Suriasumantri (2010: 234), epistemology discusses ways of obtaining knowledge through the use of scientific activities referred to as the scientific method. The epistemology approach engaged in this study is in line with the provisions of the CDA model. The area of focus is the appraisal system in the state speech from Indonesian President, and it is then narrowed down to state speeches that touched on the education policies.

Data collection techniques used in this study included documentation, literature reviews, interviews, and note-taking. Additionally, the technique used to analyze the data aimed at connecting micro, maso, and macro on three-dimensional frameworks, namely: text, discourse practice, and socio-cultural practices.

In order to ascertain the validity of the data collected, it was subjected to the following testing techniques for particular objective; triangulation to ensure credibility, systematic and objective presentation to ensure transferability, compiling the entire research process based on data to ensure dependability, as well as documenting the data in the attachment to facilitate its confirmation.

2.3 *Axiology*

Axiology, too, has its roots in Greece, its two syllables Axios and logos mean value and theory respectively, ultimately axiology is a theory of value. According to Suriasumantri (2010: 234), axiology is a value theory related to the usefulness of the knowledge obtained.

It has been previously observed on sciences' ability to change the world in both positive and negative ways, and axiology seeks to dig deeper into this analogy by asking the following questions; what is science used for and what is the relationship between science and moral ethics?

Kirom believes that science bears positive impacts necessary in solving national problems. Unlike science and technology, that is characterized by negative impacts such as the nuclear discovery that led to the Second World War, consequently acting as an indicator that science had lost its useful values for human life (2011: 116).

The axiological inquiry has led to the conclusion of science losing its meaning. Suariasumantri (2010: 239) narrates that a scientist's social attitude should be consistent with the scientific review process carried out. Science has been characterized as being free from the value system, besides being neutral in that it is the scientists that accord it value. In this study, the extent to which science is free from certain values relies entirely on the scientific steps in question and not on the scientific process as a whole. Accordingly, Rahayu (2015: 541-543) states that the issue of scientific ethics will lead to a problem of value-free. However, when the application of scientific methods by scientists must ascribe to other values such as human and religious, then, in essence, science is not value-free.

From the analysis above, we deduce the fact that science is neutral and is not confined by the common knowledge of the good or bad. Similarly, the owner of that knowledge must possess this type of attitude regarding science. This neutral nature of science is heavily dependent on epistemology, and facts ought to be stated as they are. Science does not require one to ascribe to a particular opinion; it calls for just judgment of the good and the bad accordingly. This great scientific power requires that a scientist possess a strong moral foundation.

The benefits of this research, in general, are to increase knowledge and to know the extent to which the President of the Republic of Indonesia instills ideology in every state speech. It offers a reference point instrumental in further research studies. The benefits of this research will be exclusively realized once the project has been completed.

3 CONCLUSION

Ontology answers the question of what science is, and epistemology shows us how to obtain it, while axiology teaches on its significance. The ontological inquiry is a sub-field of applied linguistics that engages Norman Fairclough's model of inquiry, Critical Discourse Analysis.

In Epistemology study, the main data is the Appraisal system in the state speech from the Indonesian president. Data collection techniques employed in this study were documentation, literature reviews, interviews, and notes taking. The data analysis aimed at connecting micro, maso, and macro on three-dimensional frameworks, namely: text, discourse practice, and socio-cultural practices.

Finally, in the axiological study, the general benefit of this research includes boosting the knowledge and understanding of the extent to which the President of the Republic of Indonesia instills ideology in every state speech. This research paper can be used as study material and reference material for further research.

REFERENCES

Asmara, Rangga. 2016. President Jokowi's Linguistic Strategy in Instilling Ideology and Government Manifesto. Litera Journal. Vol. 15. No. 2. p 379–388.

Bahrum. 2013. Ontology, Epistemology, and Axiology. Sulesena Journal Volume 8 Number 2, hh. 35–45.

Darma, Yoce Aliah. 2009. Critical Discourse Analysis. Bandung: Yrama Widya.

Djajasudarma, Fatimah. 2012. Discourse and Pragmatics, Bandung: Refika Aditama.

Ekawati, Dian. 2013. Ontology, Epistemology and Axiology Reorientation in the development of Science. Tarbawiyah Vol 10 Journal Number 2, pp. 75–84.

Eriyanto. 2005. Discourse Analysis. Yogyakarta: LKiS.

Fairclough, N. 1995. Critical Discourse Analysis: The Critical Study of Language. London: Longman.

Gie, The Liang. 1999. Introduction to the philosophy of science. Liberty. Yogyakarta.

Humaidi, Akhmad. 2016. Structure of Text, Social Sciences, and social dimensions Speeches of Susilo Bambang Yudhoyono. Journal of Language, Literature, and Learning. Journal Vol. 6. No.1.

Kirom, Syahrul. 2011. Philosophy of Science and Direction of Pancasila Development: Its relevance in overcoming Nationality Issues. Journal of Philosophy. Vol. 21, no. 2. hh 99–117.

Kristiawan, Muhammad. 2016. Philosophy of Education: The Choice is Yours. Valia Library. Yogyakarta.

Maufur. 2008. Philosophy of Science. Bintang Arli Wartika. Bandung.

Mulyana. 2005. Discourse Study. Yogyakarta: Yogyakarta: Tiara Wacana.

Rahayu, Walny Sri. 2015. The contribution of the Philosophy of the Science of Ethics Scientific Modern Society. Kanun Law Science Journal. Vol.67, Th.XVII. pp. 533–553.

Suriasumantri, Jujun. 1997. Science in Perspective A Collection of Science Articles, Cet. XIII; Indonesian Torch Foundation. Jakarta.

Suriasumantri, Jujun. 2010. Philosophy of science A Popular Introduction. Sinar Harapan Library. Jakarta.

Susanto, A. 2011. Philosophy of Science A Study in Ontological, Epistemological and Axiological Dimensions. PT Bumi Aksara. Jakarta.

Interpretation, Ahmad. 2004. Philosophy of science. Unraveling Ontology, Epistemology, and Axiology of Knowledge. Youth Rosda Karya. Bandung.

Widyawati, Setya. 2013. Philosophy of Science as the Foundation for the Development of Educational Sciences. Art and culture degree journal Vol. 11, no. 1, hh. 87–96.

Emerging Perspectives and Trends in Innovative Technology
for Quality Education 4.0 – Kusmawan et al (eds)
© 2020 Taylor & Francis Group, London, ISBN 978-0-367-25803-0

Improving early childhood education students' knowledge about culture as the local wisdom

Titi Chandrawati
Universitas Terbuka, Tangerang Selatan, Indonesia

ABSTRACT: Indonesia is a big country with many cultures; therefore, Indonesian children have to learn their local wisdom beginning with their early years. Young children are in their golden years, so they have to be helped to acquire all the knowledge they need, and parents and teachers should thus be aware of children's needs. This study was an effort to increase early childhood teachers' awareness of their own culture. Teachers have to develop some activities for kindergarten children to learn and embrace their own local wisdom, such as learning the traditional Indonesian dresses and languages. Children and teachers were asked about what they experienced in learning the Indonesian culture. The results show that both the teachers and their students were not really aware of their local wisdom, but after participating in the researchers' study, the early childhood education students and the teachers had a positive response to the experiences.

Keywords: Indonesian culture, learning activities, local wisdom

1 INTRODUCTION

Indonesia is a country in Southeast Asia, comprising around 15,000 islands spreading from the east to the west. Indonesia also has a large population of more than 200 million people, even though 70% of them live in Java Island. Given the large number of islands, Indonesia has many dialects and cultures. Even though Indonesia has many people, islands, dialects, and culture, its national motto is "Bhinneka Tunggal Ika," meaning, "unity in diversity (Admin, 2013). Today, however, Indonesian culture is not really known or understood by Indonesian people, especially the young generation. Not many Indonesians care about or try to learn about their own culture. Instead, foreigners are more interested in learning Indonesian culture than the Indonesian people themselves (Kusmiyati, 2014; Gunawan, 2016).

Not knowing their own culture makes the Indonesians lose their own identity, their own ethics, and many good qualities that should be inherited from older people. Therefore, it is our duty as Indonesian educators to teach our children about Indonesian culture, since learning the culture can also mean learning the local wisdom. In addition, there are still only few teachers in this country who have the vision and mission to maintain the nation's artistic culture. It became one of the reasons why Indonesia's young people are not so concerned about traditional culture (Republika, 2014).

These facts become the motivation for the study. The researchers tried to encourage kindergarten teachers to introduce, teach, and familiarize their early students with their own culture.

The Center for Advanced Research on Language Acquisition (Zimmermann, 2017) defines culture as the patterns of behaviors and interactions, cognitive constructs, and understanding used by a community including religion, food, clothes, language, marriage, music, religion, way of life, manner, ethics, and the way people communicate.

To promote learning about Indonesian culture, we have to introduce and teach Indonesian children from their early years. This is based on research by the American expert on development and behavior T. Berry Brazelton, who concluded that the experience of children in their first months and years of life determines whether they will be able to face the challenges in their lives and whether they will have a passion for learning and succeeding in their endeavors. It was shown that children 0–6 years of age are at an important level of human development and growth. Then, the desired character can be formed from their early years (Fitri Sahlan, 2015).

Students in Indonesia rarely learn about their own culture while foreign cultures, especially from the West, succeed in influencing Indonesian students. It is seen in the infusion of foreign culture in their private spaces, where the foreign culture is finally able to replace the traditional culture that should be held by the students (Republika, 2014). If culture could be learned by students from their early years, it is hoped that the succeeding generations of this nation can fortify themselves to love and embrace their traditional arts and culture (Republika, 2014).

One way to apply the art of culture to future populations of the nation is to reproduce lessons on this subject at school because the value of traditional arts and culture is absent in the Indonesian common curriculum. The greater focus on theory than on practice has not yielded maximum results (Republika, 2014).

Sunaryo and Laxman in Saputri (2017) stated that local wisdom is the local knowledge that is fused with a system of beliefs, norms, and culture and expressed in tradition and myth embraced over a long time. The traditional values can make human life rich and complete by respecting, maintaining, and preserving the natural environment. The characteristic of local wisdom is able to withstand, accommodate, control, and integrate elements of foreign culture into the native culture and develop the culture.

Early childhood education (ECE) is education for the first eight years in a child's life. Learning culture from the early years will help young children understand their own culture more comprehensively, since these are the child's golden age. It means children can learn to understand the culture easily and happily and retain knowledge longer in their minds.

There are some important concerns about the importance of learning the culture and the local wisdom for ECE. During this time children are in a vital phase of life in terms of their intellectual, emotional, and social development. Certainly, ECE is the key element that helps build a good foundation for a child's educational success. Every child learns habits and forms patterns that are not easily changed in later years. If parents and educators can develop productive early education patterns, these children will be on their way to achieving great educational success (Tiwari, 2009).

In Indonesia, ECE is a field where people are involved in helping young children from 0 to 6 years old (Indonesian definition of ECE). ECE teachers should always improve their knowledge and skills because they will help those young children develop their cognitive, affective, and psychomotor skills. To help young children know and learn about their own culture, ECE teachers should understand that children are not born knowing many things, even the difference between red and green, sweet and sour, rough and smooth. There is no way a young child can learn the difference between sweet and sour, rough and smooth, hot and cold without tasting, touching, or feeling something. Lev Vygotsky suggested that children cannot fully realize their abilities without the help of adults. He argued that there is a zone of proximate development that could be attained only with guidance and modeling by adults (Elkin, n.d.).

In conducting the research, the ECE students were introduced to some examples of cultures that were learned under some basic themes such as my country, house, food, clothes, or places.

3 METHOD

This study used a qualitative approach. The researcher decided to use the case study research method because the research was conducted at a certain time, in a certain program, and with a certain group of people. Rossman and Rallis (Chandrawati, 2015) explain that in conducting case studies, researchers try to understand the larger phenomenon through close examination of a specific case and focus on the particular case. This qualitative approach was used because the researcher tried to gain a deeper knowledge and understanding of the respondents. Moreover, the researchers also tried to help the kindergarten teachers and their early students learn about what culture and what local wisdom are.

This study was conducted at TK Ananda PondokCabe in September 2016. Since the researcher knew the kindergarten, the teachers were asked to help implement the study.

4 RESEARCH QUESTIONS

1. Why should we teach ECE children about some aspects of Indonesian culture that can be understood as local wisdom?
2. How can we teach ECE children about some aspects of Indonesian culture that can be understood as local wisdom?

5 RESULTS AND DISCUSSION

1. ECE teachers should teach and introduce ECE children about Indonesian local wisdom, given that many ECE children first learn about Indonesian culture in their classrooms under their teachers' guidance. It is expected that the succeeding generations of this nation fortify themselves to revive traditional arts and culture
2. To introduce ECE children to some aspects of Indonesian culture that can be understood as local wisdom, the ECE teacher can do organize activities such as making traditional Indonesian food and teach about Indonesian dress and dialects.

6 CONCLUSION

The researcher concluded that culture could be learned by ECE students by playing and learning about traditional food, games, clothes, and houses in Indonesia.

REFERENCES

Admin (2013) All about Indonesia: Indonesia fast facts. Indonesia'd.com.
Chandrawati, T. (2015) *Understanding dialogue in distance education: A case study at the Open University of Indonesia*. Simon Fraser University.
Elkin, D. (n.d.) Much too early. educationnext.
Fitri Sahlan (2015) *Penerapan Nilai Budaya untuk Anak Usia Dini, Ibu Pondation*. Available at: https://ibufoundation.or.id/penerapan-nilai-budaya-untuk-anak-usia-dini/ (accessed: August 9, 2019).
Gunawan, W. M. (2016) *Bagaimana pandangan orang-orang terhadap budaya Indonesia ?, dictio*. Available at: https://www.dictio.id/t/bagaimana-pandangan-orang-orang-terhadap-budaya-indonesia/1098 (accessed August 9, 2019).
Kusmiyati (2014) *Ini Alasan Orang Asing Tertarik Budaya Indonesia, Liputan 6*.
Republika (2014) Belajar Mencintai Budaya Sendiri Sejak Dini. *Republika*.
Saputri, H. M. (2017) Indonesian culture-based comic for teaching young learners in indonesia.
Tiwari, M. (2009) *Importance of early childhood education*. Ezine Articles.
Zimmermann, K. A. (2017) What is culture? Definition, meaning and examples, live science. Available at: https://www.livescience.com/21478-what-is-culture-definition-of-culture.html (accessed: August 9, 2019).

Emerging Perspectives and Trends in Innovative Technology for Quality Education 4.0 – Kusmawan et al (eds)
© 2020 Taylor & Francis Group, London, ISBN 978-0-367-25803-0

Role of religious value education based on IPS learning to build social intelligence

Siswandi, Suwarma Al-Muchtar, Elly Malihah & Helius Sjamsuddin
Sekolah Pascasarjana Universitas Pendidikan Indonesia

ABSTRACT: Boarding school is a place of teaching about the various sciences, as well as developing the skills, knowledge, attitude, and skills of the santri to become Islamic scholars, promoting individual piety and social piety by upholding the spirit of sincerity, simplicity, independence, brotherhood among Muslims, humility, tolerance, balance, moderation, exemplary and healthy lifestyle, and love of homeland. Islamic boarding schools have devoted all their energy and thought to the realization of the love of the homeland to educate people to have social intelligence, namely sensitivity and social skills. The values taught are inherent and embedded in the teachings of Islam.

Keywords: Islamic boarding schools, religion, social intelligence, values

1 INTRODUCTION

The Math'laul Anwar Islamic boarding school in Pontianak is one of the Islamic boarding schools that conduct religious and formal education. The formal education currently includes kindergarten (TK), Madrasah Tsanawiyah (MTs), and Madrasah Aliyah (MA). The existence of these three formal education institutions is in one area of the Islamic boarding school, which is also part of the Math'laul Anwar mass organization in the West Kalimantan Province. Math'laul Anwar is also part of the second oldest mass organization in Indonesia after the birth of Muhamadiyah in 1913, headquartered in Menes, Banten Province Pandeglang. Math'laul Anwar's orientation at that time also was toward contributing to the struggle for sovereignty and independence.

Islamic boarding schools are educational institutions whose existence has been recognized in the national education system. As an educational institution, pesantren are required to be able to be agents of change and to adapt to the changes themselves by providing intellectual and social experiences through which children develop skills, interests, and attitudes enabling them to manifest themselves as individuals and shaping their ability to play roles as adults later (Ma'arif, 2012). Islamic boarding schools can grow the seeds of tolerance and foster respect and love for other human beings of diverse cultures, religions, and ethnicities through the material provided by the pesantren. All of this is so that the young santri, as the generation of the future, get used to and skillfully communicate with other communities.

As an institution specifically engaged in education and da'wah, pesantren are able to have a positive influence on the surrounding community (Kusdiana, 2014). Islamic boarding schools develop learning of various ethics; especially important and essential are teaching about Islam, ways of life and thinking, absolute obedience to the kyai, the application of strict discipline in daily life, and equality and brotherhood among the santri, and therefore pesantren become a place for personality formation (Kartodirjo, 1984).

The existence of Islamic boarding schools has been guaranteed by Law No. 20 of 2003 concerning the National Education System, and this has implications for the demands of the community for the ongoing education process. Islamic boarding schools are required to provide educational services to santri in accordance with the objectives of national education, namely to create a learning atmosphere and learning process so that students actively develop their potential to have religious, spiritual strength, self-control, personality, intelligence, noble character, and skills needed in the society, nation, and state. The mission of the Islamic boarding school is to carry out educational services based on the religious, social, cultural, and aspirational distinctiveness and potential of society as an embodiment of education from, by, and for the community.

2 VALUE EDUCATION

There are several terms that have similar meanings to value education, namely moral education, character education, and moral education. Value education is considered more generic because, in education, values contain elements of morality, character, morals, character, and personality elements of an individual (Narmoatmojo, 2010) so that in this discussion, several terms that have been described previously will be used.

In value education it is important to implement not only the demands of the curriculum, but also to take into account that instinctively humans want to live in conditions that are full of certainty, where law, rights and obligations, propriety, appropriateness, harmony, and peace can be felt in human life. Kosasih (2010) states that value education is education that considers objects from a moral and nonmoral point of view, which includes aesthetics, namely assessing objects from the point of view of beauty and personal taste and ethics, that is, assessing right or wrong in interpersonal relationships. Gafur (2012) affirms the opinion of Muchlas Samani (2012) that various characters that must be possessed by Muslims both according to the Qur'an and hadith include:

a. Maintaining self-esteem. "Look for life needs by always maintaining your self-esteem" (HR. Asakir from Abdullah Bin Bakri)
b. Diligently working to find sustenance. "Morning in search of sustenance and necessities of life, in fact, the morning contains blessings and fortune" (narrated by Ibn Adi of Aisyah)
c. Communicating well and spreading greetings (QS.An-Nahl: 125)
d. Honest, not cheating, taking promises and trust (QS.Tathfif: 1)
e. Doing justice, helping, and loving one another (QS. An-Nahl: 90)
f. Patience and optimism (QS. Hud: 115)
g. Working hard from a halal origin (Surat al-Ankabut: 69)
h. Compassion and respect for parents is not deceptive (Surat al-Ankabut: 8)

3 SOCIAL INTELLIGENCE

Various social problems that occur require the attention of all parties. Since the beginning, adolescents must be given an understanding of the importance of the ability to communicate and understand the surrounding environment so that they are not alienated from everyday life. Our concern as members of the community is also to develop an effective way to minimize all forms of irregularities that occur so whenever there are symptoms, they will be quickly anticipated. This attests to the importance of social intelligence.

Howard Gardner suggests eight types of intelligence: linguistic, mathematical logic, visual and spatial, musical, interpersonal, intrapersonal, kinesthetic, and natural intelligence. Of the various types of intelligence proposed by Gardner, interpersonal intelligence is intelligence that has proximity to social skills. Interpersonal intelligence is concerned with the ability to understand the intentions, motivations, desires, and feelings of others. Gardner (Porath, 2009) defines *social expertise as the ability to perceive and make differences in the intention, motivation, points of view, and emotion of other people.*

4 ISLAMIC BOARDING SCHOOLS

According to Sodiq (2011), Islamic boarding schools seen from infrastructure have several variations in forms or models that are broadly grouped; there are three types:

a. *Pesantren type A* has the following characteristics: (i) the students study and settle in boarding schools; (ii) the curriculum is not explicitly written, but in the form of a hidden curriculum (hidden curriculum in the minds of clerics); (iii) learning patterns use learning belonging to pesantren (sorogan, bandongan, and others); and (iv) education is not conducted within the madrasa system.

b. *Type B boarding schools* have the following characteristics: (i) the santri live in boarding houses, (ii) there is scouting between the original learning patterns of the pesantren and the madrasa system/school system, (iii) there is a clear curriculum, and (iv) they have a special place that functions as a school.

c. *Islamic boarding schools type C* have the following characteristics: (i) pesantren is only a place of residence for santri; (ii) santri study in madrasas or schools that are located outside of pesantren; (iii) there is time to study at pesantren, usually at night or daytimes when santri do not study in schools (when they are in huts/dormitories); and (iv) they are generally not programmed in a clear and standard curriculum.

To realize this goal, each pesantren has a unique learning method. Learning in Islamic boarding schools is very special because it uses learning methods that originate from pesantren and are different from learning in formal schools. Some of these methods are:

a. *Sorogan*, also known as *the individual learning process*, emphasizes the development of individual abilities for santri, under the guidance of clerics or clerics.

b. *Bandongan*, also called *Wetonan*, is a method of learning that is delivered directly by the kiai to a group of santri. The kiai read classical religious manuscripts in Arabic (kitab kuning), while the santri listen carefully while taking notes in the book being read.

c. The recitation method *Market*, namely the learning activities of students through the study of certain material (books) on a kiai or ustadz conducted by a group of students in a continuous activity (marathon).

d. *Muhafazhah* (memorization), which is the process by memorizing a particular text under the guidance and supervision of a cleric or cleric. The students were given the task to memorize certain readings, which at the next stage were memorized periodically or incidentally in the presence of clerics.

e. *Riyādlah*, which is one of the learning methods in the pesantren that emphasizes the mental aspects of achieving the purity of heart of the santri in various ways based on clerics' guidance and guidance.

f. *Muhādatsah* or *Muhâwarah*, namely learning activities in applied sciences, such as language, by practicing conversations in Arabic that are required of the santri (Jailani, 2012).

5 CONCLUSION

One of the differences between pesantren and formal educational institutions is that santri live in a dormitory with various regulations aimed at developing their independence. Results of the study by Uci Sanusi (2012) and literature review show that education in pesantren encourages the independence of santri in the following ways: (a) A high level of self-confidence of the santri becomes the main factor in shaping independence. (b) The students have a sufficiently high level of trust in themselves and institutions. (c) Santri can control themselves both in anger and prohibition of boarding schools. (d) Santri can solve problems faced both in establishing life and study in boarding schools. (e) Santri have good responsibility for themselves and the cottage pesantren. (f) Santri help friends who are in distress. (g) Santri have high expectations about the success and manifestation of the future. (h) Santri's creativity and innovation are seen in activities outside the recitation. (i) Santri show the value of independent learning in their level of independence. (j) Santri have certain skills in managing life. (k) Motivation of the santri for learning comes most from self-encouragement.

REFERENCES

Al Muchtar, S. (2016a) Development of thinking and values in social studies education. Independent Library Degree, Bandung 2016.

Al Muchtar, S. (2016b) Innovation and transformation in learning Social Studies Education. Degree Pustaka Mandiri, Bandung 2016.

Jailani Imam Amrusi (2012) Islamic boarding school education as a portrait of cultural consistency amid the crash of modernity. *KARSA Journal*, 20(1).

Kusdiana Ading (2014) *History of Islamic boarding schools: Traces, spreads, and networks in the Priangan Region (1800–1945)*. Bandung: Humanities.

Law of the Republic of Indonesia No. 20 of 2003 concerning the National Education System.

Maarif Ahmad Syafii (2013) Religion, terrorism, and the role of the state. *Maarif Journal: The Flow of Islamic and Social Thought*, 8(1).

Maftuh Bunyamin (2010) Strengthening the role of social sciences in teaching social skills and conflict resolution. Speech of Inaugural Position of Professor in the Field of Education in Social Sciences at the Faculty of Social Sciences Education, University of Indonesia.

Mastuhu (1994) *The dynamics of Islamic boarding school education systems: A study of pesantren education system elements and values*. Jakarta: INIS.

Matondang, Zulkifli (n.d.) National character education based on Islamic boarding school traditions. Study on PP Syekh Burhanuddin Kampar Riau.

Muchlis Solichin Mohammad (2012) Reconstruction of Islamic boarding school education as character building. *KARSA Journal*, 20(1).

Narmoatmojo Winarno (2010) Value education in the global era. Paper in the Regional Seminar *Implementation of value education in the global era* on September 22, 2012, UNISRI Surakarta.

Regulation of the Minister of Religion of the Republic of Indonesia No. 3 of 2012 concerning Islamic Religious Education.

Rosana Ellya (2011) Modernization and social change. *TAPIS Journal*, 7(12).

Sanusi Uci (2012) Independence education in Islamic boarding schools: Study of the reality of the independence of Santri at the Al-Istiqlal Islamic boarding school in Cianjur and the Bahrul Ulum Tasik Malaya Islamic Boarding School. *Journal of Ta'lim Islamic Education*, 10(2).

Sapriya (2014) *IPS education: Concepts and learning*. Bandung: Rosda Karya.

Sayono (2005) Development of pesantren in East Java. *Language and Arts*, 33(1).

Shodiq M. (2011) Pesantren and social change. *Fa Lasifa*, 2(2).

Somantri Muhammad Numan (2001) *Initiating renewal of social studies education*. Dedi Supriyadi & Rohmat Mulya (Ed). Bandung: Rosda Karya.

Supardan Dadang (2015) *Social sciences learning philosophy and curriculum perspectives*. Jakarta: Bumi Aksara.

Suprayogi, et al. (2011) *Social sciences education*. Eko Handoyo (Ed). Semarang: FIS Semarang State University.

Emerging Perspectives and Trends in Innovative Technology
for Quality Education 4.0 – Kusmawan et al (eds)
© 2020 Taylor & Francis Group, London, ISBN 978-0-367-25803-0

Students' perception of Open University of Indonesia online tutorial

Prayekti
Universitas Terbuka, Tangerang Selatan, Indonesia

ABSTRACT: Open University provides online tutorial learning assistance services, yet not all physics education students follow them. This is attributed to technical problems and a lack of socialization among students. In general, several factors lead to socialization problems, including failure to understand the internet, inadequate time to subscribe to online tutorial activities, and lack of information on the significance of these services to their overall performance. Also, students perceive the responses given by tutors to be very long and discouraged them from participating in other sessions. The same tutorial assignments are given each semester, and the university has to improve the quality of services provided to students

1   INTRODUCTION

From the final semester of 2018, the open university of Indonesia began to change the form of online learning assistance service. It started to make them simpler and required all students to take online tutorials and required all who registered to be active every day. If the second week, the students do not activate their portal, they are automatically dismissed as online tutorial participants, and their scores are nullified. Likewise, tutors are required to open online tutorials at all times and fill in the attendance list and assess the results of student discussions of the material presented. In case a tutor does not positively impact the results of student discussions, the score of the tutorial assignment cannot be processed. This is an improvement from the organizer of learning assistance, and it is expected to impact performance positively. According to Susanti (2007), the ability of learners to use the internet might be improved through a new student orientation program. The material in the program provides more information about accessing the internet, discussion questions, and tutor assignments. Students' perceptions of the material of tuton initiation are high, meaning that they consider the material to be easily understood, according to the module, and has a systematic arrangement. The ease of students in understanding tuton initiation material is expected to boost their understanding of the module. Therefore the contents of the initiation material cannot deviate from the required contents. Furthermore, according to Murray (2012), students selectively access online learning aids based on their perception of how the material being studied is going to affect their achievements and learning outcomes of assignments and judgments. The purpose of this study therefore is to assess the perception of Physics Education students in the online tutorial provided by the Open University of Indonesia.

2   STUDY OF LITERATURE

Open University (UT) requires students to learn independently from the materials (modules) they receive since there is no face-to-face learning. This education approach is determined by the ability of the students to learn effectively. However, the potential to learn depends on the speed of reading and the ability to understand the contents of the material. To be able to learn independently effectively, UT students need to have self-discipline, initiative, and strong learning motivation.

Additionally, they are required to manage their time efficiently and learn regularly based on a self-determined study schedule. To make the study of UT and other educational materials easier, a tutorial is provided. This is academic assistance or tutoring to students (tutee) to facilitate smooth learning. Tutors are people who provide knowledge to students directly and understand the concepts and practices of better non-formal education.

The objectives of implementing tutorials include optimizing the use of the internet network to provide learning assistance services to students. Tuton is an internet-based tutorial service offered by UT and is followed by students through the internet. It is also meant to enabling the process of distance learning in more communicative and interactive Link design and provide choices for students who have access to the internet. To Tuton, students must activate an account on the site http://elearning.ut.ac.id and receive a password for accessing the site. Before students log in, it is recommended that they download and read the tuton guide available on the site.

An online tutorial serves as a teacher, where all interactions occur between computers and students, while lecturers are only facilitators and monitors. In this model, computer software replaces the tutor system carried out by the teachers or instructors. Learning is presented through text or graphics displayed by a computer screen. Importantly, the computer displays the questions according to the problem presented. Designing interactions in the form of tutorials is usually followed by giving questions or exercises and cases. The user's answer to the question and case is analyzed by the computer and immediately the computer, which responds and provides feedback on the learning outcomes. Also, the information and knowledge presented are very communicative, providing directions and guidance directly to students.

The concept of independent learning implies that tutorial is a learning aid that triggers and stimulate individuality, discipline, and self-initiative of students in learning by minimizing the intervention of Tutors. The main principle of the tutorial is "student independence. "This means it cannot exist without independence. If students do not study at home, then the usual "lecture" takes place, not a tutorial.

According to Muilenberg and Berge (2005), there were eight obstacles experienced by students in online learning, including 1) administrative issues; 2) social interaction; 3) academic skills; 4) technical skills; 5) learner motivation; 6) team and support for studies; 7) cost and access to the internet; 8) technical challenges. Furthermore, Gosmire, Morrison, and Osdel (2009) argue that the study of graduate student interaction patterns shows positive student perceptions of online learning on their interactions with material, tutors, and other fellow learners. Students in online discussions show significant variability and heterogeneity in cross-student and cross-module. The study also indicated that there was no relationship between the input and the further success of students. Computer-based learning has aspects that improve learning effectiveness, and the results of the student's first work are often entered into the appropriate computer language technical system for data to be accessed at any time.

3 RESEARCH METHODS

The study distributed questionnaires to students to determine their perceptions of online tutorials held by the Open University. The results of the questionnaire were then analyzed and described to answer the research questions. Importantly, the questionnaire included the age groups, area of residence, thoughts, and the difficulty level of following online tutorials and other courses.

4 RESULTS AND DISCUSSION

There were 60 respondents from Surabaya, and 80% of respondents stated that they read the learning material published. Also, 60% suggested that they have worked on the modules and participated in informative tests. Additionally, 80% of respondents also read the overall learning material, while another 60% suggested stated that tutors introduced learning materials and course concepts. In the presentation of tutor materials, 80% of respondents asked for links to web sites related to the subject matter, while 60% examined interesting instructional materials.

Moreover, the other 60% thought the images listed illustrated the contents of the learning material, while the combination of colors on display attracted the attention of 80% of the respondents. Other basics in this regard include the font size listed being large enough (80%); the type of letters on learning materials (80%); Ease of material in instructional aids (60%); Systematic demands on materials (80%); and use of illustrations/graphics help understand material (80%). The tutorial assignment helps students understand 80% of the course material. There are several advantages of Online Tutorial, such as helping students to (1) obtain learning services individually for the specific problems they face to be addressed; (2) learn at a pace that is in accordance with their abilities without being influenced by the speed of learning of other students or the "Self-Paced Learning." This helps students in carrying out tuton assignments and improve their final grades. Online tutorial task evaluation is one of the learning outcome assessment tools for examining students' mastery of the subject and the ability to apply their understanding in the relevant context. Students are increasingly able to use the internet, and therefore their perception of online tutorial questions is declining. The use of the internet by students outside online tutorial shows they are used to seeing interesting sites and read the contents there. As a result, students have a low perception of Tuton's questions. In the implementation of learning, there is always an interaction between students and resources used (Wardani, 2004). Interaction is recognized as one of the most critical components of the learning experience, both in conventional and distance education.

According to Picciano (2002), in the interaction process, there is a relationship qualitatively and quantitatively with student satisfaction. Interactions in web-based learning are often perceived as more successful, especially in analyzing student relation patterns in online tutorials and analyzing their perceptions. In general, there are three types of interactions occurring in distance education, including student dealings with material (content), tutors, and colleague learners. Interaction with the material is defined as the characteristics of education, which facilitates the learning process with the help of a tutor. Student interaction with tutors helps to translate all the content of the material being studied, demonstrate skills, and give examples of the behavior and values taught in the content of the material. The interactions between students themselves often occur in the form of discussion groups. Collaborative interactions happen in case of students discuss relevant topics using the material taught online and solve the problems related to the subject being discussed.

REFERENCES

Gosmire, D., Marison, & Osdel. (2009). Perception of interaction in online courses. Merlot Journal of Online Learning and Teaching, 5 (4), December 2009. Retrieved 10 January 2013.

Muilenburg, LY, & Berge, ZL (2005). Student barriers to online learning: A factor analytic study. Distance Education, 26 (1), 2948.

Murray, Meg. et al. (2012). Student interaction with online course content: Build It, and they might come. Kennesaw State University, USA. Journal of Information Technology Education: Research.

Open University Writers Team. (2011). Catalog of the 2011 Open University. UT. Jakarta.

Picciano, AG (2001). Beyond student perceptions: Issues of interaction, presence, and online performance. JALN, V6, (1) July 2002. New York.

Swan, K. (2002). Building communities in online learning courses: The importance of interaction. Journal of Education, Communication, and Information. Routledge.

Susanti (2007). Effect of student internal factors on student participation in online tutorials. Journal of Open and Distance Education, 8 (1), 68–82.

Wardani, IGAK. (2004). The learning process in distance education in the potpourri book: Distance Education. Jakarta: Open University.

*Emerging Perspectives and Trends in Innovative Technology
for Quality Education 4.0 – Kusmawan et al (eds)*
© *2020 Taylor & Francis Group, London, ISBN 978-0-367-25803-0*

The Baumatahutn traditional values of Dayak Kanayatn communities in implementing social studies based on ethnopedagogy

Saiful Bahri, Hemafitria & Emi Tipuk Lestari
IKIP PGRI Pontianak, Indonesia

ABSTRACT: There is an adequate need to keep up and safeguard the honorable qualities of cultures, which is slowly losing its value due to globalization. Therefore, the legacy of the Dayak Kanayatn people needs to be actualized using Ethno pedagogy-based learning method to ensure its continuous existence. The exploration strategy is subjectively utilized to analyze ethnography and its activities properly. The results obtained from this examination demonstrate that *baumatahutn*, is a type of human action with respectable qualities which ought to be safeguarded in accordance with nature. This tradition has nearby shrewdness esteems, which are actualized in Ethno pedagogy-based social investigations learning. In addition, the estimations of its astuteness consist of religious qualities, rules, obligations, social consideration, and natural qualities. Therefore, its execution in Social Studies Learning using Ethno pedagogy indicates great outcomes.

Keywords: BaumaTahutn, Values, Dayak Kanayatn, Social Studies, Ethnopedagogy

1 INTRODUCTION

Social Science Education needs to be systemic, holistic (Capra, 2002) and futuristic with a sense of sustainability (Capra, 2002; Kincheloe, 2008; Gadoti, 2010; Kahn, 2010). However, modernism has brought humans to a very complicated dilemma, with a low sense of concern for nature, thereby leading to the destruction of forests, and the natural ecosystems. According to Capra, (Capra, 2002) this hegemony is universally accepted, thereby, making it a natural law. The earth comprises humans, t animals, and plants. Harari (Harari, 2014) stated that humans, as the only species of the remaining genus of Homo, are often treated as the only creatures existing on earth.

In line with this, Supriatna (Supriatna, 2016) stated that modern philosophy placed humans as superior beings, thereby influencing the occurrence of crises and imbalances of natural elements. Nature is considered valueless due to the development of modern science and technology, which leads to destruction in the form of wars leading to suffering and poverty (Keraf, 2010). However, the current generation of the Dayak people has dropped the attitudes, norms, and values associated with the community's tradition due to the inception of modern communication technology.

Also, the community's local wisdom in practicing the *baumatahutn* system, which is a method of rice cultivation, should be the main form of agriculture in Borneo. However, this farming system is considered unfair because the local wisdom held by them is considered a forest destroyer due to the system of deforestation and burning in accordance with conservation technology in agriculture.

Therefore, in learning social studies, the school of Reconstructionism seeks to determine the agreement among all people on the goals capable of regulating the human system in overhauling the old and new arrangement of the entire environment. This, however, led to the creation of the Ethno pedagogy concept by Alwasilah, in learning the ideas and importance of local wisdom as the basis of education and civilization (Alwasilah, Suryadi and Karyono, 2009) Ethno pedagogy views local knowledge or wisdom as a source of innovation and skills that is empowered for the welfare of society. Its meaning is due to what students see, hear, feel, and experience, when taught using activities to facilitate as well as construct their knowledge by involving in daily life activities. In line with the above opinion, Mars (2008: 9) explained that "Social Studies education plays an important role in the inheritance of knowledge on the relationship between the community and its environment as a means of cultural transmission."

However, today's younger generation has no regard for local wisdom as they prefer relocating to the city, thereby leaving the *baumatahutn* tradition for parents. Based on the above explanation, Ethnogagogy tends to make students preserve the local wisdom of the bauma tradition. Therefore, it is necessary to conduct research which examines its depth values.

2 LITERATURE REVIEW

Baumatahutn, also known as a shifting field, has been carried out by the Dayak people for hundreds of years with the guidance of a natural ruler possessing good knowledge. These people have the knowledge and technology which is balanced with the ability to deal with problems related to nature and creatures around us for the sustainability of life on earth.

Scott, 2012, defined a movement of ideas that replaces those of the modern-day characterized by prioritizing ratios, objectivity, totality, structuring/systematization, single universalization, and progress of saints known as Postmodernism. According to Capra (Capra, 2002) problems in life consist of interrelated and dependent components known as systemic. It also utilizes anthropocentrism, a philosophical theory which states that moral values and principles only apply to humans, with their paramount needs and interests. Therefore, all demands regarding moral obligations and responsibilities towards the environment are regarded as excessive, irrelevant, and out of place (Keraf, 2010).

Rice cultivation is the primary form of Dayak agriculture in Borneo, and it is generally called bauma, which is local wisdom used to manage nature. This method believes in the philosophy of life, which states that Life reflects our Heaven, and depends on Jubata (God). Kahn (Kahn, 2010) stated that development had displaced many humans in the management and utilization of natural resources.

In addition, the Baumatahutn tradition adapts to the natural environment through local geniuses practiced for generations. Its ability to read the signs of nature becomes a strength used by local communities to live in harmony and balance with the environment, as expressed by Goleman (Goleman, Bennet and Barlow, 2012; Normuslim, 2018). Baumatahutn, is also defined as a rotation system that is environmentally friendly. According to Goleman (Goleman, 2009), ecological wisdom integrates understanding and cognitive skills as developing empathy for all forms of life. By combining social and cultural intelligence and empathy, ecological wisdom is built to create people that care about nature.

Based on an analysis of the dimensions of culture and education, Alwasilah et al. (Alwasilah, Suryadi and Karyono, 2009) Suratno (2010) viewed Ethno pedagogy as an educational practice based on local wisdom in various domains with emphasizes on knowledge or wisdom as a source of innovation and skills which is empowered for the welfare of society. Local wisdom is related to how knowledge is generated, stored, applied, managed, and inherited in supporting the inheritance of departing cultural values.

Furthermore, social studies learning is carried out based on the philosophy of reconstructivism. According to Somantri (Somantri, 2001), Social Studies Education discusses basic human activities, which include culture and wisdom, with learning conducted by integrating its values. Sumaatmadja (2008: 47) stated that culture is the result of learning that relies heavily on developing unique human abilities in utilizing symbols or gestures without coercion or natural connection with the values maintained.

3 RESEARCH METHODS

This study uses a multi-methodical, ethnographic qualitative approach along with, with Elliott's elaboration model (Hopkins, 2014). The elaboration model is carried out in three phases or steps, namely general idea, reconnaissance, and implementation.

4 RESULTS AND DISCUSSION

Farming, which is carried out in accordance with the values and culture, is passed down from the ancestors from generation to the present. It is a center of civilization, also known as the cycle of life consisting of traditional knowledge, solidarity, togetherness, spirituality, local wisdom, and anticipatory attitudes, which still colors the ancestral path of life in farming. This process is, however, not as simple as one might imagine, and are carried out with traditional ritual ceremonies.

The *baumatahutn* tradition has values that need to be developed and preserved. It is related to sustainable life by referring to human relations with the surrounding environment and the Creator of the realm of God Almighty. The values include developments in terms of rules, norms, sanctions, and patterns of community behavior in its implementation. The local wisdom consists of religious values, discipline, hard work, democracy, environmental care, social care, and responsibility.

This action research is conducted based on learning scenarios jointly arranged between researchers and partner teachers. However, there are some notes that must be corrected, namely:

1. Teacher creativity and innovation are expected to be improved in developing and linking teaching material to students' daily lives, from the local wisdom of the community.
2. Communication and routine meeting activities must be carried out by teachers with their colleagues through groups of social studies subjects to share and exchange experiences in classroom management; learning media, strategies, and models were more oriented to students' activeness.

The use of local wisdom values contained in the local culture of the indigenous people in Sungai Ambawang is an effort to instill knowledge, attitudes, and understanding. This enables students to interpret and follow up on the social studies learning material obtained in class; therefore, learning is not only oriented to the cognitive level but the affective and psychomotor aspects. The main goal is to form a caring attitude of students towards the surrounding environment and to make them have good skills as a provision in their daily lives.

5 CONCLUSION

Shifting cultivation known as Baumatahutn, is a farming system carried out in the fields by the Dayak Kanayatn communities hundreds of years ago. It is naturally managed by a ruler with good knowledge. The values include developments in terms of rules, norms, sanctions, and patterns of community behavior in implementing these values. Dayak Kanayatn has always been firm in maintaining the values inherited by ancestors because it is a guide in life that should not be violated. The result of the action research on the implementation of the values of the bauma tradition in social studies based on Ethno pedagogy was good. Students possess a sense of enthusiasm, and deep curiosity towards the local culture to instill local wisdom values that are developed and in their daily lives.

REFERENCES

Alwasilah, A. C., Suryadi, K. and Karyono, T. (2009) *Etnopedagogi: Landasan Praktek Pendidikan dan Pendidikan Guru.* Bandung: Kiblat Buku Utama.

Capra, F. (2002) *Jaring-jaring Kehidupan, Visi Baru Epistimologi dan Kehidupan.* Yogyakarta: Fajar Pustaka Baru.

Gadoti, M. (2010) "Reorienting Education Practice Towards Sustainability," *Journal of Education for Sustainability*, p. 203.

Goleman, D. (2009) *Ecological Intelligence: How Knowing the Hidden Impact of What We Buy Can Change Everything.* New York: Broadway Books.

Goleman, D., Bennet, L. and Barlow, Z. (2012) *Eco literate, How Educators Are Cultivating Emotional, Social, and Ecological Intelligence.* San Francisco: Jossey-Bass.

Harari, Y. N. (2014) *Sapiens: a brief history of humankind. Canada Limited, a Penguin Random House Company.* New York: Random House.

Hopkins, D. (2014) *A Teacher's Guide to Classroom Research.* 4th ed. England: Open University Press, McGraw-Hill Education.

Kahn, R. (2010) *Critical Pedagogy, Ecoliteracy& Planetary Crisis, the Ecopedagogy Movement.* New York: Peter Lang.

Keraf, A. (2010) *Etika Lingkungan Hidup.* Jakarta: Kompas.

Kincheloe, J. L. (2008) *Knowledge and Critical Pedagogy, an Introduction.* New York: Springer.

Normuslim, N. (2018) "Kerukunan Antar Umat Beragama Keluarga Suku Dayak Ngaju di Palangka Raya," *Wawasan: Jurnal Ilmiah Agama dan Sosial Budaya*, 3(1), pp. 67–90.

Somantri, M. N. (2001) *Menggagas Pembaharuan Pendidikan IPS.* Bandung: PT. Remaja Rosda Karya.

Supriatna, N. (2016) *Ecopedagogy, Membangun Kecerdasan Ekologis dalam Pembelajaran IPS.* Bandung: PT. Remaja Rosda Karya.

Emerging Perspectives and Trends in Innovative Technology
for Quality Education 4.0 – Kusmawan et al (eds)
© 2020 Taylor & Francis Group, London, ISBN 978-0-367-25803-0

Perception of kindergarten teachers on the utilization of traditional games in learning in West Nusa Tenggara, Indonesia

Sri Tatminingsih
Universitas Terbuka, Tangerang Selatan, Indonesia

ABSTRACT: Indonesia is an archipelago country consisting of 17,504 islands spread throughout 34 provinces, one of which is West Nusa Tenggara Province. Indonesia is also a country rich in traditional games because it has more than 1,340 ethnic groups. If it is assumed that one ethnic group has only two traditional games, then in Indonesia there will be around 2,680 types of traditional games. Traditional games are games that develop from generation to generation and usually contain elements of education that can stimulate aspects of child development such as motor skill, cognitive ability, language, social-emotional, creativity, and character. This article describes the perceptions of kindergarten teachers about the use of traditional games in the learning process in kindergarten. The research used a descriptive qualitative paradigm with interviews and was conducted from July to October 2018. The research subjects were 100 kindergarten teachers in West Nusa Tenggara (NTB) Province. The results of the study show that the use of traditional games in learning in kindergarten is only enriching and is not optimal and needs to be encouraged so that traditional Indonesian games, especially in NTB, can be preserved and empowered to stimulate aspects of child development.

Keywords: early childhood, kindergarten, perception, traditional games

1 INTRODUCTION

Every country certainly has a traditional game. Even one country can have more than 10 traditional games. Why is this? Traditional games are folk games that are created because processes are passed down from one generation to the next. Traditional games are also usually an illustration of local wisdom from a particular region or place. Min-Seok Kim Jin-Ah Choi Hal (Kim, 2015) states that traditional games are complex games acquired through culture. Meanwhile, Choi (Choi and Sohng, 2018) states that traditional games are a concept that contrasts with modern and folk games, and that includes foreign games traditionally adopted by a nation. Traditional games are also activities that have elements of educational value (Sarana, 2019). This is in line with the statement (Danandjaja, 2007) that traditional games are one form of local wisdom in the form of children's games that circulate verbally among certain community members, shaped traditionally with or without self-made tools and are passed down from generation to generation. Traditional games have diverse content, ranging from the social life of the community, urbanity, and cultural patterns (Alizadeh et al., 2014) to people's lives over time (Jacobs, 2019), and philosophy of life and connecting families to nature, respecting elders, and encouraging tenacity in everyday life (Batsaikhan and Kaye, 2017). In addition to varied content, traditional games can also be performed by various groups, both men and women, young and old (Gold et al., 2015) and can be used in school learning, including well-designed kindergartens (Waters et al., 2014; Waller *et al.*,

2017). Games can also generate motivation and curiosity in early childhood (Compagnoni et al., 2019).

Indonesia is an archipelago country in Southeast Asia consisting of 17,504 islands spread across 34 provinces. Indonesia's geographical position is very strategic because it is flanked by two continents (Asia and Australia) and two oceans (Hidia Ocean and the Pacific Ocean). The total area of Indonesia is almost 2 million kilometers, stretching from 6°N to 11°C and 95°–141°E longitude. This strategic location also influences various aspects of people's lives, one of which is the sociocultural. This influence can be seen from the many foreign cultures that merged and grew in Indonesia and shown in various way such as dress, musical instruments, films, dances, games, and more. With such conditions, the State of Indonesia becomes a country that is rich in culture and local wisdom, one of which is traditional. According to data from the Indonesian Central Bureau of Statistics, this country has more than 1,340 ethnic groups. If it is assumed that one ethnic group has only two traditional games, then in Indonesia there will be around 2,680 types of traditional games.

Perception is the result of evolution. Our perception is the result of natural selection (Hoffman et al.g, 2015) and is a definition that contrasts with hallucinations, imagination, and other experiences (Charles, 2017). Perception is generated from vision, hearing, and experience (Trivers, 2006) and is formed based on sensory information and the actual state of the environment (Yuille and Bülthoff, 1996). So perception is someone's opinion based on vision, hearing, and experience of something and then presented in the opinions and responses of someone. Therefore, in this article, we will discuss the perceptions of kindergarten teachers in West Nusa Tenggara (NTB), Indonesia, about traditional games and their relationship to the learning they have imparted in kindergarten. The thesis of this study is that kindergarten teachers are likely to never use traditional games in the learning process. Basically, they know that traditional games are very rich and valuable and offer enormous benefits for a child's growth and development. Based on their experience, the teachers have personally felt the benefits of traditional games. Therefore, this article specifically describes the perception of kindergarten teachers toward the use of traditional games in the learning process in kindergarten in NTB.

2 METHOD

This research applies a descriptive qualitative paradigm with surveys via interviews. Interviews were conducted with 100 kindergarten teachers in the West Nusa Tenggara region who were randomly selected. One hundred teachers represent the entire NTB region, which has 10 cities/district. Each city/district is represented by 10 teachers. The data are collected by giving five questions in writing to each respondent. These questions are as follows: (1) What game is it that you still remember? There may be more than one answer. (2) In your opinion, what traditional games do you often see played by the people in NTB? (3) What traditional games are suitable for early childhood? Give the reason! (4) What traditional games have you used in learning activities in your class? (5) What developmental aspects can be stimulated by the traditional games that you apply?

3 RESULTS AND DISCUSSION

Interviews of 100 kindergarten teachers in West Nusa Tenggara showed that many of them did not know the names of traditional games in NTB. But when one teacher mentioned how a game is played, another teacher stated that he knew the game. This shows that the teacher did not know the name of the game, but actually knew the game. After deeper analysis, it turned out that as many as 62 kindergarten teachers who were the subjects of this study came from outside NTB. Thirty-eight people from NTB came from different regions.

3.1   Traditional games that are still known to kindergarten teachers in NTB

The answers from respondents to the first question were quite surprising because there were not many traditional games still remembered and known by kindergarten teachers in NTB. The traditional games contained are outdoor games that require a fairly wide and flat playground. Although the playgrounds described by respondents are not exactly the same, the games basically have the same concept, as well as the way to play them. This is very reasonable because, in Indonesia, traditional types of games have similarities in terms of the game apparatus, the shape of the playing arena, and the rules for play. One way in which they are distinguished is in the name. The game requires cooperation, cohesiveness, and strategy that must be agreed upon by the group members. In addition, the game also requires gross motor movements. This is very suitable for children because children move a great deal or can't stay still.

3.2   Games that are still played by the community in NTB

The answers given by the respondents are the result of their observations of the environment around their residence. Twenty-four of the respondents stated that they had never seen children or people around their homes playing traditional games because they were rarely at home during the day. They worked from morning to evening and arrived home only at night. But from other respondents, data were obtained about games that were often played by the people in NTB. The games they mentioned are usually played when there are certain celebrations, for example, on the anniversary of the Independence Day of Indonesia. This game was carried out as a race. Usually, the players are adults in one village. Although this competition was held to compete and get prizes, victory was not the main goal, but the togetherness and joy that were gained by each contestant.

3.3   The type of traditional game that is suitable for early childhood

This is very reasonable because traditional games are games that are rarely performed by teachers. This game is only enrichment and is used for filling up leisure time. Therefore, teachers will look for games that are not too much trouble; for example, they do not need to prepare complicated tools and rules. They also tend to choose games that can be played by many children at once so that the teacher does not need to explain repeatedly or make complicated rules. In addition, games that are considered appropriate for early childhood are games that can stimulate many aspects of a child's development. All selected games stimulate motor development of the child, both fine and gross motor. Besides that, they also stimulate the development of children's emotional and social skills, such as patience, positivity, perseverance, concentration, and communication.

3.4   Traditional games used by teachers in learning and stimulated aspects of children's development

Teachers admit that they rarely use traditional games in their learning activities because they feel they will have sufficient time if they have to include traditional games in the learning process. In addition, the curriculum that they have to implement is quite complex, so they don't have the chance to use traditional games in their learning. However, they also admitted that during teaching they had used or applied this traditional game at least once, but only as an enrichment and not included in the actual learning plan. It appears that the type of game chosen by the teacher is one that does not require complicated equipment and preparation. They can choose simple games that involve many children at once. The developmental aspects stimulated by the game chosen by the teacher are motor skills (both gross and fine motor) and cognitive skills, especially the ability to count, and the rest are social-emotional aspects. This is in line with the opinion of Hurlock (2012), which states that playing s very important to simulate aspects of child development.

4 CONCLUSION

Traditional games are a cultural heritage that is actually loaded with noble values and contains good strategies to simulate aspects of early childhood development, but for teachers in Indonesia, traditional games still cannot be utilized optimally in their classrooms for various reasons. One reason is that the curriculum that must be applied by teachers is considered sufficiently solid so that they do not have the opportunity to enter traditional games in their learning activities. This study needs further research to prove the relationship between traditional games and early childhood development. Basically, teachers know that traditional games can stimulate children's abilities, but because they are not familiar with traditional games in the areas where they teach, they do not apply the games in their classrooms.

REFERENCES

Alizadeh, F., Hashim, M. N. Bin, and Amini, R. (2014) Place of teahouse in performance of Traditional plays on Iran. In *5th International Conference on Humanities, Geography and Economics (ICHGE'2014)*.

Batsaikhan, J. and Kaye, C. (2017) Horse racing with sheep ankle bones: The play of nomadic children in Mongolia. *Journal of Childhood Studies*, 42(3), 40.

Charles, E. P. (2017) The essential elements of an evolutionary theory of perception. *Ecological Psychology*, 29(3), 198–212.

Choi, M.-J. and Sohng, K.-Y. (2018) The effect of the intergenerational exchange program for older adults and young children in the community using the traditional play. *Journal of Korean Academy of Nursing*, 48(6), 743–753.

Compagnoni, M., Karlen, Y., and Merki, K. M. (2019) Play it safe or play to learn: Mindsets and behavioral self-regulation in kindergarten. *Metacognition and Learning*, 1–24.

Danandjaja, J. (2007) *Folklor Indonesia, Ilmu Gosip, Dongeng, dan Iain-lain*. Jakarta: Grafiti.

Gold, Z. S., et al. (2015) Preschoolers' engineering play behaviors: Differences in gender and play context. *Children, Youth and Environments*. JSTOR, 25(3), 1–21.

Hoffman, D. D., Singh, M. and Prakash, C. (2015) The interface theory of perception. *Psychonomic Bulletin & Review*, 22(6), 1480–1506.

Jacobs, J. (2019) *Jewish Tradition and the Judeo-Christian Tradition*. New York: John Jay College/City University of New York.

Kim, M.-S. (2015) [전래놀이 프로그램] 이 지역아동센터 아동의 자아존중감과 사회적 능력에 미치는 영향," 아동학회지, 36(6), 39–57.

Sarana, N. V. (2019) *The Bildungsdrama and Alexander Ostrovsky's Plays*. Russia: De Gruyter.

Trivers, R. (2006) Foreword. In R. Dawkins (ed.), *The Selfish Gene*. New York: Oxford University Press.

Waller, T., et al. (2017) *The SAGE handbook of outdoor play and learning*. Thousand Oaks, CA: SAGE.

Waters, P., Waite, S., and Frampton, I. (2014) Play frames, or framed play? The use of film cameras in visual ethnographic research with children. *Journal of Playwork Practice*, 1(1), 23–38.

Yuille, A. and Bülthoff, H. (1996) Bayesian decision theory and psychophysics. In D. Knill and W. Richards (eds.), *Perception as Bayesian inference*. New York: Cambridge University Press.

Emerging Perspectives and Trends in Innovative Technology
for Quality Education 4.0 – Kusmawan et al (eds)
© 2020 Taylor & Francis Group, London, ISBN 978-0-367-25803-0

A self-regulated learning process to enhancement of student teachers' education in Indonesia: A qualitative approach

Astri Dwi Jayanti Suhandoko
Universitas Terbuka, Tangerang Selatan, Indonesia

ABSTRACT: This qualitative study discusses the impact of the intervention associated with self-regulated learning given by the researcher to 43 undergraduate students at Universitas Negeri Jakarta (UNJ), Indonesia. The participants were from the department of primary education (PGSD) who attended the learners' development course and one-hour intervention of self-regulated learning strategies from March to May 2016. The two core discussions of this study are observations during the three months of intervention and open-ended interviews after conclusion of the entire program. Observation helped the researcher to portray a variety of students' actions during the self-regulated learning (SRL) program. The in-depth interviews were given to eight students selected directly by the researcher, including those who actively followed the SRL program both in practice in front of the class and also in the question-and-answer session. Results showed that the intervention of SRL strategies can enhance students' SRL and their academic performance.

Keywords: in-depth interview, observation, qualitative research, self-regulated learning strategies

1 INTRODUCTION

Teacher quality is obviously associated with the quality of teacher education institutions that have the main responsibility of producing prospective teachers. In Indonesia the quality of many teachers' training colleges, FKIP (The Faculty of Teacher Training and Education) and STKIP (School of Higher Learning of Teacher Education and Science Education) still do not meet the expectations of the government and students. They employ traditional teaching methods and emphasize theory more than practice. This has brought about a poor quality of the graduating student teachers. Various research have been conducted in recent years to promote and improve self-regulated learning (SRL) of student teachers (Yakar et al., 2013). SRL ensures that high-performing students are able to regulate themselves by planning, applying, and monitoring their cognitive abilities and motivation in learning processes (Patrick and Middleton, 2002).

This study is inspired by research conducted by Hofer and Yu (2003), who were also lecturers who conducted a course called "Learning to Learn to Teach Undergraduate Students to Be Self-Regulated Learners." This course is specially designed to help students to augment or upgrade their study skills for better achievement. This study aimed to assess the impact of a similar intervention on self-regulated learning given to undergraduate students of the Universitas Negeri Jakarta (UNJ) majoring in primary education who attend developmental psychology class.

2 LITERATURE REVIEW

Research conducted in the area of educational psychology experienced many changes over the since the early 1990s; self-regulation is a topic that most people discuss at the present time (Torrano Montalvo and González Torres, 2004). Self-regulation is generally defined as the basic capacity of an individual to adapt to a variety of contextual circumstances and is recognized as a crucial part of the healthy development of life (Gestsdóttir et al., 2011).

Self-regulation has evoked further discussion among the researchers in the field of psychology of education who linked SR with students' academic achievement due to self-regulated learning (SRL) (Zimmerman, 2008). SRL has to do with the ability of students to organize their cognitive, affective, behavioral and contextual processes in planning, monitoring, and evaluating their academic goals (Schwartz, 2012). Self-regulation has a crucial role to play in the students' learning process; it directs the knowledge of the students toward predicting motivation and academic achievement (Banarjee and Kumar, 2014). When students are able to organize themselves to become self-regulated learners, they can develop deep and meaningful learning skills along with significant advantages for their academic performance and achievement. An intervention carried out for this study focused on cognitive, metacognitive, motivation, behavioral, and contextual aspects. This intervention was implemented in regular classes that teach skills or strategies required to achieve good academic performance: (1) learning from a lecture (i.e., note-taking method and conditional reasoning strategy); (2) the SQ3R method (survey, question, read, recite, and review); (3) learning from the discussion; (4) self-management strategy (i.e., goal setting and setting up the study area); (5) writing strategies; and (6) the cognitive models of motivation (i.e., CRAFT method).

3 METHODOLOGY

This research indicates how the qualitative methodology can be used in a beneficial way to improve the understanding that is associated with SRL and the context that supports it. Data collection is the description of data analysis which is conducted inductively. The researcher used observation supported by field notes, interview, and memo writing. Observation helped the researcher to portray a variety of actions of students while administering the SRL program; it helped her to view students' responses while listening to SRL strategies, to observe the students' actions during the practice of SRL strategies, to discover the students' opinions on the topic every week, and to research their responses in a question-and-answer session. The in-depth interview (Stage and Mattson, 2003) explored the motivational beliefs and self-regulated learning differences among students; in addition, the researcher wanted to further describe students' notions of intervention and learning strategies.

4 ANALYSIS AND DISCUSSION

The results of the interview that was granted to eight participants and the observation that was conducted by the researcher during the intervention indicated that the students reflected on, comprehended, practiced, and evaluated the SRL strategies during their learning process.

4.1 *Combination of cognitive and metacognitive strategies*

The strategies of note-taking, conditional strategy, SQ3R (Survey, Question, Reading, Recite, and Review), discussion, and writing given by the researcher enhanced the students' academic performance. Students acknowledged the strategy of note-taking was not completely new. They practiced the strategies before the intervention began. But there were crucial points from the researcher that let the students reflect on their past experiences and thus make some progress in practice. The students were not only focusing on the technique while taking notes, but they were also doing some strategic preparation to support themselves in note-taking. As for a conditional strategy used by students when they need to interpret reasoning on a certain condition, the classroom practice of conditional strategy made the students realize the importance of reading the material before class.

The SQ3R strategy was a new reading technique for the students. They usually use the conventional way to read the material, that is, to read the full length of the book and then take notes. From the material given by the researcher and the classroom practice, the students admitted that this strategy has been very effective in saving the reading time and grasping the ideas from the course material. The reading homework given by the researcher enabled the students to fully comprehend the steps needed to perform the SQ3R strategy. They affirmed that there was a relationship between reading and note-taking strategies. The researcher's feedback to the homework gave crucial points to students as an evaluation of their understanding of reading and SQ3R strategy.

Classroom discussion not only needs the students' cognitive and metacognitive strategy for them to be able to use their ability to speak, but it also requires that they utilize contextual strategy in order to manage and control other students' opinions and feedback during the discussion. That is why from the observation it appeared that not all students were capable of participating in classroom discussion, because they needed to perform these two responsibilities at the same time. The material that was used by the researcher to explain the lecturers'/students' role in classroom discussion and the habit that can improve classroom discussion ability made the students reflect their past experience during classroom discussion.

The intervention given by the researcher was not to generate excellent writing skills in students, but rather to enhance students' understanding of writing strategies and motivate them to continue practicing. All the participants acknowledged that the intervention contributed to their comprehension of the good research paper model and the different phases in making the paper assignment. The researcher knew that it was one thing to comprehend the material clearly but another thing entirely to achieve mastery of the strategies, but she expected that this stage will help the students head to where they need to get to.

4.2 *Motivational strategies*

The SRL strategies given to the students showed that there is a connection among the strategies, and motivational strategies are also connected to them. When the students have mastered the writing strategies, possibly the others will follow, but the task may not be completed if the students lack the motivation to accomplish this. It means that the students should self-motivate themselves to improve their SRL strategies. The motivational strategies intervention given by the researcher was more in the way of explaining how the students regulate their own motivations. The process was completed with a classroom discussion practice of two motivational strategies (i.e., mastery self-talk and CRAFT method). From this intervention, the students not only try to understand and reflect but they also practice the strategies conveyed. The students had practiced a self-talk strategy when they faced learning problems in order to manage and control their scheduled time and to motivate themselves during the learning process. Meanwhile, the most practiced of the motivational strategies was self-reinforcement, whereby the students set the goals for themselves and on completing the assignment or goals, they rewarded themselves. The students tried their best on exams. The other motivational strategy that students practiced was an interesting enhancement to break the boredom in the process of information acquisition or construction of meaning while they were listening to the lecturer's teaching or reading the material.

4.3 *Behavioral strategies*

The material that the students received such as goal-setting, help-seeking, and setting up the area made them more aware of the importance of managing their time to accommodate their study environment as well as their relationships with peers and lecturers. The self-management strategies given to them by the researcher helped the students to reflect on their past learning habits such as when they set goals but did not monitor and evaluate the goals and when they were involved in unorganized activities because they did not draw up a timetable of tasks and study. The classroom practice that the students attended gave direction to them on how they should set their short-term and long-term goals, set the goals that are big and specific, make the goals align with their own values, organize and elaborate the goals, and monitor and evaluate the goals. Before the intervention began, the students were already accustomed to doing their tasks together with friends. They preferred to ask their friends for help in understanding the course material rather than asking the lecturer. One participant expressed that friendships could experience problems due to constraints of help-seeking from peers. The intervention given by researcher helped the student to realize the importance of help-seeking because the strategies taught were not only for her to carry out the tasks, but also to serve as the medium for her to attain her goals.

5 CONCLUSION

The SRL intervention strategies were designed to enhance students' cognitive, metacognitive, behavioral, motivational, and contextual strategies. They have been very beneficial to sophomore students as a directional guide for them to face academic problems and difficulties. These strategies were inserted into the teaching program for the purpose of enhancing the students' abilities to achieve a good academic performance by using the SRL strategies. The researcher believes that the SRL intervention strategies are designed as a constructive program to help students realize the importance of being good prospective teachers and also to motivate them to master the SRL skill as a way to achieve better academic performance. She also believes that if the Teacher's Education Institution opens more opportunities to student teachers to learn and apply SRL skills on a structured program through the four-year undergraduate program, this institution in Indonesia will become known for producing good prospective teachers.

REFERENCES

Gestsdóttir, S., Urban, J. B., Bowers, E. P., Lerner, J. V., and Lerner, R. M. (2011) Intentional self-regulation, ecological assets, and thriving in adolescence: A developmental systems model. *New Directions for Child and Adolescent Development*, 2011(133), 61–76.

Hofer, B. K. and Yu, S. L. (2003) Teaching self-regulated learning through a "Learning to Learn" course. Teaching of Psychology, 30(1), 30–33.

Patrick, H. and Middleton, M. J. (2002) 'Turning the kaleidoscope: What we see when self-regulated learning is viewed with a qualitative lens', Educational psychologist, 37(1), pp. 27–39.

Schwartz, S. S. (2012) Self-regulation, motivational conflict, and values. College of Saint Elizabeth.

Stage, C. W. and Mattson, M. (2003) 'Ethnographic interviewing as contextualized conversation', Expressions of ethnography: Novel approaches to qualitative methods, pp. 97–105.

Torrano Montalvo, F. and González Torres, M. (2004) Self-regulated learning: Current and future directions.

Yakar, Z., Can, B., and Besler, H. (2013) Does the teaching program effect on pre-service teacher's self-regulation? *International Journal of Academic Research*, 5(3).

Zimmerman, B. J. (2008) Investigating self-regulation and motivation: Historical background, methodological developments, and future prospects. *American Educational Research Journal*, 45(1), 166–183.

Emerging Perspectives and Trends in Innovative Technology
for Quality Education 4.0 – Kusmawan et al (eds)
© 2020 Taylor & Francis Group, London, ISBN 978-0-367-25803-0

To reach the unreached: Online learning to improve early childhood education teachers' competencies in the digital Era

Della Raymena Jovanka
Universitas Terbuka, Tangerang Selatan, Indonesia

ABSTRACT: This research aimed to describe the quality of online learning during 2018/19.1 and 2018/19.2 registration to improve teachers' competencies by using technology. Universitas Terbuka (UT), as an online learning service provider, tries to accommodate these needs. A survey was conducted in the form of online questionnaires distributed to 208 respondents who registered for online tutorials on practical courses and teaching practice of the Early Childhood Teacher Education (ECTE) Study Program in the 2018/19.1 semester. The results describe the experiences of students who engaged in online learning in the Program as required. The factors that support is learning through online. The ability to assisting students in the process of learning and completing assignments is a factor in support of the implementation of the online learning program. Obstacles to the implementation of online learning are a limited internet network, especially in the island areas; limited ability of students to access the internet; etc.

Keywords: online learning, online tutorial, practice course

1 INTRODUCTION

The Early Childhood Teacher Education (ECTE) Study Program was established in 2007 to provide assistance to teachers in ECTE institutions in Indonesia in meeting the requirements of the applicable National Education System Law. This ECTE program is a continuation of Universitas Terbuka (UT)'s Kindergarten Teacher Education program. The ECTE of UT study program aims to (1) provide access to world-class quality education for ECTE educators through the implementation of various Open and Distance Learning (ODL) programs; (2) produce professional ECTE educators who are capable of acting globally; (3) expand access to community participation in sustainable education in order to realize a knowledge-based society; (4) produce academic products in the education and scientific fields of ECTE, as well as ODL; (5) improve the quality and quantity of ECTE's educational and scientific research and development, as well as ODLs; (6) utilize and disseminate the results of ECTE's education and scientific studies and ODL; and (7) enhance partnerships and cross-cultural cooperation networks to strengthen national unity.

One of the learning outcomes expected from S1 ECTE graduates is the ability to use science and technology in solving education and teacher problems in the ECTE field and to adapt to the situation at hand. The Regulation of the Minister of National Education No. 16 of 2007 concerning Academic Qualification Standards and Teacher Competence states that one of the competencies that must be developed in accordance with the profession relates to mastering information and communication technology (ICT), namely utilizing ICT in the interest of organizing educational development activities (pedagogical competencies). On an individual basis, the goal of ICT mastery is to utilize ICT to communicate and develop professional competence.

UT implements distance education to facilitate various ways in which its students can learn independently. One form of learning assistance services provided to UT ECTE S1 students is face-to-face tutorials. Face-to-face tutorials are held for learning locations that allow students to gather in one group (approximately 25 people) to study with tutors. They are designed and focused on specific subjects that have high substance complexity as well as practice/practicum. The tutorial value obtained by students contributes to the final grade of the course. Given the large number and distribution of students, the peculiarities of the program, and the elements involved in the face-to-face tutorials (for example, S1 ECTE is a bachelor input), cost assistance for the Outermost, Leading and Disadvantaged Regions (3T), and other ECTEs, special efforts carry out organizing procedures effectively.

At this time, students who can take face-to-face tutorials study assistance are enrolled in a learning group with a maximum number of 25 people in one class. However, currently, in several places it is difficult to fill a classroom with 25 students. The location of UT ECTE students at this time is increasingly scattered, with faraway geographical locations, so service should be provided for them.

In addition to providing face-to-face tutorial assistance services, UT has been conducting learning assistance services in the form of online tutorials, which serve students who miss the opportunity to take part in face-to-face tutorials. The online tutorial was also given for courses that were not given face-to-face tutorial service assistance.

To overcome the problems in the face-to-face tutorial, learning assistance has been provided in the form of the online tutorial called the non-face-to-face tutorials program. Starting in 2017/208.1, UT, especially in the Department of Basic Education, also launched the non-face-to-face tutorials program, namely the Semester Package System Service (SPSS). This service is designed for students who cannot attend lectures to receive learning assistance without having to be in class (online), except for practicing courses that still require cooperation with ECTE classes and supervisors in locations around students.

Non-face-to-face tutorial services implementation for one prior semester has been conducted for 18 regional offices in Indonesia. The non-face-to-face implementation needs to be evaluated for its effectiveness in UT'hs ECTE S1 Study Program, especially for practical subjects. Therefore it is necessary to conduct in-depth studies through research.

Based on the background of the aforementioned problems, the main objective of this research is to evaluate the effectiveness of implementing online learning for practical courses in the UT ECTE Program. Based on the aforementioned problems, several research questions are formulated as follows. What are the general requirements of the non-face-to-face S1 UT ECTE Program compared to Non-face-to-face SPSS Guidelines? What factors support the implementation of UT's ECTE S1 Online Learning Program? What factors are obstacles to the implementation of UT's ECTE S1 Online Learning Program? How does the online learning service registration compare to the Non-face-to-face Code Guidelines? What is the suitability of academic guidance services for UT's ECTE practice courses in online learning at UT Regional Offices? What is the effectiveness of the non-face-to-face program for UT ECTE practice courses?

2 THEORETIC STUDY

Electronic learning, or e-learning, began in the 1970s (V Waller, 2001), so the concept of learning that utilizes web and internet technology is not a new idea. . But for the Indonesian context, new e-learning has been widely used since the 2000s.

At present, the phenomenon seen in developing countries including Indonesia is the rapid penetration of mobile phones in everyday life. Mobile devices such as internet-enabled phones are very popular and increasingly used for blogs, social networking, and e-learning. This program is very helpful in improving users' attitudes toward e-learning.

But according to Hazarika et al. (2011), physical separation between students and instructors feels isolated, which leads to negative attitudes, especially for students who do not master ICT. They experience anxiety when using computers, have low expectations for learning technology, and e-learning is surely useless for them (Vrana et al., 2005).

Factors that influence student attitudes toward online learning determine the success of learning online (Zhang and Bhattacharyya, 2008). Significantly influential factors are associated with increased awareness of using technology in learning online, enriching basic technical knowledge and skills, improving learning materials, providing computer training, motivating users to use learning systems online, and full support from universities (Bhuasiri et al., 2012). Other factors are characteristics of residential areas or demographics (Roca et al., 2006; Bertea, 2009; Chen and Huang, 2012). Also, access to technology affects student attitudes toward learning online.

The expected results of e-learning are substantially influenced by the way students understand the advantages and disadvantages of online learning. Therefore, a scale is needed to measure student attitudes toward e-learning (Musawi, 2014). The attitude of students toward learning online is defined as a positive or negative feeling when doing something to achieve a target or goal (Palmer and Holt, 2009). This means that students' positive or negative feelings about their participation in e-learning activities through the use of computers will directly influence their behavior when participating in online learning for different purposes. Understanding student attitudes toward e-learning can help determine the extent to which students utilize the e-learning system provided by universities and online courses and the effects on quality of education (Ong and Lai, 2006). This is supported by (Cereijo, 2006), who said that student attitudes toward e-learning are useful for predicting learning outcomes.

2.1 Online learning at S1 ECTE UT study program

This non-face-to-face Semester Package System Service (SPSS) was prepared by the S1 ECTE Study Program for prospective students who have a background education non-graduate. Non-face-to-face services do not require the fulfillment of learning groups. The main requirements that must be met to obtain non-face-to-face tutorials are being able to operate computers and access the internet and have an active email address/account.

Learning is done independently by students through the Basic Material Book and Digital Teaching Materials, which can be downloaded on two electronic devices. Students can also obtain other distance learning services including Virtual Reading Room, Independent Training Online, Smart Online Teacher, Online Tutorial, TV Tutorial, Online Based Enrichment Material, and guidance for Practical/Practicum courses/Practicing with supervisors.

Online learning has been running for one semester in the S1 UT study program, even though it has been implemented only on a limited basis at 18 UPBJJ. Before being carried out thoroughly in all regional offices, it is necessary to have an in-depth study of the effectiveness of non-face-to-face tutorial administration implementation for practical subjects in UT's ECTE S1 Study Program. Therefore this research has become very relevant.

3 RESULTS

Research data were obtained by distributing questionnaires through Microsoft 365 forms to 208 students (1 man and 207 women) who took part in the online tutorial on practicing courses at ECTE Study Program. Even though there was only one man who answered the questionnaire, it has no major effect on the results because this research did not measure learning by gender.

The following results of the questionnaires were obtained: there are still many students who do not understand the location in which they were registered. This is evidenced by the fact that there are still many students, as many as 62, who answer "others" or are not following their regional office. Examples of answers given by students are "Rimbo Bujang," one area at Sumatra Island, while the correct response is that they are UT Jambi regional office's students. A factors that supports online learning is that it helps students in the learning process and work assignments. Aligned with this advantage is that students taking online courses need to be particularly disciplined, especially in what they are learning and where and when they are doing this (Boettcher and Conrad, 2016). On the other hand, factors that constrain the implementation of online learning are limited internet networks, especially in the islands; the ability of students to access the internet; and others.

4   CONCLUSIONS

Based on the results of data processing, the following conclusions were reached in this study:

General requirements of the non-face-to-face Civil Engineering UT ECTE S1 Program are consistent with the requirements of the non-face-to-face service guidelines.

A factor that supports the implementation of online learning in the UT ECTE Program is that online learning helps students in the learning process and work assignments.

Factors that become obstacles to the implementation of online learning in the ECTE S1 Program UT is a limited internet network, especially in the island areas; the ability of students to access the internet; and others.

The registration for online learning services follows the non-face-to-face service guidelines.

Academic guidance services for S1 ECTE practice subjects UT on online learning at the regional office are carried out during education in independent learning activities, namely during the orientation of new student study and independent learning skills training.

The non-face-to-face tutorial for UT ECTE practicing subjects is believed to be quite effective for students as long as students can access learning assistance services in the form of online tutorials.

5   SUGGESTIONS

Based on the foregoing conclusions, we propose that UT expand the online services center to areas that still have limited internet access.

REFERENCES

Bertea, P. 'Measuring students' attitude towards e-learning: A case study. In Conference proceedings, "eLearning and Software for Education (eLSE): Carol I." National Defence University Publishing House, pp. 417–424.

Bhuasiri, W., Xaymoungkhoun, O., Zo, H., Rho, J. J., and Ciganek, A. P. (2012) Critical success factors for e-learning in developing countries: A comparative analysis between ICT experts and faculty. *Computers & Education*, 58(2), 843–855.

Boettcher, J. V. and Conrad, R.-M. (2016) The online teaching survival guide: Simple and practical pedagogical tips. Hoboken, NJ: John Wiley & Sons.

Cereijo, M. V. P. (2006) Attitude as predictor of success in online training. *International Journal on E-Learning*, 5(4), 623–639.

Chen, H.-R. and Huang, J.-G. (2012) Exploring learner attitudes toward web-based recommendation learning service system for interdisciplinary applications. *Journal of Educational Technology & Society*, 15(2), 89–100.

Hazarika, M., Deb, S., Dixit, U. S., and Davim, J. P. (2011) Fuzzy set-based set-up planning system with the ability for online learning. *Proceedings of the Institution of Mechanical Engineers, Part B: Journal of Engineering Manufacture*, 225(2), 247–263.

Musawi, N. M. A. (2014) Development and validation of a scale to measure student attitudes towards E-learning. *Journal of Teaching and Teaching Education*, 2(1), 1–12.

Ong, C.-S. and Lai, J.-Y. (2006) Gender differences in perceptions and relationships among dominants of e-learning acceptance. *Computers in Human Behavior*, 22(5), 816–829.

Palmer, S. R. and Holt, D. M. (2009) 'Examining student satisfaction with wholly online learning. *Journal of Computer Assisted Learning*, 25(2), 101–113.

Roca, J. C., Chiu, C.-M., and Martínez, F. J. (2006) Understanding e-learning continuance intention: An extension of the Technology Acceptance Model. *International Journal of Human-Computer Studies*, 64(8), 683–696.

V Waller, J. W. (2001) A definition for e-learning. Open Learning Today.

Vrana, V., Fragidis, G., Zafiropoulos, C., and Paschaloudis, D. Analyzing academic staff and students' attitudes towards the adoption of e-learning. In ICDE International Conference.

Zhang, P. and Bhattacharyya, S. (2008) Students' views of a learning management system: A longitudinal qualitative study. *Communications of the Association for Information Systems*, 23, 351–374.

Emerging Perspectives and Trends in Innovative Technology
for Quality Education 4.0 – Kusmawan et al (eds)
© 2020 Taylor & Francis Group, London, ISBN 978-0-367-25803-0

Reflective practice teacher education: Constraints at Universitas Terbuka

Isti Rokhiyah
Universitas Terbuka, Tangerang Selatan, Indonesia

ABSTRACT: Reflective practice has been mentioned extensively in teacher education practices. Schön's book (1983) on reflective practice initiated interest in the application of reflective practice in teacher education programs. Universitas Terbuka (UT/Indonesia Open University) offers a teacher education program for primary school teachers in order to improve teachers' competencies and qualifications. This concept paper aims to review the theoretical and practical application of reflective practice in teacher education, as well as possibilities, constraints, and challenges faced in the application of reflective practice in the Primary Teacher Education Program (PTEP) at UT.

1 INTRODUCTION

Universitas Terbuka (UT), the Indonesia Open University, has been mandated by the government to educate under-qualified and unqualified teachers since its establishment in 1984. The 2005 law number 14 requires teachers at all levels of education to possess a bachelor's degree. One feasible way for teachers to upgrade their competencies and qualifications is through open and distance education system, which does not require teachers to go to campus regularly. They teach while studying at UT. UT offers Primary Teacher Education Program (PTEP), which is designed for primary school classroom teachers who do not possess bachelor degrees.

There are two courses in PTEP which are related to improving teaching competencies through teaching practice courses. One of the teaching practice courses requires student teachers to conduct action research in their teaching. Reflection is one of the steps in action research (Kemmis and Taggart, 2005).

This paper presents the theory of reflective practice in teacher education and provides examples of its application in teacher education. Furthermore, it describes primary school teacher education as it exists at UT currently and the degree of reflective practice in the institution. In this respect, it proposes the implementation of reflective practice at UT and enumerates the constraints and challenges faced.

2 THE ROOTS OF REFLECTIVE PRACTICE IN TEACHER EDUCATION

Reflective practice is derived from the concept of reflective thinking. According to Dewey (2007), to reflect means to find proof and other sources that help to bring about suggestions of further thought or actions. He defines reflective thought as "active, persistent and careful consideration of any belief or supposed form of knowledge in the light of the grounds that support it, and the further conclusions to which it tends (p. 7)".Reflective thinking, in short, means judgment suspended during the further inquiry; and suspense is likely to be somewhat painful. (p. 10). Dewey (2007) further argues that reflective thinking emerges as a result of a phase of perplexity, hesitation, and doubt. In other words, reflective thinking is needed when we are in a state of uncertainty because of the problem we face. As the process of reflective thinking unfolds, it follows "an act of search or investigation directed toward bringing to light further facts which serve to corroborate or nullify the suggested belief" (p. 9).

Schön's (1983) further differentiates reflection into "reflection-in-action" and "reflection-on -action". Reflection on action is conducted after-action, while reflection in action takes place at the same time as practitioners are conducting the action. Reflection on action is common with other names such as "thinking on your feet" or "learning by doing," and he asserts that "... we can think about doing something while doing it" (p. 54). In the teaching practices, teachers often find unintended and unique results of teaching strategies. Teachers become researchers because they often cannot rely solely on established theories and techniques to guide and 'predict' activities in the classroom; they cannot separate thinking from doing, and means from ends of the planned action. Good teachers, according to Schön, are teachers who are capable of finding their students' difficulties from their teaching, not from their Students' deficit. Reflective teachers construct and reconstruct the materials of the situation, including their initial understanding of situations and events in their practice. Schön refers to this activity as 'problem-setting.'

Another view on reflective practice, especially in reflective teaching, is proposed by Stuart Parker (1997). His views on reflective practices are similar to Schön's reflection-in and on-action. Schön mentions "reflection in and on action," and Parker adds "reflection before action." Parker argues that reflection can be conducted before action–which means scrutinizing the plan in advance and after the action–by evaluating the action and reflection in action–which means during the action.

Reflective teachers are the ones who are not only concerned about classroom problems but are also concerned with wider issues on education, such as the aim of education, the social and personal impacts, the ethics, the rationale of their method and the curriculum being implemented, and the close relationship between these issues and their direct classroom practices. The first characteristic indicates that reflective teaching is part of the tradition of action research; there is something that needs to be improved based on the situated practice. Teachers themselves conduct action research on their classroom situation, which is unique for each teacher. The second characteristic implies that reflective teachers attain the quality of open-mindedness in engaging various views or theories, responsibility for their duty as related to authority and rationality, and wholeheartedness of dedication. These characteristics of reflective teachers differ from the positivistic point of view in the sense that reflective practitioners become aware of the social implication of their educational practices and policy, and they become adaptive in an effective way, welcoming change and taking part actively in bringing about change. This notion is in line with Kincheloe's (1998) views on empowering teachers to become knowledge generators and the agents of transformation, not the agent of knowledge transfer.

Rose (2013) views reflection as a mode of thought that entails mulling over ideas that have no necessary connection and eventually producing from them, and the perceived interconnection among them, new meanings and ideas. (p. 19). She defines reflection as the separation between thought and action, or as reflection-then-action, which involves a process of synthesis to generate new ideas.

3 REFLECTIVE PRACTICE IN TEACHER EDUCATION

Reflection or reflective practice has become an accepted component of teacher education program (Beauchamp, 2015). Schön (1987) views educating teachers as educating practitioners. He delineates three ways in the education of practitioners. Practitioners may learn the practice by themselves, through apprenticeship, and practicum. Prospective practitioners learn "a kind of reflection in action that goes beyond stateable rules–not only by devising the new method of reasoning, as above, but also by constructing and testing new categories of understanding, strategies of action, and ways of framing problems" (p. 39). The third type of practicum is the combination of the first and the second, together with apprenticeship: "These practicum are reflective in that they aim at helping students learn to become proficient at a kind of reflection-in-action. They are reflective, as well as see, in the further sense that they depend for their effectiveness on a reciprocally reflective dialogue of coach and student" (p. 40).

Research on reflective practice in teacher education has been conducted extensively since the concept of reflective practice evolved. MacKinnon (1987) developed a scheme for detecting reflection-in-action in student teachers in science teaching based on the vignette in a clinical model of supervision. The reflection in action detected through dialogue between the coach and the student. Maarof (2007) researched to investigate the type of reflections, strategies, and perceptions of student teachers through reflective journal writing. The co-researchers were student teachers conducting teaching practice in Malaysia. Khoury-Bowers (2005) studied reflective dialogues in a virtual learning environment by means of analyzing student teachers' postings, threaded discussions, and students' essays using Pathwise's criteria of reflection. This study involved 22 student teachers in 4 consecutive weeks during the field study. Ostorga (2006) conducted a study on reflective practice in primary teacher education. She employed an interview protocol and reflective journals as part of the teaching processes. From this research, it can be seen that the application of reflective practice in teacher education varies, such as in the form of a reflective dialogue between the coach and student, and the form of narrative through reflective journals. Reflective dialogue can be conducted through direct conversation or online.

4 EXISTING PTEP AT UNIVERSITAS TERBUKA, CONSTRAINTS AND CHALLENGES FACED IN IMPLEMENTING REFLECTIVE PRACTICE IN THE PROGRAM

At UT, PTEP is under the management of the Faculty of Education and Educational Studies among 10 study programs offered by the faculty. PTEP is intended to improve the quality and qualification of primary school classroom teachers with a competence-based curriculum. The curriculum covers four basic competencies based on Law Number 14, 2005, namely pedagogical, personality, professional, and social competencies (Zuhairi, 2010). As mentioned, there are two teaching practices in PTEP, the first and the second. The first teaching practice is conducted during the fifth semester, which focuses on improving teaching competencies based on the knowledge and skills acquired. The second teaching practice is conducted during the tenth semester, and it focuses on improving teachers' professional competencies. The main characteristic of the second teaching practice is the application of action research theory in the real classroom context. Student teachers conduct cycles of action research on their teaching practices and write a research report.

One of the steps of the action research cycle is reflection. Theoretically, students of PTEP learn what, how, and why think reflectively, because they should record their reflection as part of their research report. In reality, this concept has not been fruitfully developed. Research conducted by Wardani et al. (2002) showed that only a small number of teachers practice reflective thinking in their teaching. Besides, in my experience as a tutor for the 'second' teaching practice, I found out that student teachers often consider their unsuccessful teaching-learning process to be a result of their students' deficit, rather than questioning themselves and evaluating their teaching methods and practices. Teachers often do not evaluate their practices.

If student teachers implement the concept of reflective practice, especially during these tutorials and supervisions (teaching practice), I believe they will develop the habit of questioning their practices rather than placing the blame on their students. They will be sensitive to their students' needs and give space for their students to learn.

In conducting its programs, UT faces difficulties and constraints. The geographical barrier is the most apparent constraint; students of UT spread across Indonesia. This imposes UT to provide support for all students regardless of their location. For example, for the face-to-face tutorials, UT collaborates with higher education institutions in providing tutors. In remote areas, tutors need to travel long hours to reach the tutorial sites. Thus, developing reflective practice becomes tricky. Computer literacy is also another obstacle, especially for students of PTEP. When I conducted face-to-face tutorials, 4 among 10 students had computer phobia. Internet infrastructure is also another issue as students in remote areas often face difficulty getting access to the internet.

Aside from technical constraints, there are several other challenges faced in the application of reflective practice in the PTEP program. PTEP curriculum is a competency-based curriculum derived from the central government. In order to implement the reflective practice, UT needs to review its curriculum. The curriculum should be more flexible and should provide more space for students to develop their reflective thinking, especially considering the Thomson and Pascal (2011) concept of reflective practice. Indonesia has a strong religious basis as the deliberation of the first principle of the 'way of life (Pancasila),' which is to believe in God. When UT applies critical reflective practices, there will be room for students to query on all facets of life, including religious dogma. UT needs to provide quality teaching-learning materials, and train developers/tutors who are capable of facilitating this concept and inquiry into delicate subjects such as religion. Special attention needs to be given to the 'second' teaching practice. UT needs to revise its related module as the basis for conducting teaching practice. It needs to incorporate more elaboration on reflective thinking and reflective practice. As a consequence, UT also needs to train supervisors who will share the same views and apply the concept of reflective practice in PTEP.

5 CONCLUSION

As a public university, UT's responsibility is to provide quality education for everyone in Indonesians. Since its establishment, UT has been providing education for in-service teachers. Moreover, since the enactment of Law number 14, 2005, UT has made considerable strides in improving teachers' qualifications, especially the primary school classroom teachers. The university is obligated to comply with government policy on curriculum and ensure that the PTEP curriculum is competence-based. In order to improve its teaching and learning quality, UT is continuously seeking innovations in the areas of computer and communication technology and analyzing the latest theories and research results.

Reflective practice in teacher education has been in existence and extensively applied by teacher education institutions worldwide. Even though UT has been adopting the reflective practice in its teacher education, especially at PTEP, UT still needs to improve on its application. UT needs to revise the teaching practice module with more focus on deliberation reflective thinking and practice, as well as increase the application of computer and communication technology. UT also needs to train qualified tutors and teaching supervisors so that they share the same perception and apply the knowledge of the reflective practice.

REFERENCES

Beauchamp, C. (2015) 'Reflection in teacher education: issues emerging from a review of current literature', Reflective Practice, 16(1), pp. 123–141.
Dewey, J. (2007). How we think. Stilwell: Digireads.com.
Kemmis, S., & McTaggart, R. (2005). Participatory Action Research: Communicative Action and the Public Sphere. In N. K. Denzin & Y. S. Lincoln (Eds.), The Sage handbook of qualitative research (pp. 559–603). Thousand Oaks, CA,: Sage Publications Ltd.
Maarof, N. (2007). Telling his or her stories through reflective journals. International Education Journal (8)1, 205–220.
MacKinnon, A. M. (1987). Detecting reflection-in-action among pre service elementary science teachers. Teaching and Teacher Education (3)2, 135–145.
MacKinnon, A. M. (1989). Conceptualizing a "reflective practicum" in constructivist science teaching. Unpublished dissertation. The University of British Columbia.
Parker, S. (1997). Reflective teaching in the postmodern world. a manifesto for education in postmodernity. Buckingham: Open University Press.

Schön, D. A. (1983). The reflective practitioner. How professionals think in action. New York: Basic Book.

Schön, D. A. (1987). Educating the reflective practitioner. Oxford: Jossey-Bass.

Undang-Undang Republik Indonesia Nomor 14 Tahun 2005 tentang Guru dan Dosen [Law of the Republic of Indonesia Number 14 Year 2005 on Teacher and Lecturer].

Wardani, IG.A.K., Andayani, Julaeha, S., Sugilar, & Arismanti, Y. (2002). Kiner¬ja Guru Lulusan Program Penyetaraan D-II PGSD Guru Kelas Kurikulum 1996 (Laporan Penelitian). Jakarta: Pusat Penelitian Kelembagaan, Lembaga Pe¬nelitian, Universitas Terbuka.

Zuhairi, A. et.al. (2010). Using Educational Technology to Enhance Learning for In-Service Primary Teacher Education Students at Universitas Terbuka, Indonesia. Paper presented to IPTPI-APPJJI International Seminar Integrating Technology into Education. Jakarta, Indonesia, 17-18 May 2010.

Emerging Perspectives and Trends in Innovative Technology for Quality Education 4.0 – Kusmawan et al (eds)
© 2020 Taylor & Francis Group, London, ISBN 978-0-367-25803-0

Student satisfaction in academic services, teaching materials, online tutorials, and practicum at the department of biology education in Open University

Gusti Nurdin, Krisna Iryani & Anna Ratnaningsih
Universitas Terbuka, Tangerang Selatan, Indonesia

ABSTRACT: The purpose of this study was to determine and improve student satisfaction in the academic services of the Undergraduate Study Program in Biology Education, Faculty of Teacher Training, Open University Education. This study focused on the persistent influence of students on various academic services to better the undergraduate program. The research was descriptively conducted at the Open University of Pontianak, Surabaya, Jakarta, Bandar Lampung, Kendari, and Kupang. The research duration was from April to October 2016, with students contacted through mail for the entire unit distance learning program. The closed and open response data from the survey were quantitatively and qualitatively analyzed to triangulate the findings. The evaluation is limited to the "output" stage of students on their experiences and perceptions of academic services, lab work, etc. The report contains statistical descriptions and syntheses on the utilization of academic services, practicums, online tutorials, and principal material books on the level of student satisfaction in gaps, problem findings, and recommendations for improvement.

Keywords: academic services, online tutorials, practicum, student satisfaction, study programs, teaching materials , undergraduate biology education

1 INTRODUCTION

Biology education for undergraduates in Open University (OU) began in 1986 to help students willing to deepen their knowledge. The objectives were related to education, research, and community service, which are used to determine its vision. It aimed to determine the extent to which students feel satisfied with the services provided by the department to capture students' opinions on the needed service. Its purpose in obtaining information and analyzing student satisfaction was to provide adequate knowledge.

The Teaching and Education Faculty in OU is a center of excellence in the implementation of teacher education service and the provision of certificates through distance learning. Based on this vision, its main objective is to maintain sustainable paths, types, levels, and various forms of education and play an active role in the research, development, and dissemination of teacher science and culture. Its academic services refer to the learning resources in the form of printed and non-print material.

The printed teaching materials packaged in the basic material book or module are the main independent learning resources for students because they are self-instructional, contained, and assessed. Furthermore, the principal material is an evaluation of the learning outcomes, which is sourced in the book. However, this does not mean it overrides the role and benefits of other learning resources, but instead the university's policy provides online enrichment resources and tutorials through television and radio broadcast, which reflect the seriousness and recognition of the available learning modes. Services are also provided in the form of information, learning assistance, academic guidance, administration, student complaints, and library to help students to overcome academic and administrative problems while studying at the university.

Tutorial services include all aspects developed online with questions explored on academic satisfaction, including mastery of tutors' material used in the classroom, helping students to understand the subject, feedback, and assignments. In addition, practical services are defined as questions on academic satisfaction, which consists of understanding instructor material, and provision of feedback during the implementation phase in order to obtain an adequate analysis.

The teaching material used for academic satisfaction is easy to obtain through online educational stores. Teaching materials were received before the first meeting, and in addition to the service in academic administration, they contained the period of study, student code of ethics, and credit transfer numbers. The study period of the OU depends on student ability. Academic ethics includes attitudes, responsibilities, scientific professionals, and collegial behavior of students to the community. Its main foundation is the values of scientific honesty, fair behavior, openness, and respect for the opinions of others. When students register for four consecutive times, their data tend to indicate inactive with unfinished credit transfer. The populations in this study were undergraduate biology education students from the entire Distance Learning Program Unit in 2016. Data samples were randomly obtained from 2012 to 2016, using 200 students based on geographical distribution.

In the general services section, respondents were asked to submit their opinions on information clarity of the university, in line with its tuition fees compared to services provided. Also, students are able to contact staff and tutors easily for assistance. In the general services section, respondents were asked to submit their opinions on information clarity of the university, in line with its tuition fees compared to services provided.

In the online tutorial, practical, teaching ability, and exam implementation services, the most prevalent student opinion indicated satisfaction.

2 CONCLUSION

The number of respondents in this study was limited. Therefore, it was difficult to draw conclusions on a real picture of the level of need associated with student satisfaction using online materials, as well as administrative and academic services. The data obtained show students were in need of a basic material book but were unsatisfied with its availability owing to the difficulty associated with its purchase. The quality of online tutors is considered satisfactory, although the results obtained were unexpected because of the small number of respondents. However, the biology education study program received good satisfaction ratings by providing for the academic services needs of student. The results of the interviews provide additional information on the survey obtained, which are expected to help improve the provision of material books and the quality of online tutors.

REFERENCES

Decree of the Minister of National Education Number 045/U/2002, concerning Core Curriculum of Higher Education.

Educational Science Consortium (1993) *Professionalization of teacher's position: Offer and challenge.* Jakarta: Educational Science Consortium.

Law of the Republic of Indonesia No. 20 of 2003 concerning the National Education System. Jakarta: Ministry of National Education.

Law of the Republic of Indonesia No. 14 of 2005 concerning Teachers and Lecturers. Jakarta: Ministry of National Education.

Ministry of National Education Open University LPPM. (2008) Research report: Search Study of S1 FKIP Graduates, UT.

Open University Catalog (2014) The Non-Public Program Implementation System, the Open University Ministry of Education and Culture.

Raka Joni, T. (1989) They are future, now: Challenges for education in welcoming information. Scientific lecture delivered at the ceremony Anniversary XXXV, Lustrum VII IKIP Malang, October 18, 1989.

Republic of Indonesia Government Regulation No. 19 of the Year 2005 concerning Standards of National Education. Jakarta: Ministry of National Education.

SMART-UT Mapping Team, PMIPA Department FKIP-UT (2010) Market mapping and curriculum mapping as development of results of tracer study for Open University graduates of PMIPA - FKIP in 1990–2004.

Tilaar, HAR (1995) National education development 1945–1995: A policy analysis. Jakarta: PT Gramedia Widiasarana Indonesia.

Emerging Perspectives and Trends in Innovative Technology
for Quality Education 4.0 – Kusmawan et al (eds)
© 2020 Taylor & Francis Group, London, ISBN 978-0-367-25803-0

Measuring a culture-based teachers education and training program's effectiveness: Implications for physics teachers' PCK

Imelda Paulina Soko & Yos Sudarso
Universitas Terbuka, Tangerang Selatan, Indonesia

ABSTRACT: Professional development of Pedagogical Content Knowledge (PCK) is the key to success in improving teaching and learning quality. With the view of indigenous science at the forefront, the diversity of culture and local science should be considered for any teachers' education and training program. This study investigates the effectiveness of a culture-based physics teachers' education and training program. This quasi-experimental research using one pretest-posttest design group involved twenty physics teachers in the conference of physics teachers (MGMP) in one of the regencies in the province of East Nusa Tenggara (NTT). The content, pedagogy, and cultural knowledge data were collected using an observation sheet focused on the participant's learning activities and an observation sheet focused on training model implementation. The data of content, pedagogy, and cultural knowledge were analyzed using descriptive and inferential statistics from the one-sample t-test and the Wilcoxon test. When the data of the observations were analyzed using descriptive statistics, it showed that physics teachers' content knowledge was increased by 0,57, pedagogy knowledge was increased by 0,54, and cultural knowledge was increased by 0,38 of n-gain. The results of statistical tests showed that there were significant differences in the achievement of content, pedagogy, and cultural knowledge before and after the implementation of the culture-based physics teachers' education and training program.

1 INTRODUCTION

PCK is essential to career development for teachers and they can develop their own PCK by research-based activities such as action research and lesson study, with a particular emphasis on the employment of classroom practice, information technology, and collaborative learning (National Research Council, 2003; Juang et al., 2008; Shulman, 2015). Carlson et al. (2013) suggest that the use of educational curriculum materials and the transformation of experience in professional development programs are an effective way to strengthen the teachers' PCK. The program design based on material characteristics, the nature of the professional development program, and the changes or ultimate goal to be achieved, resulted in significant improvements in academic content knowledge, general pedagogical knowledge, the content knowledge component of PCK, the pedagogical knowledge component of PCK, and changes in teaching practices based in inquiry. Teachers' professional development can also be done in teacher education and training. Based on the Development Guidelines for KKG and MGMP Activities (Kementerian Pendidikan Nasional, 2010), the content of education and training programs should be designed with an eye to local culture and wisdom based on the essential needs of physics teachers, in terms of content, learning strategies based on the 2013 curriculum, and teacher competency standards established (Peraturan Menteri Pendidikan Nasional Nomor 16 Tahun, 2007) on the rationale of the diversity of local wisdom associated with content, especially physics, that has never been used in educational activities.

NTT is rich in cultural tradition and diversity, such as traditional houses, local arts, ikat weaving, custom rituals, various beliefs related to natural phenomena, and so forth. The cultural treasures of NTT are also full of moral and educational values, such as house building, carving motifs, crafting and playing traditional musical instruments, traditional texts, community habits, and so forth. In the district of Southwest Sumba, they practice Pasola, a customary ritual celebrating the rice-planting season. Pasola is a war game where two groups of equestrian troops throw wooden spears at each other in a savanna. The ritual of throwing a Pasola while riding toward the opponent is closely related to the concept of particle dynamics regarding the laws of motion. Science is a subculture of the Western environment; consequently, teaching science to students in non-Western countries or students within a multicultural environment who straddle these cultural borders can experience significant problems with the material and must deal with cognitive conflicts between those two worlds (Jegede, 1995; Subagia, 1999). Students' prior knowledge can strongly influence the formation of new concepts on an individual level. They thrive in a world with other human beings, so social relationships have a major influence on what students do and how they think (Siegler & Alibali, 2005). Their cultural background has a stronger influence on the learning of Science than the content of the subject matter (Baker & Taylor, 1995). To learn science is to acquire the culture of science; students must travel from their everyday lifeworld to the world of science found in their science classroom (Van Driel et al., 2001; Janik et al., 2009; Gay, 2013; Chinn, 2014). Based on the description above, this research would like to apply and measure the effectiveness of a culture-based physics teachers' education and training program at the conference of physics teachers in a regency of NTT.

2 METHODS

This is quasi-experimental research with a one-group pretest-posttest design. The research design is presented as follows:

$$O_1 \qquad X \qquad O_2$$

Where O1 = Content, pedagogy, and cultural knowledge pretest; X = Culture-based teachers' education and training program, and O2 = Content, pedagogy, and cultural knowledge posttest.

The research population included only physics teachers from MGMP and involved twenty physics teachers in one city as a research sample. The research instruments are (1) physics content knowledge test, (2) pedagogy knowledge test, (3) cultural knowledge test, (4) observation sheets of the participants learning activities, and (5) observation sheets of education and training implementation. Pretest and posttest data of content, pedagogy, and cultural knowledge were analyzed by descriptive and inferential statistics using the one-sample t-test, the Wilcoxon test, and the Kruskal-Wallis test. The data from observations and response questionnaires were analyzed using descriptive statistics and presented in the diagram. The flowchart of culture-based teachers' education and training is presented in Figure 1.

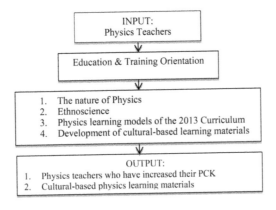

Figure 1.   Culture-based teachers' education and training flowchart.

3 RESULTS AND DISCUSSION

Teachers' education and training of culture-based physics to improve senior high school teachers' PCK were conducted in 43 lesson hours. This number qualifies the calculation of credit points of Peraturan Menteri Negara Pendayagunaan Aparatur Negara dan Reformasi Birokrasi nomor 16 Tahun 2009 on the functional position of teachers and credit numbers. These activities involved physics teachers at MGMP. The education and training facilitator was a researcher herself, along with two supervisors. The structures of this program were systematically arranged based on the description of training materials with respect to the achievements and time they require. Activities were introduced first with the material, followed by a discussion, and ended with the preparation of learning tools, peer teaching, and evaluation.

The effectiveness of education and training with a cultural approach is proven by the increase of the content, pedagogy, and cultural knowledge on the pretest and posttest vectors by calculating the n-gain. An overview of the development of the three components of PCK: content, pedagogy, and cultural knowledge can be seen in Figure 2.

The lowest average pretest value is pedagogy knowledge, and the highest average is cultural knowledge. In the middle category, there was an increase in the value of content knowledge by 0.57e, pedagogy knowledge increased by 0.54, and cultural knowledge increased by 0.38.

The increase in content knowledge was obtained from group discussion activities in cycle I to cycle IV; each group discussed the difficulties encountered in teaching the specified physics topic, the strategic solution to the difficulties faced, as well as the cultural elements associated with the physics topic. The heterogeneous group of junior and senior teachers facilitating the discussion of content knowledge developed well. With the discussions in each new cycle, there were sharing experiences, growth in the types of discussions, and work towards developing a lesson plan; they realized it would be beneficial to master all the topics because they are all interrelated. Gibbs and Coffey (2004) revealed that education and training activities could improve a teacher's ability to attract and retain students' interest in the material, improve some aspects of teaching including content knowledge, and change teachers' cognitive outlook and motivation to improve students' learning ability.

The increase of pedagogy knowledge was caused by a series of training activities that provided new insights on the characteristics of Physics, learning models and strategies, discussions on how to tackle specific topics, and peer teaching. Postareff et al. (2007) and Van Driel (2010) showed that education and training activities consisting of pedagogy knowledge and insights would help awaken effective teaching methods. Pedagogy knowledge, as the knowledge of the art of teaching, cannot be separated from content knowledge. The training

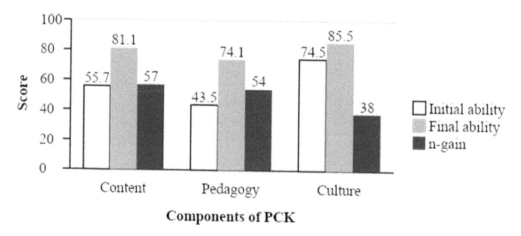

Figure 2. The increasing of participants' content, pedagogy, and culture knowledge.

activities provided a comprehensive mix of content, pedagogy, and cultural knowledge so that teachers have the experience and skills to put what they've learned into practice.

3.1 The achievement of content, pedagogy, and cultural knowledge

The analysis of differences in the attainment of content, pedagogy, and cultural knowledge using paired sample tests.

Table 1. Hypotheses test results of content, pedagogy, and cultural knowledge.

Data	Statistics Test	Sig. Value	Decision
Pretest-posttest of content	Wilcoxon	0,000	Significant differences exist
Pretest-posttest of pedagogy	Wilcoxon	0,000	Significant differences exist
Pretest-posttest of culture	t-test	0,000	Significant differences exist

The probability value (sig.) < 0.05, H0 (the sample is from a normally distributed population) is rejected, the implementation of the culture-based education and training effectively improves physics teachers' content, pedagogy, and cultural knowledge.

3.2 Differences in content, pedagogy, and cultural knowledge

For listing facts, use either the style tag List summary signs or the style tag List number signs.

By criterion, if the significance > 0.05 is obtained, then the variant of each sample is the same (homogeneous).

Table 2. The result of the homogeneity test of content, pedagogy, and cultural knowledge n-gain.

Levene Statistics	df1	df2	Sig.	Decision
3,786	2	57	0,029	Not homogeneous

The test results obtained the significance of 0.029 < 0.05, H0 rejected. By criterion, if the probability value (sig.) < 0.05, then H0 (there is no difference in the average increase of content, pedagogy, and culture knowledge) is rejected.

Table 3. The result of n-gain differences test of content, pedagogy, and cultural knowledge.

Statistics	Sig.	Decision
Kruskall-Wallis	0.018	Significant differences exist

Significance value < 0.05 or H0 is rejected, the education and training activities of culture-based physics give a different and significant influence on the improvement of the content, pedagogy, and cultural knowledge.

3.3 Observation results of physics teachers' learning activities

The observation of physics teachers' learning activities aimed to describe the relevant activities shown to the participants that took place during the program. The observed learning activities were analyzed and presented in Figure 3.

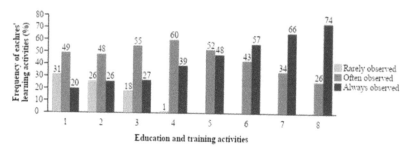

Figure 3.   Observation results of physics teachers' learning activities.

The observers assessed there were positive activities in the participants' actions. This condition describes the improved relevancy of activities relating to participation and motivation as well. The improvement in physics teachers' learning activities in each training activity was caused by; (1) the implementation of a participatory training model, one that encourages active participation of facilitators and participants alike; (2) each cycle being designed to include an open discussion involving the participants to deepen the ideas and understanding of the training materials; (3) discussions on the preparation of tools, on the teaching material, and on cultural exploration; (4) peer teaching supporting the active participation of physics teachers in the training.

3.4 Observation results of the education and training implementation

The observation aimed to assess the effectiveness of the program in two major parts: the implementation of activities and evaluation. In the implementation, observed aspects included facilitator's delivery of the materials, guided discussions, and monitoring participants. At the evaluation phase, the observed aspects included reflection at the end of the program and the preparation of the program follow-up. Implementation of the training is in two categories: often observed and always observed. In activities, I to III, observation results from both observers indicate that the implementation of the training model in the often-observed category he observers saw that the activities undertaken by facilitators and participants are often seen based on the observation items in their observation guidelines. Activities IV through VIII are activities of the cycle I to cycle four, which included group discussions, lesson plan preparation, and peer teaching. It was found that the training participants were more interested in the discussions and presentations.

4   CONCLUSION

Teachers' professional development through education and training has become a prevailing theme in the quest for improving education quality. It has been assumed to refer to qualified physics teachers, and, in this research, it refers to a culture-based approach. Culture-based education and training that recognizes and connects modern and indigenous or local science can improve access to education and validate the worth of individuals from the local community. The culture-based education and training program is a participatory model developed by taking into account the perspectives, customs, beliefs and values held by the people of NTT, applying andragogy principles, meeting physics teachers' learning needs as adult learners, and equipping the teachers of the participants of the education and training program with the ability to develop and implement activities for learning physics based on NTT's local culture. The culture-based physics teachers' education and training were very effective in increasing the physics teachers' PCK. The program meets the effectiveness aspects measured based on the improvement of content knowledge and pedagogy, the observation of the learning activities completed by the participants, and the implementation of the training activities. As a follow-up, the participants agreed to develop culture-based physics learning activities as a means of cultural preservation and to show the interconnection of culture and science.

REFERENCES

Baker, D. & Taylor, P. C. S. (1995). The effect of culture on the learning of science in non-western countries: The results of an integrated research review. International Journal of Science Education, 17(1): 695–704.

Carlson, J., Gess-Newsome, J., Gardner, A., Taylor, J. A. (2013). A framework for developing pedagogical content knowledge: the role of transformative professional development and educative curriculum materials. Paper presented at European Science Education Research Association Conference Nicosia, Cyprus.

Chinn, P.W.U. (2014). Place and culture-based professional development: Cross-hybrid learning and the construction of ecological mindfulness. Cultural Studies of Science Education, 10(1): 121–134.

Gay, G 2013, 'Teaching to and through cultural diversity', Curriculum Inquiry, vol. 43, no. 1, pp. 48–70.

Gibbs, G. & Coffey, M. (2004). The impact of training of university teachers on their teaching skills, their approach to teaching and the approach to learning of their students. Active Learning in Higher Education, 5(1): 87–100.

Janík, T., Najvar, P., Slavík, J. & Trna, J. (2009). On the dynamic nature of physics teachers' pedagogical content knowledge. Orbis Scholae, 3(2): 47–62.

Jegede, O.J. (1995). Collateral learning and the eco-cultural paradigm in science and mathematics education in Africa. Studies in Science Education, 25(1): 97–137.

Juang, Y.R., Liu, T.C. & Chan, T.W. (2008). Computer-supported teacher development of pedagogical content knowledge through developing school-based curriculum. Educational Technology & Society, 11(2): 149–170.

Kementerian Pendidikan Nasional. (2010). Rambu-rambu pengembangan kegiatan KKG dan MGMP. Jakarta: Direktorat Jenderal Peningkatan Mutu Pendidik dan Tenaga Kependidikan.

National Research Council. (2003). National Science Education Standards. Washington, DC: National Academy Press.

Postareff, L., Lindblom-Ylänne, S. & Nevgi, A. (2007). The effect of pedagogical training o teaching in higher education. Teaching and Teacher Education, 23(1): 557–571.

Shulman, L. S. (2015). Knowledge and teaching: foundations of the new reform. Harvard Educational Review, 57(1): 1–22.

Siegler, R. S. & Alibali, M. W. (2005). Children's thinking. Upper Saddle River, NJ: Prentice Hall.

Subagia, I.W. (1999). Science education and cultural clash: a case analysis of science education in Bali. Prosiding Postgraduate Student Conference, Graduate Schools of Education (430-436). Melbourne: La Trobe University.

Van Driel, J.H., Beijaard, D. & Verloop, N. (2001). Professional Development and Reform in Science Education: The Role of Teachers' Practical Knowledge. Journal of Research in Science Teaching, 38 (1): 137–171.

Van Driel, V. (2010). Model-based development of science teachers' pedagogical content knowledge. Paper Presented at international Seminar, Professional Reflections, National learning centre, New York. February 2010.

Emerging Perspectives and Trends in Innovative Technology
for Quality Education 4.0 – Kusmawan et al (eds)
© 2020 Taylor & Francis Group, London, ISBN 978-0-367-25803-0

Developing mathematical disposition of 8th grade students through MEAs strategy

E. Wahyuningrum
Universitas Terbuka, Tangerang Selatan, Indonesia

Y.S. Kusumah
Universitas Pendidikan Indonesia

ABSTRACT: This research examines the mathematical disposition of students after experiencing Model Eliciting Activities (MEAs) strategy. This Quasi-static research with comparison group design was conducted involving 122 eighth grade students. The students are from Junior high school from high and medium school level category. The data was collected by mathematical disposition scale instruments. Furthermore, they were analyzed by using the average differential test and two-way ANOVA. The results of the combination of the schools' high and medium category found that students' mathematical disposition engaged in learning with MEAs was better than that of engaging in the conventional one. The effect of MEAs on the mathematical disposition was not affected by the school category and students' prior knowledge of mathematics. These results were indicated by no significant effects occurred between the school category and learning category, and among the students' prior knowledge of mathematics, learning category, and mathematical disposition.

1 INTRODUCTION

Following characteristics of mathematics as a science that has a pattern of regularity, a logical sequence, and abstract object, it is often seen as a difficult science. Also, considering its hierarchical characteristics, mathematics is hard for students to master. This phenomenon becomes noteworthy, both in terms of students' mastery of the subject and students' mathematical abilities. Due to mathematical characteristics, students' awareness related to the understanding that mathematics is a common human activity is necessary because it has the potential to motivate students to build skills in using mathematics to understand the situation and solve daily life problems.

Activity habituation should contain the mathematical ability of students' development sustainably, which can be aggravated as well established. The habit of thinking and developing mathematical ability will be successful and become a joyful process if students possess mathematical awareness, desire, and a strong dedication to study mathematical concepts to communicate ideas openly. Their desire, awareness, and dedication to study mathematics or perform various mathematical activities are effective forces (NCTM, 2000), which appear along with growing cognitive powers.

The concepts of awareness, desire, and a strong dedication (Sumarmo, 2008) to study mathematics and perform a variety of mathematical activities show dispositions of mathematics in students, which can be developed through problem-solving and discussion. Several studies (Saija, 2008; Shafridla, 2012) revealed that teachers' learning innovation has a potency to build mathematical disposition capabilities. One of the selected learning strategies that allegedly explore the potency to develop a mathematical disposition is Model Eliciting Activities (MEAs).

This study chose and tested the strategy of MEAs with the consideration of MEAs learning (Erick, 2008, Hamilton 2009, Yildirim, 2010), including the activities of habituation to think and work mathematically based on six principles. They are issues of real (reality), enabling students to develop mathematical models (modeling), motivating students to produce an effective model which can be used for other relevant issues (prototype model), encouraging students to judge their ability and work (self-assessment), motivating students to dare to express mathematical ideas during discussion (sharing), and inviting students to document the work (documentation). Ekmekci and Krause (2011) suggested that MEAs motivate students to describe, retest, and refine their mathematical thinking.

2   METHODS AND RESEARCH SUBJECTS

Quasi research (Quasi-Experimental) with a static group comparison was conducted in two junior high schools with middle and high school levels in Depok, West Java. It involved eighth-grade students as research subjects. The availability of the population in the form of units did not allow the use of random samples (Setyosari, 2012) so that the sample selection was performed by employing purposive sampling.

Every research subject was divided into two separate groups: experimental and control group. The experimental groups obtained MEAs learning within a certain time, while the control groups learned the usual lesson conventionally at the same time. After their study ended, both groups were given the same tests, which assessed mathematical disposition. This study will analyze students' mathematical disposition based on their Prior Knowledge of Mathematics (PAM) and level of school groups after experiencing the MEAs initial mathematical knowledge.

3   RESULTS AND DISCUSSION

Results of the study show that the disposition of students' mathematical ability receiving MEAs treatment was better than that of receiving conventional learning, as shown in Table 1.

Table 1 demonstrates that overall, the experimental group learning math by MEAs received a higher average score in mathematical disposition (130.83> 124.98) compared to those studying math by conventional learning. The standard deviation score of the experimental group

Table 1. Description of students' mathematical disposition score based on PAM, learning strategy, and schools' level.

Level of School	PAM	MEAs			Conventional		
		Σ Student	$\bar{x}$	sd	Σ Student	$\bar{x}$	sd
High	High	13	145.54	15.00	12	132.00	13.18
	Medium	16	126.69	14.43	15	124.93	10.04
	Low	5	128.20	8.58	8	121.88	13.73
	Sub Total	34	134.12	16.40	35	126.66	12.38
Medium	High	6	128.67	18.96	9	129.67	12.22
	Medium	10	124.60	12.89	11	121.64	10.82
	Low	10	127.20	14.46	7	115.86	9.04
	Sub Total	26	126.54	14.48	27	122.81	11.82
Total	High	19	140.21	17.74	21	131.00	12.51
	Medium	26	125.88	13.63	26	123.54	10.30
	Low	15	127.53	12,.48	15	119.07	11.79
	Sub Total	60	130.83	15.93	62	124.98	12.19

was also higher (15.93> 12.19) compared to the control one, which indicates that scores of the experimental group in mathematical disposition are more heterogeneous than the control group learning mathematics conventionally.

Based on the school level aspect, students who obtained top-level mathematics learning through MEAs show higher mathematical disposition (134.12 > 126.66) compared to those learning math conventionally. The standard deviation score of students who learned with MEAs was higher (16.40 > 12.38) than those learned in a conventional setting. These results indicated that students who studied mathematics through MEAs had highly heterogeneous diversity compared to those who studied this subject through conventional learning.

Furthermore, students with medium school level demonstrate mathematical disposition scores between students who learned mathematics with MEAs, which was higher (126.54 > 122.81) than those attending in the conventional setting. The standard deviation score was higher in mathematical disposition (14.48 > 11.82) compared to those who studied mathematics conventionally. This result indicates that the mathematical disposition test results of the MEAs students had higher diversity than the conventional ones.

From the overall categories of PAM, it was found that the highest score of mathematical disposition was obtained by the students who received the learning of mathematics through MEAs (140.21 > 130.00) than those who learned it conventionally. The standard deviation also revealed that the students learning through MEAs received a higher score (17.74 > 12.51) compared to those studying in a conventional setting. These results indicate that the MEAs students obtained more heterogeneous mathematical disposition results than conventional ones.

Based on the PAM group, it was found that mathematical disposition and standard deviation scores of students receiving MEAs treatment were higher (125.88 > 123.54) and (13.63 > 10.30) respectively, compared to those studying in a conventional setting. These results reveal that the students who learned through MEAs received more heterogeneous mathematical disposition results than those who studied conventionally.

Furthermore, students at medium school level who studied conventionally with high PAM scores reveal slightly lower mathematical disposition scores (128.67 < 129.67) compared to those who received treatment with MEAs. Likewise, the standard deviation shows students receiving MEAs treatment obtained higher mathematical disposition (18.96 > 12.22) than those who studied the same subject in a conventional setting. These results indicate that the mathematical disposition of a group of students studying math through MEAs received higher diversity than those studying math conventionally.

3.1 The effect of interaction learning and school level on students' mathematical disposition

The test results showed no influence between learning and school level on students' mathematical disposition indicated by the value of α = 0.232 in Table 2.

Another analysis that can be obtained from the results in Table 2. is the ANOVA analysis. It reveals that each of the main sources influences students' learning and school level significantly, which then affects students' mathematical disposition, as shown in the value of Sig. The values demonstrate the influence of the level of school and teaching and learning accounted for 0.027 and 0.030, respectively.

Table 2. Test of the effect of interaction learning and school level on mathematical disposition of students.

Source	Type III Sum of Squares	df	Mean Square	F	Sig.
Level of School	977.294	1	977.294	5.023	0.027
Teaching & Learning	937.096	1	937.096	4.816	0.030
Interaction	104.614	1	104.614	0.538	0.465
Total	2,019,573	122			

3.2 The relations between learning and PAM test to students' mathematical disposition

The test results showed no relations between learning and school level to influence students' abilities related to mathematical disposition as shown in Table 3.

Table 3. ANOVA test results showing the relations between learning and PAM on students' mathematical disposition.

Source	Type III Sum of Squares	df	Mean Square	F	Sig.
Level of School	1291.225	1	1291.225	7.407	0.007
Teaching&Learning	3532.244	2	1766.122	10.132	0.000
Interaction	321.708	2	160.854	0.923	0.400
Total	2,019,573	122			

The absence of relations between learning and students' PAM indicated that no integration between learning and PAM students occurred in shaping students' mathematical disposition variations. The largest difference is shown by the students with a high level of PAM compared to students with medium and low PAM, which indicates that MEAs learning is more appropriately applied to students with a high level of PAM.

4 CONCLUSION

Following the research questions which have been formulated and based on analysis of the data, it can be concluded as follows. Disposition students receiving mathematical learning strategy through MEAs obtain a higher score than those studying conventionally, either student from the combined school (high and medium level) or the top-level school. There is no relation between learning strategies and school level to students' mathematical disposition, as well as among learning strategy, students' prior knowledge of mathematics, and students' mathematical disposition. The absence of relations between learning strategies and school level and among learning strategies and prior knowledge of mathematics to mathematical disposition explains that learning is the most influential factor in students' mathematical disposition compared to their prior mathematical knowledge and school level.

REFERENCES

Ekmekci, A. dan Krause, G. (2011). Model Eliciting Activities (MEAs). 5thAnnual UTeach Institute-NMSI Conference, May 26, 2011. [Online]. Tersedia: http://www.uteach-institute.org/images/uploads/2011_ekmekci_model_eliciting_activities (11 Maret 2013).

Eric, C.C.M. (2008). Using Model Eliciting Activities for Primary Mathematics Classrooms. The Mathematics Educator, Vol. 11No. 1/2, 47–66. Singapore: Association of Mathematics Educators. [Online]. Tersedia: http://repository.nie.edu.sg/jspui/bitstream/10497/135/1/ME-11-1-47.pdf (11 Agustus 2011)

Hamilton, E. (2009). Modeling and Model-Eliciting Activities (MEAs) as Foundational to Future Engineering Curricula. 20th Australasian Association for Engineering Education Conference University of Adelaide, 6–9</pg> December <yr>2009. Adelaide: University of Adelaide. [Online].Tersedia: http://aaee.com.au/conferences/AAEE2009/PDF/AUTHOR/AE090130.PDF (11 September 2011)

National Council of Teachers of Mathematics, (2000). Principles and Standards for School Mathematics. Reston, VA: NCTM. [Online]. Tersedia: http://www.fayar.net/east/teacher.web/Math/Standards/index.htm (13 December 2010)

Saija, L.M. (2012). Analyzing The Mathematical Disposition and Its Correlation with Mathematics Achievement of Senior High School Students. Jurnal Ilmiah Program Studi Matematika STKIP Siliwangi Bandung, Vol I, No. 2, September 2012. [Online]. Tersedia: http://e-journal.stkipsiliwangi.ac.id/index.php/infinity/article/.../48/23 (15 April 2013)

Setyosari, P. (2012). Metode Penelitian Pendidikan Dan Pengembangannya. Jakarta: Kencana Prenada Media Group.

Shafridla (2012). Peningkatan Kemampuan Komunikasi dan Disposisi Matematis Siswa melalui Pendekatan Matematik Realistik. Tesis, Universitas Negeri Medan. [Online]. Tersedia: http://digilib.unimed.ac.id/peningkatan- kemampuan-komunikasi-dan-disposisi-matematis-siswa-melalui-pendekatan-matematik- realistic-22886.html (28 Maret 2013)

Sumarmo, U. (2008). Berpikir matematik: Apa, Mengapa, dan Bagaimana cara mempelajarinya. Universitas Pendidikan Indonesia, Bandung: tidak diterbitkan.

Yildirim, T.P., et al. (2010). Model Eliciting Activities: Assessing Engineering Student Problem Solving and Skill Integration Processes. Int.J. Engng Ed. Vol. 26, No.4, pp.831–845. Great Britain: TEMPUS Publications. Online]. Tersedia: http://www.modelsandmodeling.pitt.edu/Publications_ files/MEA_Ije e2332_1.pdf (24 Maret 2012)

*Emerging Perspectives and Trends in Innovative Technology
for Quality Education 4.0 – Kusmawan et al (eds)*
© *2020 Taylor & Francis Group, London, ISBN 978-0-367-25803-0*

Writing language ability of Indonesian education program student in Universitas Terbuka

Enang Rusyana
Universitas Terbuka, Tangerang Selatan, Indonesia

ABSTRACT: In the Universitas Terbuka (UT), Indonesian Language Education is actively studied in junior and senior high schools. While teachers are expected to speak Indonesian appropriately, their language skills do not meet the expected standards. This was evident after they were taken through the Literature Appreciation Teaching online tutorial. Generally, their written language skills were incorrect. This paper, therefore, describes the use of the Indonesian language in writing. It uses the descriptive method in analyzing the writing skills of students in a discussion forum based on spelling and sentence structures. The inaccuracies in the spelling were of four types and three errors in sentence writing.

Keywords: Teacher, Language Skills

1   INTRODUCTION

Indonesian Language Education in Universitas Terbuka (UT) is included in the Teacher Training and Education Faculty (TTEF). Every TTEF UT student needs to have an active education study program in schools (UT, 2018). According to PP Number 74 of 2008 Chapter I General Provisions, teachers are professional instructors tasked with the responsibility of educating, teaching, guiding, directing, training, and evaluating students. These duties are discharged during early childhood education in formal, basic, and secondary schooling. Moreover, in Chapter II, Competence and Certification, teachers are expected to have academic qualifications, competencies, and certificates, be physically and mentally healthy and have the ability to foster national education goals. The competencies are portrayed in terms of pedagogy, personality, social interaction, and professionalism obtained through relevant education. In terms of educating, teaching, guiding, directing, and training, teachers should be able to speak appropriately and correctly. This relates to the quality of the content delivered and the proper use of the linguistic rules.

For the Indonesian Language Education to be used in writing appropriately, UT provides teaching materials Writing I. The goals which students need to achieve in these courses include (1) applying diction, spelling, and punctuation, and (2) making correct and effective sentences (UT, 2014: xi). The purpose of Writing 2 is to ensure students are able to write journalism, scientific work, short stories, etc. (UT, 2014: i). Additionally, the Indonesian Language Education students are also introduced to the General Guidelines for Improved Indonesian Spelling (Indonesian Language Development Center Committee, 2004), which explains punctuation, spelling, and words. Learning the two Writing courses and paying attention to the General Indonesian Language Spelling Guidelines (PUED), students should be in a position to express their ideas in written form during discussions on the online tutorial Literature Appreciation Teaching.

In the online tutorial discussion forum for Literature Appreciation Teaching, students provide written comments on topics provided by the tutor. There are eight discussion forums with questions to be responded to by students. The responses should be written in Indonesian and in a formal, detailed, sequential manner depending on the topic of discussion. The debates are expected to help students develop a broader viewpoint on useful topics.

However, students' responses in several forums examined are still far from the expected level. Moreover, their comments from students are also written in various forms of social media, which do not heed the rules for punctuation, writing letters, words, and standard sentences.

In this paper, the problem is formulated based on two questions. First, what is the use of spelling in the responses of students of Indonesian Language Education TTEF UT in Literature Appreciation Teaching online tutorial discussion forum associated with PUED? Second, what is the effectiveness of the sentences given?

The purpose of this paper is to describe the uses of spelling in the responses or comments of students of Indonesian Language Education TTEF UT in the Literature Appreciation Teaching online tutorial discussion forum PUED. It also aims to examine the sentence effectiveness of responses or comments of students.

The study used a descriptive method, in which the author examines spelling and sentences in responses from students. The comments studied are from the 6th, 7th, and 8th discussion forums. The reason for collecting data from these forums is to determine whether the suggestions and improvements from the tutor on the previous forums were adopted. In all the three panels, students who commented were mostly the same. To improve the efficiency of the discussion, each comment is numbered according to the forum and the specific student giving it. Those commenting on one or two forums are given serial numbers after all students have participated.

2 THEORETICAL FRAMEWORK

Data from the online tutorial discussion forum Literature Appreciation Teaching should be rated according to the rules of the *General Improved Indonesian Spelling Guidelines* (PUED), *General Guidelines for Formation of Terms* (PUPI), *Indonesian Language Grammar* (TBBBI), and *Large Indonesian Language Dictionary* (KBBI).

3 RESULT AND DISCUSSION

The inaccuracy of spelling usage in student responses is described as follows.

a) Punctuation
The inaccuracy of the use of punctuation in the responses or comments of the students can be described as follows.

The comma (,) is affixed without spaces between the words before and after it. Example:, Moody, ie, within, ie, 1

Punctuation comma (,) shall be shown in a space with the word preceding it. Example: Moody, Schuman, Moody, following,

Comma punctuation (,) is affixed without spaces with the word preceding it or following it. Examples: Gordon, Moody's creativity, how problems,(2) problems,(3) practical,3. Introduction,4. Presentation,5. discussion

The punctuation point (.) It is affixed to using spaces in tang words preceding it. Example: students. Found. Increase.

The punctuation mark (.) It is affixed without spaces with the word that follows it. Examples: language.Literary literature.Principles 1.Preliminary tracking.2. Determination

There are also some inaccuracies in the use of punctuation marks. For instance, commas are not used where the clause precedes the main clause, colon (:) use a space after the word before it, a period (.) is used even though the sentence is not finished, etc.

The inaccuracies in the use of punctuations by all participants in the 6th forum amounted to 115 cases, 51 in the 7th, and 34 in the 8th. Evidently, there was a decrease in the number of inaccuracies.

b) Capital Letters

Inaccuracies in the use of capital letters included the use of lowercase letters in a word at the beginning of a sentence and the first letter of a person's name. The overall Inaccuracies in the 6th forum totaled to 12 cases, 17 in the 7th, and 7 in the 8th, showing a significant decrease.

c) Word Writing

The inaccuracies in this regard include writing affixation (prefix, infix, suffix, confix) separately. Example: done, given, etc. Also, the preposition is written in a series with the words that follow it — example: above, below, including, besides, etc. Additionally, there were incomplete writing words, for example, evaluation, remediation, knowledge, creativity, etc.

The inaccuracies in word writing by all participants in the six discussions were 29, the 7th had 34, and 17 on the 8th. There are fluctuations in the number of inaccuracies in the writing of words.

d) Writing up the absorption element.

The inaccuracies in all participants in the six discussion forums were 2, the 7th had none, while the 8th had 1 case. Generally, there were fluctuations in the number of inaccuracies in writing the absorption elements in the responses or comments of the students.

e) Ineffective sentences

There was an absence of a clear sentence subject. Who is meant by "we" in that sentence, is it bad, bro? Not clear. Also, who should carry out the evaluation, teacher or student? It is not clear either.

The ineffectiveness in the discussion forum 6 amounted to 20 cases, 18 in the 7th, and 4 in the 8th. There is a decrease in the inefficiency in writing sentences in the responses given by the students.

4 CONCLUSION

4.1 *Conclusion*

Based on the results, the students were unable to write their responses according to the *General Guidelines for Improved Indonesian Spelling*. However, there are positive changes, along with a consistent reminder by the tutor. The 7th and 8th discussion forums had better results than the 6th based on the spelling and sentence inaccuracies.

4.2 *Suggestions*

This paper only discusses the accuracy of the use of spelling and sentences in the comments of students in the Indonesian Language Education online tutorial participants' Literature Appreciation Teaching. Further research is needed to find out why the students' language skills are low and how to improve.

REFERENCES

Alwi et al. (1993). Indonesian grammar. Jakarta: Balai Pustaka.
Department of Education and Culture. (1991). General guidelines formation term. Jakarta: Ministry of Education and Culture.
Open University. (2018). Catalog of Open Universities in 2018. Jakarta: Open University.
Open University. (2014). PBIN 4109 writes I. Jakarta: Open University.
Open University. (2014). PBIN 4433 writes II. Jakarta: Open University.
Open University. (2014). PBIN 4219 teaching literary appreciation. Jakarta: Open University.
Indonesian Language Development Committee. (2004). General spelling guidelines Improved Indonesian. Jakarta: Ministry of education National.
PP No. 74 of 2008.

Emerging Perspectives and Trends in Innovative Technology for Quality Education 4.0 – Kusmawan et al (eds)

The relationship of parents' psychosocial stimulation with multiple intelligence level of kindergarten children, age 5–6 years

Dian Novita
Universitas Terbuka, Tangerang Selatan, Indonesia

ABSTRACT: Preschoolers are creative individuals, though most parents and teachers are less aware of or cannot appreciate their abilities. Children who are always obedient and do things parents want or behave like other children are always preferred. Originality is less acceptable, considered severe, and might even be viewed as dangerous. Without realizing it, well-intentioned adults, in an attempt to instill discipline and obedience, do not provide the opportunity for children's creativity to grow and develop. Offering educational services in early childhood is an influential basis for their development. According to Hurlock, the early years of a child's life are the foundation that tends to survive and influence the attitudes and behavior of children throughout their lives. Every child has the creative talent, and in terms of education, it is possible to develop and foster it early on. Based on the importance of parenting patterns, it is always challenging to understand the effectiveness of parents in increasing children's creativity. This is a cross-sectional study conducted at Ananda Kindergarten Open University with a sample of 30 children utilizing primary and secondary data types. Importantly, data were obtained through observation, interviews, and documentation studies. The results of this study show that the sex and birth order of children does not correlate with the parenting style applied at home. Family characteristics do not correlate with parenting style, nor with children's characteristics for creativity. There is a significant relationship between the education of fathers and children's creativity in dealing with boredom. In addition, there is a significant relationship between the opinions of mothers and answering children's questions in play activities that enhances creativity. Moreover, there is a significant relationship between the care given by parents and children's imagination, knowing the surrounding environment, and answering questions raised, thus increasing the child's ability to experiment, giving rise to a new stimulus for children to overcome boredom.

1 INTRODUCTION

Children are a gift and trust given by Allah SWT to each parent, accompanied by the responsibility to develop their personality and virtue. An important task as parents, educators, and caregivers is to provide the best possible education for all students. According to Pestalozzi, education for children in kindergarten should be fun, meaningful, and warm (Masitoh, 2003). This is in line with Solehuddin (1997), who stated that preschool education is intended to facilitate the growth and development of children following the norms and values of life. Rather than being academically oriented, it needs to provide learning experiences for children. In addition, preschool education programs must be tailored to the needs, interests, and development of learners.

Apart from learning in schools, family education also has a vital role in increasing children's intelligence. For parents and learning institutions not to make mistakes in educating children, there is a need to establish harmony and good cooperation between the two parties. Parents educate their children at home while teachers take that responsibility in educational institutions. The learning process, after an early age, needs to provide basic concepts that enable children to be optimally active and curious. For this reason, teachers are companions, guides, and facilitators for children.

Gardner (Armstrong, 1996) states that everyone has different forms of intelligence, not just one. A person's intelligence consists of language/linguistics, mathematical logic, visual-spatial, kinesthetic, interpersonal, musical, and naturalist. Every child has these forms of intelligence, though they are often not honed appropriately by parents, educators, or the education system, and therefore are underdeveloped.

The general purpose of this study was to determine the effect of parenting and learning processes on the creativity of kindergarten age children in TK Ananda UT Pondok Cabe, South Tangerang City. It also aimed to (1) determine the influence of individual and family characteristics on the level of creativity of kindergarten age children; (2) define the influence of parenting on the level of creativity; (3) determine the effect of the learning process at school on the level of creativity; (4) know the difference due to parenting or learning processes in school in the level of creativity of children and the effect of the combination of parenting and learning processes at school on the level of creativity of children; and (5) determine whether there is a relationship between sex, age of the child, and parents' work and income and the level of creativity of kindergarten-age children.

2 THEORETIC STUDY

Parents consist of a father and mother and result from a legal marriage relationship that can form a family. Parents have the responsibility to educate, nurture, and guide their children to reach certain stages at which they are ready for social life. The definition of parent cannot be separated from that of the family for a valid reason. Parents are part of a large family that has been principally replaced by the nuclear household consisting of father, mother, and children. Traditionally, the family is defined as two or more people connected with blood ties, marriage, or adoption (law) and have a place to live together. According to Morgan in Sitorus (1988: 45), a family is a primary social group based on marriage (husband–wife relationship) and kinship ties (intergenerational relations, parent–child) at the same time. However, dynamically, the individuals who make up a family are described as members of the most essential groups of people who live together and interact to meet their individual and interpersonal needs.

According to Suparlan (1993: 76), a family is a social group consisting of father, mother, and child. It is a social association among family members, relatively fixed and based on the bond of marriage, blood, or adoption. Relationships between family members are imbued with affection and a sense of responsibility. Generally, childcare has an essential role in forming personality in a family. This is reinforced by the opinion of Brown (1961: 76) that the family is the environment that first receives children. Parents have many roles in this regard, such as caring for their children.

The pattern of care, according to Stewart and Koch (1983: 178), consists of three tendencies for parenting, which include (1) an authoritarian approach that sets absolute standards to be followed and is usually accompanied by threats; (b) democratic parenting that prioritizes the interests of children but does not hesitate to control them, adopted by parents who are rational, always basing their actions on reason or thought; and (c) permissive parenting, which is commonly loose supervision, where a child is allowed to do something without enough supervision.

2.1 Creativity

The term creativity comes from the English phrases *to create*, that is, to compose or make something different from the shape, arrangement, or style of what is commonly known by people. The whole product does not need to be new; it might also be a combination of preexisting elements (Semiawan, 1999). Creativity is defined not only as the ability to create something new or a combination. Munadi (1987) also defines creativity as a thought process that makes someone try to determine new methods and ways of solving a problem and then stresses that what is important is not what is produced from the process but rather the pleasure and preoccupation that is seen in carrying out creative activities.

2.2 Results

Based on the distribution of children, the majority are girls, precisely 69 in number. This is 57.5% of the total sample, followed by boys, 42.5%, or 51 in number. The average age of the children in the sample was 69.42 months, and almost half of them were first-borns (49.2%). Family characteristics show that three-quarters of fathers are at an average age of 38.45 years, and half of the mothers are about 34.45 years of age. The average family income is IDR 11,717,500 per month. According to Gunarsa (2000) in Rahmaulina (2007), the sufficient economic situation of the family results in parents having more time to guide their children. This is because they no longer think about economic conditions and what they are lacking. At the level of parental education, the results of the study showed that more than half of the fathers (55%) and nearly half of the mothers (41.7%) had college degrees.

The results showed that the average late score stimulation psychosocial family total is 47.15. This means 85.7% of families have provided the psychosocial stimulation needed by children. Even an SI-1 of 88% had provided psychosocial stimulation.

Intelligence shows no real difference between examples in public and religion-based schools. The average score of achievement for each intelligence is relatively the same, but the examples of public schools have an average score of excellent intelligence in terms of motor, interpersonal, intrapersonal, and spatial skills in comparison with religious schools. In contrast, examples in schools based on religion have higher average scores on gross motor intelligence, language, mathematics, and music than those in public schools. The lack of differences in terms of multiple intelligences might be due to the relatively similar sample and family characteristics.

The essence of multiple intelligences theory, according to Gardner (Armstrong, 1994), is to appreciate the uniqueness of each individual, with a variety of ways of learning and realizing several models for valuing and almost unlimited ways of self-actualizing in this world. According to Gardner, there is no intelligence that is better or more important than the other. In this study, it was found that the sample had an average score of compound intelligence achievement of 71.94, or fulfilled 89.9% of the total compound intelligence. Intrapersonal intelligence has the highest achievement score of 9.35. This means there is an awareness of feelings within oneself, intuition, motivation, temperament and desire, ability to self-discipline, self-understanding, and self-confidence (Moleong, 2004).

The characteristics of children, both the age and the birth order, are not significantly related to psychosocial stimulation received. This means the psychosocial stimulation in this study is not significantly related to age and birth order, but in contrast to family characteristics, several variables are significantly related to psychosocial stimulation.

Family characteristics significantly related to psychosocial stimulation are the father's and mother's education and the mother's and family's income. This shows that the higher the education of parents, the better the psychosocial stimulation given to children.

2.3 Relations between psychosocial stimulation and multiple intelligence

The test results showed that the overall correlation of psychosocial stimulation is not related to multiple intelligence. However, if multiple intelligence is perceived in its individual aspects, a significant positive relationship between language stimulation and excellent motor intelligence is found, with a correlation coefficient of 0.226. Similarly, there is a relationship between the physical environment and excellent motor intelligence, with a correlation coefficient of 0.281.

3 CONCLUSIONS

From the study, parents are in the productive age range of 25–54 years, with the highest proportion holding undergraduate degrees. Additionally, there were no significant differences in psychosocial stimulation based on school type except in the exemplary dimension. On this basis, the religion-based schools had better scores. There is no significant difference in multiple intelligence based on school type. Importantly, there is a positive correlation between family characteristics (father's and mother's education and their income as well as that of the family) and psychosocial stimulation. However, there is no significant relationship between psychosocial stimulation and multiple intelligences.

4 SUGGESTIONS

This study shows a positive correlation between the learning process and multiple intelligences. Nevertheless, several suggestions are presented to researchers, parents, schools, and government and related institutions to improve the multiple intelligences during early childhood. For instance, schools should strengthen the institutional system that leads to the creation of a comfortable, safe, and mutually supportive atmosphere between their fraternal members. The government and related institutions need to provide support for improving the quality of teachers and schools to optimize multiple intelligence in early childhood.

REFERENCES

Armstrong, T. (1996) *Multiple intelligences in the classroom.* Alexandria, VA: Association for Supervision and Curriculum Development.
Armstrong, T. (2003) *Sekolah Para Juara. Terjemahan.* Yudhi Mrtanto. Bandung: Kaifa.
Beceren, B. O. (2010) Determining multiple intelligences preschool children (4–6 age) in learning process. *Procedia Social and Behavioral Sciences,* 2(2010), 2473–2480.
Giyarti (2008) *Pengaruh Stimulasi Psikososial, Perkembangan Kognitif, dan Perkembangan Sosial Emosi terhadap Perkembangan Bahasa Anak Usia Prasekolah di Kabupaten Bogor.* [Skripsi]. Bogor. Fakultas Pertanian. Institut Pertanian Bogor.
Hastuti, D. (2006) *Analisis Pengaruh Model Pendidikan Prasekolah pada Pembentukkan Anak Sehat, Cerdas, dan Berkarakter* [Disertasi]. Bogor: Fakultas Pascasarjana. Institut Pertanian Bogor.
Masitoh. Dkk. (2003) *Pendekatan Belajar Aktif di Taman Kanak-Kanak.* Jakarta: Depdiknas. Dirjen Dikti. Bagian Proyek Peningkatan Pendidikan Tenaga Kependidikan.
Moleong, J. (2004) *Teori Aplikasi Kecerdasan Jamak pada PAUD*, seminar dan lokakarya nasional pendidikan anak usia dini di Jakarta.
Rahmaulina, N. (2007) *Hubungan Pengetahuan Ibu tentang Gizi dan Tumbuh Kembang Anak serta Stimulasi Psikososial dengan Perkembangan Kognitif Anak Usia 2.5-5 Tahun* [Skripsi]. Bogor: Fakultas Pertanian. Institut Pertanian Bogor.
Solehuddin, M. (1997) *Konsep Dasar Pendidikan Prasekolah.* Bandung: FIP UPI.
Turner, J. S. and Helms, B. D. (1991) *Lifespan development*, 4th ed. Philadelphia: Saunders College Publishing.
Xie, J. C. and Lin, R. (2009) *Research on multiple intelligences teaching and assessment. Asian Journal of Management and Humanity Sciences*, 4(2–3), 106–124.

Emerging Perspectives and Trends in Innovative Technology
for Quality Education 4.0 – Kusmawan et al (eds)
© *2020 Taylor & Francis Group, London, ISBN 978-0-367-25803-0*

Using GeoEnZo software in geometry to teach Primary School Teacher (PST) students

Idha Novianti
Universitas Terbuka, Tangerang Selatan, Indonesia

ABSTRACT: The use of appropriate learning aids makes it easier for students to understand the material. There are various kinds of learning software used to build space. However, GeoEnZo makes it easier for students to understand the material and become confident as individuals and capable of sharing their knowledge in their classes.

Keywords: GeoEnZo, mathematics learning, primary school teacher students

1 INTRODUCTION

According to Robert S. Zais, the Curriculum and Learning modules of an educational institution are based on five foundations: (1) philosophical assumptions, (2) epistemology (the nature of knowledge), (3) society/culture, (4) the individual, and (5) learning theory. Creative learning activities effectively increase students' knowledge (Khuziakhmetov and Gorev, 2017). Furthermore, Kuboja and Ngussa stated that there are four essential foundations in developing a curriculum, namely the philosophical, psychological, sociocultural, and the development of science/technology (Kuboja and Ngussa, 2015). The philosophical foundation becomes the main foundation compared to other aspects, with different views resulting in varying curriculum development applications. The philosophical foundation produces national education, institutional, field, and instructional goals. The psychological foundation is related to learning theory and developmental psychology, which is used to determine the contents of the curriculum presented to students to ensure the level of breadth and depth is in accordance with their development. The curriculum needs to be able to meet the needs of every student (Ornstein, 1990), and changes need to consider various mathematical aspects and education systems (Robitaille and Dirks, 1982). It is a strategic and crucial aspect of education because of its ability to lead learners in their lives in the future.

2 USE OF LEARNING MEDIA

Three things need to be considered in the use of tools in the form of software to achieve learning needs. These include the context in which the software is used to ensure students learn and to ease their ability to access information (Brown, 1990). The visual appearance significantly increases students' understanding (Clements, 1982), with computers used to make the atmosphere in the classroom conducive to learning, thereby increasing students' enthusiasm to learn (Kumar and Kumaresan, n. d.). The use of information and communication technology (ICT) in learning mathematics tends to improve students' mathematical literacy (Rahmawati, 2018).

GeoEnZo is a free math application created to quickly draw geometrical shapes such as cones, triangles, circles, cubes, and lines. It is a useful tool that allows students to turn the

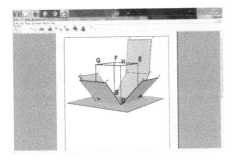

Figure 1

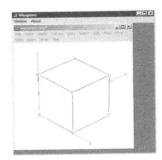

Figure 2

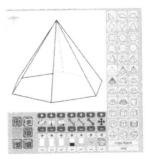

Figure 3

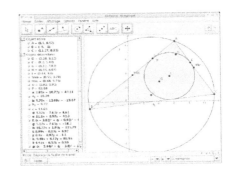

Figure 4

screen into a standard blackboard, thereby displaying its ability. Besides the ability to draw freely on the board, one of the most exciting features of this program is its ability to create all kinds of simple geometric shapes. The use of GeoEnZo in geometry lessons with contextual models significantly increases student understanding (Mauladaniyati and Kurniawan, 2018). Furthermore, its use in elementary schools, such as in learning mathematics, is the right step in technology introduction. In building space, many graphic presentations are needed to explain the material. Therefore, the use of media is very appropriate to facilitate teaching.

The software products considered are Cabri (Figure 1), Wingeom (Figure 2), GeoEnZo (Figure 3), and GeoGebra (Figure 4).

In addition, there is other software products capable of displaying three-dimensional geometric images. However, the selection of GeoEnZo software in mathematics learning courses in elementary schools is based on its ease of use compared to other software such as Cabri, GeoGebra, or Wingeom. The use of these latter software products makes students depressed and think of ways to utilize it rather than paying attention in class. Also, based on the results of interviews with students, they do not want to use these because they are complicated.

3 RESEARCH METHODS

This research was conducted on students of the teacher education study program during the seventh semester, with two classes sampled in the pretest–posttest group. The design was as follows:

Experimental class:

AO X1 O
AO X2 O

Specification:

    A = Samples randomly selected according to class
    O = Initial test and final test
    X1 = Learning with GeoEnZo
    X2 = Learning with GeoGebra

Data used in the study were obtained from the results of the pretest–posttest, interviews, and observations during the learning process. Similarly, data obtained from the results of the initial and final tests were analyzed via descriptive analysis.

4 DISCUSSION

The primary school teacher study program comprises teachers with at least two years of teaching experience. The purpose of establishing this program is to improve teacher competencies and make them professional. Students who work daily as educators or teachers are trained to improve their various teaching abilities using this software. Based on the detailed test results, classes that utilized the GeoEnZo software performed better, as shown in Table 1.

Table 1. Descriptive test results.

Descriptive statistics					
	N	Minimum	Maximum	Sum	Mean
PRE	25	.00	20.00	230.00	9.2000
POST	25	60.00	100.00	1980.00	79.2000
PRE1	25	.00	20.00	320.00	12.8000
POST1	25	40.00	60.00	1320.00	52.000
Valid N (listwise)	25				

In Table 1, the pre- and posttest results in class A, which uses GeoEnZo, with a mean of 79.2, were higher than in class B, which does not use the software, with a 52.8 mean. However, both classes have increased learning outcomes using the learning media, with more improvement in class A. Mauladaniyati and Kurniawan (2018) stated that the use of this software increases understanding.

The results of interviews with all students proved that those who used GeoEnZo felt happy and confident of their knowledge, while others failed to understand, were not satisfied, and the class was not conducive (crowded) to learning.

Students need always to prepare to learn something new to be able to adapt in the future. The instructor needs to analyze their characters and prepare and motivate them to keep learning with perseverance.

5 CONCLUSION

The use of GeoEnZo in learning geometry is essential in improving students ability to understand the material. Furthermore, it makes the classroom conducive to learning and increases students ability to pay attention, enthusiasm, and confidence. There is much free software available on the internet that can be used to learn mathematics. However, it is necessary to study their usefulness in order to attract students.

REFERENCES

Brown, L. (1990) *The right to be wrong.*

Clements, K. (1982) *Visual imagery and school mathematics.*

Khuziakhmetov, A. and Gorev, P. (2017) Rio Claro. *SP*, 31(58), 58. doi: doi.org/10.1590/1980-4415v31n58a06.

Kuboja, J. and Ngussa, B. (2015) Conceptualizing the place of technology in curriculum formation: A view of the four pillars of curriculum foundations. *International Journal of Academic Research in Progressive Education and Development*, 4(2). doi: doi.org/10.6007/IJARPED/v4-i2/1728.

Kumar, A. and Kumaresan, S. (n. d.) *Use of mathematical software for teaching and learning mathematics.*

Mauladaniyati, R. and Kurniawan, D. (2018) GeoEnZo utilization as mathematics learning media with contextual approach to increase geometry understanding. In *3rd International Conference on Mathematical Sciences and Statistics*. doi: doi.org/doi:10.1088/1.

Ornstein, A. (1990) Philosophy as a base for curriculum decisions. *The High School Journal*, 74(2), 102–109. doi: doi.org/10.1007/978-90-481-2804-4.

Rahmawati, N. (2018) Utilization of ICT in improving mathematical literacy ability. *PRISMA*, 1, 381–387.

Robitaille, D. and Dirks, M. (1982) *Models for the mathematics curriculum.*

Emerging Perspectives and Trends in Innovative Technology
for Quality Education 4.0 – Kusmawan et al (eds)
© 2020 Taylor & Francis Group, London, ISBN 978-0-367-25803-0

Development of mobile learning with constructive models in trigonometry material at senior high school 8 in Semarang

Nurmawati & Pukky Tetralian Bantining Ngastiti
Universitas Terbuka, Tangerang Selatan, Indonesia

ABSTRACT: This research is motivated by the non-optimal increasing use of smartphones among students with the aim of developing a mobile learning media on the subject of trigonometry using a decent constructive model for an interactive and effective mathematical teaching process. Data were effectively analyzed using the Research and Development (R&D) methods with the Borg and Gall model. The product showed that the effective criteria of the experimental class learning achievement were better than the control class using the t-test. The obtained t count> t table is 2,09 > 1.67. Therefore, Ho is rejected, which means learning using media mobile with the constructivist model is better than the conventional. Therefore, it is concluded that mobile learning media with the constructivist model is an effective learning process for senior high school students.

Keywords: Constructive Model, Conventional Model, Mobile Learning

1 INTRODUCTION

1.1 *Background*

In the globalization era, the development of Science and Technology is rapidly developing. According to data from International Data Corporation (IDC), a market research institute, sales growth in 2014 for smartphones and tablets grew by 12% and 18%, respectively, compared to 2013. In Southeast Asia, Indonesia is the biggest contributor to its sale with a 30% increment. Sophisticated technology tends to encourage its use in the teaching and learning process, to maximize student outcomes.

Mathematics is a subject taught at every educational level in Indonesia, starting from elementary to college. It needs to be given to equipping them to think logically, analytically, systematically, critically, and creatively, with the adequate corporation. Many students still find it difficult to absorb the materials taught by teachers in accordance with concepts and formulas. This tends to be an obstacle, which makes them lose interest in studying the subject. Students seem quiet and lack interaction with teachers during the learning process, thereby making it difficult to use the material to achieve the required objectives. One way to overcome the increasing problems is by improving the delivery method using media containing text, images, videos, animations, and sounds, which tends to foster their interest in mathematics.

The teacher plays an important role in the learning process. Therefore, they need to be innovative and creative in using technology to facilitate their understanding of mathematics. Teachers as facilitators are required to be able to present current technological material to make learning more effective and efficient using the media, which are now developing along with the era of globalization.

The smartphone is a learning media, with the ability to facilitate and increase students' interest in learning through characters and visuals designed. According to Kumar, in 2013, mobile learning, also known as m-learning, is conducted using a small or portable computing device, which enables users' to access learning content anywhere, irrespective of the time and place (Kumar, 2013).

Abu bakar and Rahmat syah (bakar and Syah, 2012) argued that learning is more directed at the formation of meaning based on prior knowledge and understanding. In this process, students are actively involved in discovering the meaning of what they are learning in order to create a direct impact on the growth and development of their thinking skills. This is capable of building cognitive knowledge in students independently to enable critical thinking in solving problems related to mathematics.

Based on the research of constructivist models conducted by Cakir (Cakir, 2008) entitled "Constructivist Approaches to Learning in Science and Their Implications for Science Pedagogy: A Literature Review," it is concluded that students need to have the basic knowledge to enable improve the learning process. This contradicts the research conducted by Khalid and Muhammad (Khalid and Azeem, 2012) entitled "Constructivist Vs. Traditional: Effective Instructional Approach in Teacher Education," which concludes that constructivist groups mark higher satisfaction and increase students' participation in learning. They tend to be more willing to answer and ask questions in order to clarify the material and group discussion, which results in the introduction of many new points.

A research conducted by B-Abee Toperesu and Jean-Paul Van Belle: 2018 (Abee and Jean, 2018) titled "Mobile Learning Considerations In Higher Education: Potential Benefits And Challenges For Students And Institutions," showed that the use of mobile learning in higher institutions, as acts as a reference. The research carried out by Jin Xue et al., 2018, (Jin, Xue and Heng, 2018) titled "Effects of Mobile Learning on Academic Performance and Learning Attitude in a College Classroom" showed that mobile learning makes students independent and more active.

The above descriptions illustrated the importance of developing interesting, innovative, effective, and efficient learning media, capable of fostering students' interest.

1.2 Problem formulation

Based on the background described above, the subjective problems associated with this study lies on the question: "Is the learning outcomes using mobile media with a constructivist model better than the conventional trigonometry in class X high school?"

1.3 Research objectives

The purpose of this research and development is to determine if the learning outcomes of mathematics using constructivist models are better than the conventional method using trigonometry material.

2 RESEARCH METHODS

This research was conducted at Senior High School 8 in Semarang from February to June 2018, with data collected from class XA and class XB. The instruments used were 20 multiple choice test questions with a 90-minute time allocation. Before evaluation, the test questions were tested in the trial class (X C) to ensure the test questions met the requirements of validity, reliability, level of difficulty, and good distinguishing power. In addition, the questionnaires were given to material and media experts.

Furthermore, this study utilized the Borg and Gall model comprising of 10 stages as its research design by using only 6 stages, namely (1) Potential and problems, (2) Data collection, (3) Product design, (4) Design validation, (5) Design revisions, (6) Product testing. The test technique was in the form of multiple-choice Trigonometry materials using sine and cosine rules, which are then analyzed using the normality (Lilliefors test), homogeneity test (two variance similarity test), and the right t-test samples. The effectiveness of the results determines the indicator and completeness of the value of each individual learning outcome to analyze the results of the average experimental and control class.

3 RESEARCH RESULTS AND DISCUSSION

Based on the research conducted, the results of the evaluation on media experts are based on 85% learning presentation, 80% language, and 80% graphic feasibility aspects. While the results of evaluations on material experts are 100% general aspects assessment, 100% material substance, and 95% learning design aspects. From the evaluation results on material and validation experts, the average score was 88.57%; therefore, it is concluded that mobile learning media with valid constructivist models is used in the classroom.

Furthermore, it is tested based on effectiveness, using the hypothesis test t one side of the right, the following an average evaluation table of the control and experimental classes.

Based on the table above it is known that the average results of the experimental and control classes are 79,13 and 75,25, respectively, with n1 = 40, n2 = 40 and s-total

Table 1. Average results evaluation of the control and experimental classes.

Class	Average	Total of students	Lcount	Fcount	Varians total	tcount
Control	75,25	40	0,1379			
Experiment	79,13	40	0,1061	1,28	8,17	2,12

= 8,17, while the normality test in the control class is obtained L count = 0,1379 < L table 0,1401 and normally distributed. The experimental class obtained L count = 0,1061 < L table 0,1401 therefore, the experimental class is normally distributed. Furthermore, to test the homogeneity of the two classes using the F test, F count = 1,28 <Ftable = 1,76; therefore, it is concluded that the two classes are homogeneous. The last step is to test the hypothesis using the right-hand t-test, from the two classes t count = 2,09. The results from the distribution list t and dk = 78 and when t 0,95 (78) is 1,67. Therefore, from the calculation t count and t table equals 2,09 and 1,67. Ho is rejected because t count > t table which is 2,09> 1,67.

Based on the calculation of effective criteria using SPSS, it is concluded that Ho is rejected with t count> t table which is 2,09 > 1,67 and the experimental class > control by 79,13 > 75,25, with a Minimum completeness criteria value of more than 75 the learning outcomes using mobile media with constructivist models are better than conventional with trigonometry in class X high school.

4 CONCLUSION

Based on the discussion of the problem, the following is concluded:

1. Mobile learning media developed products with constructivist models in the form of e-book applications using trigonometric for X grade students of high schools.
2. Learning outcomes using mobile media with constructivist models are better than conventional models using trigonometry.

REFERENCES

Abee, T. B. and Jean, P. V. B. (2018) "Mobile Learning Considerations In Higher Education: Potential Benefits And Challenges For Students And Institutions," *14th International Conference Mobile Learning*, pp. 31–38.

bakar, A. and Syah, R. (2012) "Menerapkan Model Konstruktivis untuk Meningkatkan Hasil Belajar Fisika Umum I Mahasiswa Semester I Jurusan Fisika FMIPA Unimed TA 2012/2013," *Jurnal Pendidikan Fisika*, 1(2), pp. 49–54.

Cakir, M. (2008) "Constructivist Approaches to Learning in Science and Their Implications for Science Pedagogy: A Literature Review," *International Journal of Environmental & Science Education*, 3(4), pp. 193–206.

Jin, X., Xue, Z. and Heng, L. (2018) "Effects of Mobile Learning on Academic Performance and Learning Attitude in a College Classroom," *4th International Conference on Advanced Education and Management*, pp. 307–311.

Khalid, A. and Azeem, M. (2012) "Constructivist Vs Traditional: Effective Instructional Approach in Teacher Education," *International Journal of Humanities and Social Science*, 2(5), pp. 170–177.

Kumar, S. (2013) "M-Learning: A New Learning Paradigm," *International Journal on New Trends in Education and Their Implications*, 4(2), pp. 24–34.

Emerging Perspectives and Trends in Innovative Technology
for Quality Education 4.0 – Kusmawan et al (eds)
© 2020 Taylor & Francis Group, London, ISBN 978-0-367-25803-0

Assessment of critical education concepts in the perspective of Islamic education

Saifullah Idris
Ar-Raniry State Islamic University, Banda Aceh, Indonesia

Z.A. Tabrani
Serambi Mekkah University, Banda Aceh, Indonesia

Fikri Sulaiman Ismail
Senior Researcher at SCAD Independent, Indonesia

Amsori
Sekolah Tinggi Ilmu Hukum IBLAM Jakarta, Indonesia

ABSTRACT: This article is an assessment of the concept of critical education that is reviewed in a paradigmatic frame and sees it from the perspective of Islamic education. The basic problems addressed in this article include the relevance, elaboration, and application in Islamic education. Education as a center for the advancement of a nation presents several paradigms that influence and represent the world of education. One such paradigm is the critical education paradigm. As for the concept of Islamic education, it basically emphasizes humanization and liberation as educational orientations and places students and educators as subjects in the learning process. Islamic education combines vertical (spirituality) and horizontal (social) aspects in one educational orientation. This is different from the paradigm of critical education, which places greater emphasis on materialistic matters and less on spiritual aspects. The Islamic education paradigm emphasizes social and togetherness aspects. The orientation of education in Islamic education is to awaken and actualize all the potential possessed by humans holistically. Because education aims to work on the reality of human consciousness, free will, critical reasoning, and creativity, critical education must be methodologically based on the principles of reflection and total action.

1 INTRODUCTION

The problem of education is a very important one and cannot be separated from the whole series of human life. The view that education as a very sacred and noble activity has long been believed by humans (Fakih, 2002: x). Education is not directed to education itself but rather to the achievement of goals and directions for the future. Thus, the time dimension in education is not limited only to the present time, namely when education takes place.

The critical educational paradigm applies critical, creative, and active patterns for students in the learning process. In a critical paradigm, the main tasks of education are to conduct a critical reflection on unjust systems, deconstruct their structures, and advocate for a more just social system (Fakih, in William F. O'Neil, 2002: xv).

The critical education paradigm proposed by Freire contains a very fundamental criticism of the liberal and conservative education paradigm, which has failed to project education as a humanizing process. An implication that results from such a dominant educational paradigm is that its outputs are unable to bring constructive changes to human reality (Fakih, in William F. O'Neil, 2002: xvi).

Such failure also attracts interest from some figures of contemporary Islamic education. Muhammad Iqbal (in Solikin and Anwar, 2005: 110–111), in an effort to formulate an Islamic educational paradigm, criticizes modern Western education for being the culprit behind the prolonged humanitarian crisis. According to Iqbal, the failure occurring in modern Western education is due to the fact that such education only focuses on the aspect of knowledge transformation, without any regard for the aspect of 'isyq or love. He believes that Islamic education includes not only the process of teaching and learning, which transfers knowledge, but also the aspect of integral self-fostering in order to bring humans to their true sense and to the highest level of their humanity.

2 METHOD

This research was a descriptive analysis that employed a qualitative method. A philosophical-pedagogical approach was used. In addition, an interdisciplinary approach was also used so that the research problems could be understood more deeply. The data analysis employed in this research was the content analysis technique. It was presented in the form of descriptive analysis; i.e., the interpretation of the content was made and arranged systematically and holistically by using two methods—deduction and induction.

3 RESEARCH FINDINGS AND DISCUSSION

3.1 *Relevance of the critical paradigm to Islamic education*

According to Muthahhari (2005: 14), the main objective of Islamic education is to create a better society. This is in line with Freire, with his concept of critical awareness, which states that education should make humans become aware of all aspects of their social life and that these aspects should be closely related. Muthahhari (2005: 17) states that one objective of education and teaching is to shape human personality and regulations in the field of law, economy, and politics, which are directly related to the field of education.

Al-Toumy al-Syaibany (1983: 47) also agrees with the concept that Islamic education should be neither untouchable nor alienated from social and cultural reality. Islamic education should be in harmony with the existing cultural, social, economic, and political systems in a society. Such education should not only adjust to what is happening in society, but also position itself as a pioneer, advisor, leader, and critic to such dominant systems.

Education based on the Islamic perspective functions more as a mechanism to maintain, make use of, and develop humans' natural characters (Solikin and Anwar, 2005: 114). Such education contradicts the paradigm of conservative education, which tends to position education as a means of legitimizing social, political, and cultural systems (dominant ideology) that exist in a society. Islamic education, according to Muthahhari (2005: 39), should liberate a human's mind from the limitation of traditions. In other words, it should direct humans to the freedom of thinking.

Muthahhari (2005: 25-26) believes that Islamic education should be aimed at maximizing students' potential for thinking. Teachers, for instance, should work hard to orient their students toward developing the ability to research and analyze, instead of only giving nonmeaningful instructions. Afterward, they should teach them how to make a very crucial conclusion (ijtihad) by referring to relevant sources.

Another important goal of Islamic education is to make humans understand that whatever becomes the decision of common people does not always mean it is the truth to uphold or follow. This is also in line with the characteristics of critical education, which rejects the hegemony of dominant ideologies as an authoritative source of knowledge, norms, and values whose truth is to be absolutely upheld by common people (Fakih, in William F. O'Neil, 2002: xvii).

The paradigm of critical education is also in line with the paradigm of Islamic education in terms of its way of seeing humans and the world. The paradigm of critical education rejects the liberal education paradigm, which believes in separation between humans and the world. In the paradigm of Islamic education, al-Taomy al-Syaibany (1983: 76–77) explains that nature is humans' partner in developing all potentials to improve the humans themselves. Humans and nature are not two contradictory entities. Nature is a source of inspiration and signs that help direct humans to find truth and virtue. Therefore, the paradigm of Islamic education clearly rejects the dichotomy between human beings and nature, which liberal education paradigms uphold.

Finally, both critical and Islamic education position education as a conscientization (developing awareness) process, which will make humans possess critical, reflective, and holistic awareness in interpreting, facing, and solving all problems of life.

3.2 *Critical education and actualization of potential in education*

Islam believes that humans have natural tendencies to recognize religion; have desires, conscience, and self-awareness; conduct self-reflection; and have free will (Tabrani. ZA, 2014: 129). From an Islamic perspective, humans are creatures who have dignity in front of God because they have been blessed with two basic potentials that other God's creatures have not: mind and conscience. Therefore, developing these two potentials may result in intellectual and spiritual dimensions (knowledge and faith) (Al-Farisi, 1982: 12).

Islamic education should finally aim at the shaping of humans based on their natural character including immanent (horizontal) dimension, i.e., the dimension about worldly or physical matters as well as transcendent (vertical) dimension, i.e., the dimension related to faith and spiritual matters that are related to humans' responsibility to the Creator (Ma'arif, et al., 1991: 29–31).

Furthermore, the Islamic education paradigm focuses on social and communal aspects. Although education in Islam also speaks to individuals, that Islamic education is based on the principle of Ibda' bi nafsih (begin with yourself); such a principle does not refer to selfishness but focuses on one's self as the beginning point of an education process. Ibda' bi nafsih has positive consequences in relation to communal interests (Solikin and Anwar, 2005: 115).

In relation to critical awareness proposed by Paulo Freire, Laleh Bakhtiar (2002: 45) maintains that from the Islamic perspective, humans' critical awareness is universal because it is related to three fundamental aspects: humans as servants of Allah (theethics). humans as individuals (psycho-ethics), and humans' awareness of their relations to others (socio-ethics). The balance of these three aspects is centered on theo-ethic awareness, which is the center of humans' awareness.

According to Muhammad Iqbal (2003: 96), Islam is a "religion of action," which directs humans to

1. Critical attitude toward traditions
2. Active attitude toward changing a given reality with full acceptance of their position as Gods co-creator
3. Respectful attitude toward the world as a place for the realization of humans' creativity in performing their duties as God's co-creators

Based on the elaboration given by Iqbal, it can be concluded that critical awareness in Islam is based on the foundation of faith in Allah, the Almighty, as the Creator and goal (end) of life. This is different from the paradigm of critical education, which only stresses the shaping of humans' critical awareness, taking into account only aspects of psycho-ethics and socio-ethics, but not the aspect of theo-ethics.

Muhammad Iqbal often criticizes the system of modern Western education, which tends to ignore the aspect of human faith and spirituality, often resulting in moral and spiritual crises, especially among Moslems. That is why Iqbal proposed an Islamic education system to address the gap left by the Western education system.

Sociologically, the main point of liberating humans through the paradigm of critical education is based on the assumption that humans have an equal position. It is based on the assumption that there is no dichotomy or hierarchical difference between students and teachers in this paradigm (see Freire, 2004: 176). Seen from a sociopolitical perspective, the paradigm of critical education believes that education should be oriented toward ensuring the equality of rights and the absence of oppression among humans. This way, the objective of education in the paradigm of critical education is to liberate humans from oppression and hegemony.

In Islam, the principle of social equality is a fundamental one. Horizontally, social stratification is largely unacceptable, which means that sociologically, no one's dignity is above others'. This way, there should be no reason for anyone to oppress others. Human stratification in Islam is only acceptable in the spiritual (vertical) perspective, with righteousness determining one's dignity in front of Allah. Humans' dignity seen from such a spiritual perspective does not give them any authority to oppress and discriminate others because righteousness is an abstract variable, the quality of which is only known to Allah the Almighty.

3.3 Critical education paradigm as inspiration for development of Islamic education

The implementation of an education paradigm in the teaching–learning process is the main requirement in achieving the desired education objective. Islamic education does not always reject any ideas that come from outside Islam. Therefore, the Islamic education paradigm is not a paradigm that should be contrasted to other secular education paradigms (Solikin and Anwar, 2005: 128).

Although the paradigm of critical education belongs somewhat to the paradigm of secular education, the learning process portrayed in critical education can still be a methodological reference for Islamic education in formulating a humanistic learning process. It can support Islamic education in achieving its educational objectives.

The paradigm of Islamic education also clearly rejects the patterns of liberal and conservative education, which Freire refers to as education with "bank style." In the paradigm of Islamic education, students are not a means of investment, which can be described as fruitful later. The paradigm holds that education should not be an arena of indoctrination to legitimize and prolong any social, political, and economic structures that are oppressing. However, one thing to pay attention to in Islamic education is tarbiyah al-muslimin (education for all Moslems) and tarbiyah 'inda al-muslimin (education among Moslems) (Muhaimin et al., 2002: 36).

4 CONCLUSION

Both critical and Islamic education paradigms have relevance in their educational orientation and process. They both focus on humanization and liberation as educational orientations. Both also regard students and teachers as a subject in the teaching and learning process.

Islamic education paradigm bases its ideas, objectives, and process on spiritual foundation and faith in Allah and its prophet; it combines vertical (spiritual) and horizontal (social) aspects as its educational orientation. This is in contrast to the critical education paradigm, which only focuses on material matters. This paradigm does not pay much attention to the spiritual aspect, which is actually the most sublime aspect of human beings.

The method of critical education can be implemented in Islamic education as an effort to improve Islamic education in general. It is hoped that this will result in education outputs that are able to bring improvement to Islamic civilization. Critical-constructive contents in the critical education paradigm can also act as a methodological reference in implementing Islamic education.

REFERENCES

Al-Toumy al-Syaibany, Omar Muhammad (1983) *Falsafaut al-Tarbiyah al-Islamiyah*, translated by Hasan Langgulung with the title of *FalsafahPendidikan Islam*. Jakarta: BulanBintang.

Bakhtiar, L. (2002) *Meneladani Akhlak Allah: MelaluiAsma al-Husna*. Bandung: Mizan.

Fakih, M. (2002) Ideologi-ideologiPendidikan. In William F. O'Neil, *Educational Ideologies: Contemporary Expression of Educational Philosophies*, translated by Omi Intan Naomi with the title of *Ideologi-Ideologi Pendidikan*, 2nd ed. Yogyakarta: Pustaka Pelajar.

Freire, P. (2004) *The Political of Education: Culture, Power, and Liberation*, translated by Agung Prihantoro and Arif Yudi Hartanto with the title of *Politik Pendidikan: Kebudayaan, Kekuasaan, dan Pembebasan*. Yogyakarta: PustakaPelajar, Cet. V.

Iqbal, M. (2002) *The Reconstruction of Religion Thought in Islam*, translated by Ali Audah, et. al with the title of *Rekonstruksi Pemikiran Agama dalam Islam*. Yogyakarta: Jalasutra.

Ma'arif, Ahmad Syafi'i, et al. (1991) *Pendidikan Islam di Indonesia; Antara CitadanFakta*. Yogyakarta: Tiara Wacana.

Muhaimin, et al. (2002) *Paradigma Pendidikan Islam*, 2nd ed. Bandung: Remaja Rosdakarya.

Muthahhari, M. (2002) *al-Fitrah*, translated by Muhammad Jawad Bafaqih with the title of *Fitrah*. Jakarta: Lentera Basritama.

Muthahhari, M. (2005) *Tarbiyatul Islam*, translated by Muhammad Bahruddin with the title of *Konsep Pendidikan Islami*. Depok: Iqra Kurnia Gumilang.

Saifullah. (2017) Learning by conscience as a new paradigm in education. *Advanced Science Letters*, 23(2), 853–856. doi: 10.1166/asl.2017.7447.

Solikin, M and Anwar, R. (2005) *Hakekat Manusia: Menggali Potensi Pendidikan Kesadaran Diri dalam Psikologi Islam*, 1st ed. Bandung: PustakaSetia.

Tabrani ZA. (2014). Islamic Studies dalam Pendekatan Multidisipliner (SuatuKajian Gradual Menuju Paradigma Global). *Jurnal Ilmiah Peuradeun*, 2(2), 127–144.

Emerging Perspectives and Trends in Innovative Technology for Quality Education 4.0 – Kusmawan et al (eds)
© *2020 Taylor & Francis Group, London, ISBN 978-0-367-25803-0*

Enhancing learners' speaking skills by using multimedia: An online learning perspective

Lidwina Sri Ardiasih & Juhana
Universitas Terbuka, Tangerang Selatan, Indonesia

ABSTRACT: English speaking is one of the four language skills categorized productive for EFL learners. Some effective techniques are important to be considered to support the online speaking learning process. Therefore, the use of multimedia is expected to be an effective technique to support EFL learners' learning process and give them meaningful learning experiences. PBIS4306/Speaking III is one of the compulsory courses offered to EFL learners at Universitas Terbuka (UT). Moreover, online learning is provided to support learners practicing English speaking. A descriptive analysis was conducted by using an observation sheet to describe the process of online English speaking online tutorials. This paper is aimed at discussing the importance of various multimedia to enhance EFL learners' speaking skills and describing how far the theory of using multimedia has been implemented. The results show that the use of multimedia, particularly through open learning sources (OER), has proven to help participants improve their speaking skills.

1 INTRODUCTION

The development of information and communication technology (ICT) requires educators to be more creative and innovative, but selective in providing an effective and enjoyable learning atmosphere so that students can gain meaningful learning experiences (Simonson *et al.*, 2011). Online learning is one the embodiment of the principles borderless educating people that can be taken by educators to enrich teaching materials as well as learning activities. Besides, it enables to provide clear course structure, various presentation of material, collaboration and interaction, and timely feedback (Lister, 2014). Therefore, the use of online platforms enables teachers to provide various kinds of online learning materials, including audios, videos, or combinations of different media known as multimedia.

PBIS4306/Speaking III is a compulsory subject for English Education Study Program at Universitas Terbuka (PBIS-UT) that utilizes Moodle Learning Management System (LMS) services for online tutorials (tuton)(Universitas Terbuka, 2010). This program allows tutors to provide materials with a variety of media such as text, video, audio, animation, or a combination of several media known as multimedia. This study is important for finding out how far the use of multimedia can boost EFL learners' practicing language skills, particularly speaking skills. Based on the significance, this paper is aimed at discussing the importance of various multimedia to enhance EFL learners' speaking skills and describing how far the theory of using multimedia has been implemented.

2 SPEAKING SKILL AND ITS IMPROVEMENT STRATEGIES

Speaking is one of the four important skills that people have to master when they learn a language. Speaking is considered an active skill, or in other words, it is the action of conveying information or expressing one's thoughts and feelings in spoken language (Oxford, 2009).

In learning a language, especially in terms of speaking skills, the learners have to be active in practicing to use the language. Hurd (Hurd, 2000) suggests that concerning strategies for developing speaking skills, the learners have to take the initiative, through becoming involved in self-help groups, attending conversation classes, going to the movies or recording them off-air, and other activities that enable them to enhance their speaking abilities(p. 74).

Delivering some learning strategies in enhancing speaking skills, with many activities teachers do as part of their course, are intended to prepare EFL for communication by focusing on specific aspects of the foreign language, i.e. how to use words or phrases appropriately, the applied grammar or structure, or their pronunciation (Adinolfi et.al., as cited by (Hurd and Murphy, 2005)). Moreover, they emphasize that to communicate effectively in the language, and the learners have to use contextual situations. It means that learners also need practice in coping with real-life tasks, that is, situations that involve using language in context. Another speaking proficiency, to be practiced by EFL learners, offered when learning a language is pragmatic ability. Moreover, Cohen, as cited by Hurd (Hurd and Lewis, 2008), explains that pragmatic ability is one skill needed by students and that it enables people to carry out specific social functions in speaking, such as apologizing, complaining, making requests, etc. He offers strategies for learning and performing EFL Pragmatics by giving a taxonomy of speech act strategies, for example (1) taking practical steps in order to gain knowledge about how specific speech acts work, (2) asking native-speakers to model performance of the speech acts as they might be realized under various conditions, and practicing the speech act's aspects of performance that have been learned. Using technology in education is both demanded and expected that the schools are developing and applying curriculum and media that enable its learners to study a language by using technology, including for a speaking course.

Based on the use of multimedia to enhance learners' speaking skills, there are two benefits of using a video program for students (Secule, Herron and Tomasello, 1992). The first video permits EFL learners to witness the dynamics of interaction as they observe native speakers through authentic settings speaking and using different accents, registers, and paralinguistic cues, such as posture and gestures). Second, video allows students to gain knowledge from native speakers through their authentic conversation backgrounds, i.e., the dynamics of interaction such as different accents and various languages. Moreover, the benefits of learning a foreign language via video are to engage a native speaker in real conversations in order to create a greater ability to understand native speakers in real conversations and to enhance the students' ability to speak (Terrel, as cited by (Coniam, 2001)). It means that the influence of using video media to involve students in real situations in a conversation is significant in improving students' speaking skills.

3 THE IMPLEMENTATION OF USING MULTIMEDIA FOR ENHANCING LEARNERS' SPEAKING SKILLS

This research applied a qualitative approach by using descriptive analysis in order to gather some theories and information about the use of multimedia to enhance EFL learners' speaking skills. Moreover, an observation was conducted in order to capture the process of online learning for speaking class, i.e., how far the tutor applied using multimedia in her class and what are the impacts of using multimedia in enhancing EFL learners' speaking skills.

PBIS4306/Speaking III is one of the compulsory courses in the PBIS-UT Study Program. An online tutorial (tuton) is one form of teaching assistance provided by UT in order for students to obtain learning assistance services through the use of information technology. Tuton allows students to interact with tutors and other students online. Moreover, the aim of the tuton service that utilizes the Moodle service is to facilitate students with the help of learning or tutorials in which tutors can provide additional material in the forms of various media, such as PowerPoint or Word, links from open source resources (Open Educational Resources/ OER) such as Youtube videos, articles from online journals, etc.) that support students' learning the material discussed in tuton. In other words, this program enables the tutors to provide material with a variety of media genres, such as text, video, audio, animation, or a combination of several media known as multimedia(Universitas Terbuka, 2010).

Based on the observation, the learning activities at tuton of the Speaking III Course, conducted from February 23, 2016, up to April 24, 2016, were carried out for eight weeks of asynchronous meetings. This means that when the tutor uploads material, discussion forums, and assignments along with the reference assessment (band-descriptor), students do not have to respond directly but can prepare themselves in advance either by reading the material or discussing with other participants. Regarding teaching material, tutors have compiled a tutorial activity design with the topics as described in the following table.

Table 1. Discussion topics of PBIS4306/Speaking III.

Weeks	Time	Materials/Topics
	23/2/16 - 28/2/16	Introduction
I.	29/2/16 - 6/3/16	Formal and informal communication
II.	7/3/16 - 13/3/16	Public Speaking
III.	14/3/16 - 20/3/16	Group discussion
IV.	21/3/16 - 27/3/16	Panel Discussion
V.	28/3/16 - 3/4/16	Debate
VI.	4/4/16 - 10/4/16	Seminar
VII.	11/4/16 - 17/4/16	Symposium
VIII.	18/4/16 - 24/4/16	Another communication forum

These topics are the essence taken from the whole subject matter of the subject matter discussed in the printed teaching material, which is equipped with an integrated video program. Regarding the importance of material on public speaking such as seminars, symposiums, or in other formal academic settings, the PBIS4306/Speaking III course tutor provides theoretical knowledge supplemented with examples through varied multimedia programs.

Based on the multimedia presentation function on the tuton, Simonson, Smaldino, and Svacek (Simonson et al., 2011) explain four key stages of success in online classes that use technology in learning, specifically in the process of selecting the right type of technology, namely 1) testing appropriate instructional technology, 2) determining learning outcomes (competencies to be achieved), 3) identifying learning experiences and adapt it to available technology, and 4) prepare learning experiences to be presented online. The four stages can be a reference for tutors to provide appropriate material with meaningful activities for participants. Instructional technology should support the learning process of students so that the understanding of the material being studied can be truly maximized. Furthermore, if the technology is available but without clear objectives or competencies, the learning process is not successful. Tutors should refer to the main outcomes or competencies that must be achieved by students. The learning activities offered must also be accessible and acceptable. The choice of material or activities using an internal network needs to consider whether the source is easily accessible and acceptable for students so that the learning experience of students is meaningful.

4 CONCLUSION

The use of multimedia genres becomes an alternative that can be taken by tutors in online learning in terms of delivering the presentation and its flexibility. However, the use of multimedia needs deep considerations in order to give students meaningful learning experiences. In speaking skills, videos that are included in the multimedia category are the right media to help students practice speaking. Multimedia-Based online learning can be a reference in providing students with interesting and meaningful learning activities. The results of the observation on PBIS4306/Speaking III tuton have shown that the use of multimedia, particularly through open educational resources (OER), has proven to help participants improve their speaking skills.

REFERENCES

Coniam, D. (2001) "The use of audio or video comprehension as an assessment instrument in the certification of English language teachers: a case study," *System*, 29(1), pp. 1–14. doi: doi.org/10.1016/s0346-251x(00)00057-9.

Hurd, S. (2000) *Distance language learners and learner support: beliefs, difficulties, and use of strategies.*

Hurd, S. and Lewis, T. (2008) *Language learning strategies in independent settings.* Clevedon: Multilingual Matters.

Hurd, S. and Murphy, L. (2005) *Success with language.* New York: Routledge.

Lister, M. (2014) "Trends in the design of e-learning and online learning," *Journal of Online Learning and Teaching*, 10(4), pp. 671–681.

Oxford, T. (2009) *The Oxford pocket dictionary of current English.*

Secule, T., Herron, C. and Tomasello, M. (1992) *The effect of video context on foreign language learning.* doi: doi/10.1111/j.1540-4781.1992.tb 05396.x/.

Simonson, M. *et al.* (2011) "Teaching and learning at a distance," in *Foundation of distance education.* 5th ed. Boston: Allyn and Bacon.

Universitas Terbuka (2010) *Rencana Strategis Universitas Terbuka 2010-2021.*

Emerging Perspectives and Trends in Innovative Technology
for Quality Education 4.0 – Kusmawan et al (eds)
© 2020 Taylor & Francis Group, London, ISBN 978-0-367-25803-0

The impact of online games on language aggression behavior of adolescents in warnet

Rika Sa'diyah & Ati Kusmawati
University of Muhammadiyah Jakarta, Indonesia

ABSTRACT: This study aimed to determine the impact of online games on language aggression behaviors in adolescents at an internet cafe, using observation of four internet cafes in South Tangerang. The results indicated the main languages and dirty words that are used and appear as mockeries among internet cafe users, including "monkey," "dog," "pig," "turmeric," and others. This demonstration is known to highly influence the development of adolescent language at school and at home, as it promotes the habit of directing dirty and unethical words both to fellow friends and older people if these characteristics are left unchecked. Thus the family, especially the parents, has a major role to play, through control attempts, while the school as an educational institution should assist the community to instill character and morals in society.

Keywords: adolescents, language aggression behavior, online games

1 INTRODUCTION

There is an upsurge in the development of information technology, leading to a limitation in the ability to control everything. This is evidenced by the use of mobile phones from an early to older age, as well as the installation of internet cafes in the neighborhood, enabling patrons to surf cyberspace and therefore see anything that is not widely known. In addition, the negative and positive effects resulting from the development of electronic media, encompassing cellphones and internet cafes, ought to be understood, as these tend to be a big concern for teenagers, especially in terms of instability as indicated by the impact on society. Also important in this consideration are online games, which are highly loved by children and adults and very interesting for public consumption, with a strong capacity to lead to addiction. Furthermore, they are also perceived as an interesting form of entertainment during holidays and at school time, as children tend to forget the time needed to pray, eat, and go home. Moreover, the avidness of the consumers for the games represents a soft consumption for businesspeople who are least concerned with moral values, education, and culture. Also, the reduced capacity to provide really educative games is seen in the promotion of violence (physical or verbal), and pornography; thus, a majority of adults, owing to a lack of understanding or care, tend to finally "let" their children play. In addition, there is also a provision of more varied types, through the payment of about Rp. 3,-000–10,000/day, which accords the privilege to a broader access. Meanwhile, an addiction to these games tends to disrupt the nerves and brains of children, leading to a reduction in the will to learn, a quickness to anger, difficulty in managing, and also the use of bad language, which is a major concern for educators and parents.

At the present time, teens have been identified as the objects of the social media world, based on the fact that the internet is an attractive medium for the age group, possessing a great deal of information. This is perceived as an interesting thing, and playing an online game, unfortunately, is conducive to the emergence of impolite and unpleasant language. In addition, the internet cafe environment tends to promote this demeanor and has also become accustomed to mention of such words; thus, aggression is developed as a habit that is usually conveyed into the environment outside the cafe. This particular characteristic involves the act of a creature attacking and hurting others, and for humans, this trait is exhibited verbally, although attention is focused on the physical/nonverbal form. According to Zakiah Darajat (1990: 23), adolescence is a transitional period between childhood and adulthood, during which the child experiences growth and physical and psychological development. Thus, based on physical features, thoughts, and behaviors, adolescents are referred to neither as children nor as mature adults. Further, the obstacles experienced by older people today are enormous, especially with the advent of technology, encompassing online games, as adolescent language development greatly influences daily behavior, both at home and within the environment. These characteristics prompted an interest to study the problem of language aggression in adolescents at internet cafes after playing online games.

2 AGGRESSION AS A RESPONSE AND AS AN EMOTIONAL REACTION

The study of aggression is possibly obtained through observation and imitation, which is increased by an enhancement in reinforcement. This behavior tends to be a dysfunction at extreme points, and Bandura (in Fiest, 2009) reported that it is adopted through the observation of others, alongside a direct experience with negative and positive reinforcement, practice or instruction, and abstract beliefs.

Five reasons have been identified for initiation of aggression: (1) pleasure in hurting victims (positive reinforcement); (2) the unintended consequences of aggression by others in response to avoidance or fighting (negative reinforcement); (3) becoming hurt or receiving an injury in order not to commit aggressive behavior (punishment); (4) setting the behavior as a personal standard for one's actions (self-reinforcement); and (5) seeing others receive rewards for aggressive actions or penalties for nonaggressive behavior.

Bandura (in Feist, 2009) believes these actions have the ability to result in the continuity of aggressiveness, proven by the statement that children observing the behavior of others show a higher tendency to act likewise than those in the control group that is not exposed to such demeanor.

According to Ana Paula et al. (2013), these traits affect the psychosocial needs caused by low education, reduced income, lack of information regarding prevention and promotion of health, and difficulties within the family and society. The subjects in this investigation included teenage girls and nurses at Campo Grande Mato Hospital Grosso do Sul state, Brazil. Ruphina Anyaegbu et al. (2012) explained that "games are often used for motivational or fun purposes." This statement, therefore, reinforces the fact that games are rarely applied to motivate or achieve goals.

Aggression, according to Freud (in Atkinson, 2000), is a basic instinct and is encouraged by frustration (particularly assumed to be the obstacles in an individual's efforts toward achieving a goal). These aggressive impulses are the dominant response that motivates the behavior designed to hurt people or objects that are responsible for frustration, although possible reactions, including emotion, always require punishment. Furthermore, the psychoanalytic theory proposed by Freud (in Atkinson, 2000) views aggression as the impetus, which is interpreted as a learned response according to social learning theory. In addition, it has been explained as encompassing behaviors that are intentionally aimed at hurting others (physically and verbally) or destroying property, involving the following as an emotional reaction.

2.1 Aggression as an impulse

According to Freud's psychoanalytic theory, numerous actions are determined by instincts, especially those that are sexually associated, thus raising the drive in instances in which the expression is not satisfied (frustration). In addition, experts extend the hypothesis with the following statement: Blocking an individual's effort to achieve a goal triggers the generation of an aggressive impulse, which motivates the behavior to break down causative barriers (people or things) (Atkinson, 2000).

2.2 Aggression and social information processing

What makes children behave aggressively? One of the reasons is related to the pattern with which social information is processed, influenced by the characteristics of the environment, and the means through which feelings are interpreted (Erick and Dodge, 1994, 1995).

Teenagers who play online games in internet cafes follow and imitate the environment, including spoken language. The perception of their new environment provides satisfaction, subsequently encouraging the incidence of aggressiveness, alongside the behavior of continuously using dirty words when at home.

Therefore, it is possible to interpret aggression as the behavior resulting from observations and direct experiences, with the aim of hurting others. This occurrence has been attributed to a low level of education, minimal income, lack of awareness of information, family difficulties, and also the nature of the social environment. There is also negative and positive reinforcement from others associated with the acceptance of these aggressive expressions. This study therefore establishes the existence of a high environmental impact on teenagers playing online games in internet cafes. This location is known to be a center for entertainment, especially for those who don't receive enough attention or interaction with friends, and also individuals who are bored with the unattractive conditions at home, which prompts the visit and subsequent invitation of peers to the café, followed by addiction. Meanwhile, the center of attention in this investigation is the amount of abusive and unfavorable language practiced by the participating individuals, also known as language aggression. This has been identified to confer a negative impact in environments outside the cafe through the use of abusive words, including "monkey," "dog," "idiot," "stupid," "asu," etc., which ought to be studied deeper.

The language raised by adolescents at the internet cafe was terrible, as seen in the aforementioned example. Thus, expert opinion on language and its acquisition, especially in adolescents, stipulates the need to focus on the attention felt directly on the behavior and the relationship between stimulus and the child's response, creating the expectation of behavior each time there is a trigger. In addition, adolescent languages tend to change after frequent hangouts in places like the internet cafes, which is a very prominent environment for teens' development, especially for those who do not obtain sufficient parental attention. This is, however, a very bad atmosphere for language growth, and also teenagers who recurrently visit to play online games tend to develop an addiction. In addition, offering online gaming facilities demands the presence of other teens, which fosters comfort, subsequently promoting the tendency to forget duties and responsibilities. This is a distinctive concern for parents with teenage children; hence, there is a need for special attention because adolescence is a period during which an individual seeks an identity and is also in need of assistance from parents, teachers, and the community in order to initiate a successful life.

The results obtained, using observations in four internet cafes located in South Tangerang, show the alarming impact of online games on aggression behavior in terms of language, especially dirty words, in adolescents, which is very worrisome. A proposed solution involves controlling the adolescents, especially with regard to dirty utterances, through the assistance of the family and school.

REFERENCES

Ali, Muhammad, Moh. Asrori (2010) *Adolescent psychology development of students*. Jakarta: Earth Literacy.

Atkinson, R. L. (2000) *Introduction to psychology*, Vol. 2. Batam: Interaction.

Crick, N. R. and Grotpeter, J. K. (1995) Relation aggression, gender, and Social-Psychological adjustment. *Child Development*, 66, 710–722.

Deswita (2008) *Child development*. Bandung: Rosda Karya.

Erik and Dodge. (1995) Review and reformulation of social information-processing mechanisms in children's social adjustments. *Psychological Bulletin*, 115, 74–101.

Feist, J (2009) *Personality theory*. Jakarta: Salemba Huanika.

Gleason, J. B. and Bernstein Ratner, N., eds. (1998) *Psycholinguistics*, 2nd ed. New York: Harcourt Brace College Publishers.

Gunarsah, S. D. and Gunarsah, Y. S. D. (1995) *Child and adolescent psychology*. Jakarta: Mr. Gunung Mulia.

Moleong, L. J. (2006) *Qualitative research methods*, revised edition. Bandung: Teen Rosdakarya.

Monks, F. J., Knoers, A. M. P., and Siti Rahayu Haditomo (1998) *Developmental psychology: Introduction to the various parts*. Yogyakarta: Gadjahmada University Press.

Motivation in an EFL classroom in Chinese Primary School (2012) *TOJET The Turkish Online Journal of Educational Technology*, 11 (1).

Paula de Assis Sales da Silva, A (2013) Aggression inflicted on an adolescent resultant: Imbalance in basic human needs. *Revista de Pesquisa Cuidado'e* (Fundamental Online. doi: 10.9789/2175-5361. 2013v5n2p3749.

Ruphina Anyaegbu, Wei Ting, Yi Li. Serious game. Nanjing Normal University, China.

Sugiyono (2007) *Educational research methods: Quantitative, qualitative, and R & D approaches*. Bandung: Alfabeta.

Emerging Perspectives and Trends in Innovative Technology
for Quality Education 4.0 – Kusmawan et al (eds)
© *2020 Taylor & Francis Group, London, ISBN 978-0-367-25803-0*

Gender differences in academic stress among psychology students

D.K. Dewi & I.E.P. Ningsih
Universitas Negeri Surabaya, Indonesia

ABSTRACT: Learning activities bring much pressure for psychology students. This study therefore aimed to analyze differences in academic stress experienced by psychology students of both sexes at Surabaya State University. The method of data collection used in this study is a questionnaire. There were a total of 116 participants in this study, including 17 male students and 99 female students. The data collected in this study were analyzed using the variance (ANOVA) 1 path. The results showed that there was no difference in academic stress experienced between the sexes. Furthermore, this research aimed to identify the level of stress in students and help overcome the effects of academic stress.

1 INTRODUCTION

1.1 *Academic stress*

Stress can happen to everyone, regardless of age, condition, and socioeconomic status (American Occupational Therapy Association, 2007), including students. Based on research, students are a group of individuals who have a higher risk of stress (Pelletier and Laska, 2013); first-year students are the most vulnerable to stress (Wintre and Yaffe, 1998), and stress conditions increase as a result of the transition to campus life (Towbes and Cohen, 1996).

Stress is a negative or destructive reaction to a pressure (Pandya et al.,, 2012). Stress occurs when individuals feel there is a difference in perception between environmental demands (stressors) and individual capacity to meet these demands (Pierceall and Kim, 2007; Hamaideh, 2009; Al-Samadani and Al-Dharrab, 2013), and individuals feel unable to cope with it (Nkem, 2015).

Stress levels vary, depending on socioeconomic status. Examples include age, life stage, gender, race, ethnicity, employment status, and economic class. (Pearl in, 1989; Keady, 1999). Stress that occurs in an academic environment and relates specifically to academic demands is called academic stress (Dussellier et al., 2005). Academic stress is a unique condition explicitly experienced by students.

Some students experience stress due to a shift in school cultures and the necessity to adapt quickly to a new environment. If unhandled, it could very possibly result in a deterioration in academic performance (Brough, 2015; Dwyer and Cummings, 2007), and even a decrease in achievement index (Wintre and Yaffe, 1998). Besides the aforementioned conditions, stress for students impacts on depression (Dyson and Renk, 2006), eating disorders (Costarelli and Patsai, 2012), and obesity (Nelson et al., 2008).

This study therefore aimed to determine differences in academic stress experienced by psychology students of both genders.

2 METHOD

2.1 *Participants*

The research subjects were students majoring in psychology, class of 2017. A total of 116 students who came from 5 classes were involved in the study. The details are shown in Table 1.

Table 1. Descriptive statistics.

Gender	Class	Mean	Std. deviation	N
Male	A	95.75	13,500	4
	B	99.33	14,503	33
	C	84.60	7,701	5
	D	101.00		1
	E	98.00	9,201	4
	Total	93.94	11,486	17
Female	A	97.04	9,906	24
	B	99.79	9,004	19
	C	91.94	11,532	18
	D	91.13	12,691	16
	E	98.14	14,167	33
	Total	95.93	11,816	99
Total	A	96.86	10,201	28
	B	99.73	9,463	22
	C	90.35	11,097	23
	D	91.71	12,519	17
	E	98.12	13,370	26
	Total	95.64	11,740	116

Based on the descriptive data in Table 1, there is a significant difference between the number of male and female students. Therefore, homogeneity tests need to be conducted to observe any overlapping characteristics between the two groups.

2.2 Homogeneity test

Table 2. Homogeneity test results.

	Fcount	df1	df2	Sig.	Description
Stress results	0.861	9	106	0.563	Homogeneous

Based on the homogeneity test, it can be seen that the significant value in stress data is 0.563, and the conformity data is 0.383. Because the significance value produced in each value group is greater than 0.05 ($p > 0.05$), it can be concluded that there is little variance between the data of each value group (homogeneous result).

2.3 Data

Descriptive statistics describe the basic features of the data collected in a study. Data are analyzed and summarized in a meaningful way such that significant patterns might emerge from the data. This statistics method does not explicitly seek or explain interrelationships, test hypotheses, make predictions, or come to conclusions before the analysis of data collected.

3 RESULTS AND DISCUSSION

Based on the results shown in Table 3, the significance value obtained was 0.970, and the F value calculated was 0.001. The Ftable value for df1 = 1 and df2 = 106 with a significance

Table 3. Data analysis results.

Source	Type III of sum of squares	df	Mean square	F	Sig.
Corrected model	1785,775	9	198,308	1,494	,159
Intercept	399501,133	1	399501,133	3010,598	,000
Gender	,182	1	,182	,001	,0970
Corrected total	15850,793	115			,094

level of 5% of 3,931. As the significance value obtained is greater than 0.05 ($p > 0.05$) and the F value calculated <Ftable, it can be concluded that there is no difference in the average stress value based on sex.

This study was conducted in 2018. The subjects of research, class of 2017, had already experienced a year of university education at the time of their completion of the questionnaire. Students in the early stage of university education experience higher stress than students in subsequent stages (Rafidah et al., 2009). New students experience higher stress than returning students (Amponsah and Owolabi, 2011).

The study is in line with research conducted by Yumba (2008), which states that there is no difference in the perception of stress among male and female students. Differences in stress perception only occur in stressors. Female students experience higher stress levels when dealing with increasing assignments, increased study hours, financial depletion, quarrels with lovers, a perceived lack of social and institutional support, a decrease in self-discipline, and a perceived deterioration in health and well-being (Yumba, 2008).

3.1 Conclusions

This study aimed to investigate the differences in levels of academic stress experienced by both genders. Based on data analysis using an ANOVA 1 path, no difference in stress levels was found between male and female students. A suggestion for further research are to equalize the number of subjects between male and female students.

REFERENCES

Amponsah, M., Owolabi, H., and Brough, K. (2015) Factors associated with college students' perceived stress, 153–169. Available at: www.sciencedomain.org.
Brough, K. (2015) Factors associated with college students' perceived stress.
Dwyer, A. L. and Cummings, A. L. (2007) Stress, self-efficacy, social support, and coping strategies in university students. *Canadian Journal of Counselling and Psychotherapy/Revue canadienne de counseling et de psychothérapie*, 35(3), 208–220.
Nkem, D. (2015) Academic stress among home economics students in higher education: A case of colleges of education in Nigeria. *IOSR Journal of Research & Method in Education Ver. II*, 5(6), 2320–7388. doi: 10.9790/7388-05624958.
Pelletier, J. E. and Laska, M. (2013) Balancing healthy meals and busy lives: Associations between work, school and family responsibilities and perceived time constraints among young adults. *Bone*, 23(1), 1–7. doi: 10.1038/jid.2014.371.
Kamarudin, R., et al. (2009) The impact of perceived stress and stress factors on academic performance of pre-diploma science students: A Malaysian study. (January).
Wintre, M. and Yaffe, M. (1998) First-year students adjustment to university life as a function of relationship with parents.pdf.
Yumba, W. (2008) Academic stress: A case of the undergraduate students. Available at: http://www.diva-portal.org/smash/get/diva2:556335/fulltext01.pdf

Conceptual framework used to improve the problem solving integration for kindergarten children

Nuraida
Universitas Negeri Jakarta (UNJ), Indonesia
Universitas Islam Negeri (UIN) Syahid Jakarta, Indonesia

Fasli Jalal & Rusmono
Universitas Negeri Jakarta, Indonesia

ABSTRACT: The aim of this research was to design an instructional problem-solving model to increase the moral development of kindergarten students. Many studies have been previously conducted on instructional models; however, none have been carried out for kindergarten. The David Jonassen 1997 design model was adopted, with the problem-solving steps as follows: (1) problem identification, (2) clarification of an alternative perspective, (3) generation of possible solutions, (4) assessment of its validity, (5) monitoring of the problem space, (6) implementation and monitoring of the solution, and (7) adaptation. These steps are adopted in Indonesia to design the right teaching curriculum for kindergarten children. The result showed that proper implementation of the problem-solving models has the ability to increase moral development.

Keywords: instructional design, kindergarten, moral development, problem solving

1 INTRODUCTION

Problem-solving is a fun learning technique for students because they are accustomed to finding solutions to problems with their friends and trying out ideas. In a pleasant atmosphere, it is very important to insert moral messages to ensure their impact is felt by the children. This needs to be carried out considering the results of a study that explains that many moral problems occur in early childhood (Counseling et al., n.d.).

According to some studies, problem-solving has been developed by researchers using questioning techniques that are proven to be effective in obtaining answers to the various problems faced by children (Lagare, 2013), improving their social competence and developing language skills (Dennis and Stockall, 2014), enhancing analytical and design skills (Akcaoglu, 2014), improving problem solving abilities (Yu et al., 2015), and enabling critical thinking ability through challenging questions and meaningful experiences in environments that challenge their safe zones (Beavers et al., 2017).

Research on problem-solving has been carried out by several researchers. For instance, Mete Akcaoglu (2014), conducted a study on learning problem-solving by designing games and a learning summer program. Students are assessed according to analysis, design, decision-making, and problem-solving at the beginning and end of the program. Mete's results indicate a significant increase in problem-solving ability after following the program conducted during the summer. In contrast to the aforementioned analysis, this study develops a module for religious and moral values based on research and development (Borg and Gall, 1983). Research based on problem-solving is very useful for kindergarten students to train creative thinking at an early age. Furthermore, it is integrated with morals, thereby making it useful for good character formation at an early age (Counseling et al., n. d.).

According to Kim, Yeon Ha (2016), the problem-solving process consists of five steps: (1) freedom of opinion, (2) analogy or parable, (3) step analysis, (4) description, and (5) continuation. David H. Jansen in *Instructional design models for well and ill-structured problem-solving learning outcomes,* explained it as follows: (1) review several required components, rules, and principles; (2) prepare a concept model; (3) give examples of job performance, (4) present practical problems, and (5) provide supportive and resolution ways (Jonassen, 2000).

In Kathryn Warner's (2016) book titled *Math problem-solving prompts,* the use of problem-solving was explained based on the following: (1) children's ability to work together respective of their strategies, (2) student-centered, (3) challenges associated with assignments, (4) empowerment, and (5) fun (KindergartenKindergarten.com)

James F. Voos and Terry R explained the usage of problem-solving in recent research in the fields of physics and social science (such as Larkin, McDemont). There are several reasons associated with its study, such as (1) to examine its expansion in social science, and (2) the reason for its social science relationship, which is an irregular problem, thereby leaving few people interested (Ritman, 1965; Simon, 1973). Currently, problem-solving is associated with economics, political science, and social science, with several domains used to improve the teaching method.

The following steps were utilized in making a general framework of the information process, describing the problem-solving model clearly, and discussing the characteristics of social science (Vast)

2 RESEARCH METHODOLOGY

The method used to develop the learning model is a modification of the Borg and Gall research model as well as the Dick and Carey instructional design in four development steps as follows: (1) the preliminary study in the form of needs analysis; (2) the planning stage used to develop the learning model; (3) testing, expert evaluation, and product revision; and (4) model implementation as shown in Figure 1.

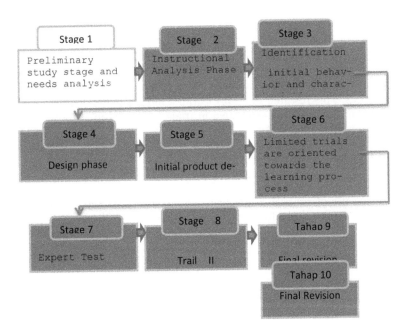

Figure 1.   Development step.

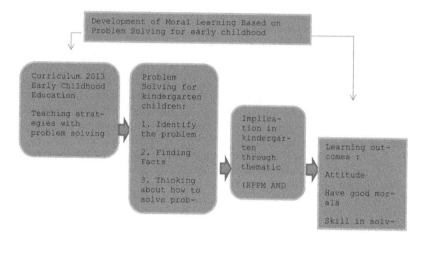

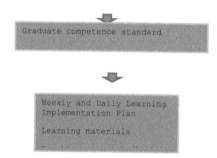

Figure 2.   Problem solving moral integration.

3   RESULTS AND DISCUSSION

Over the past few years, problem solving has been widely used to teach kindergarten children to improve their cognitive abilities; however, there were no studies on its usage to improve religion and morals. This research develops learning models integrated with morals in early childhood based on the problems associated in trials conducted in schools. Problem-solving moral integration is shown in Figure 2.

4   CONCLUSION

In conclusion, the use of problem-solving procedures has the ability to promote the moral development of kindergarten children.

ACKNOWLEDGMENTS

The authors are grateful to PUSLITPEN UIN Jakarta for their moral and material support in ensuring that this research was properly conducted.

REFERENCES

Akcaoglu, M. (2014) Learning problem-solving through making games at the game design and learning summer program. *Educational Technology Research and Development*, 62(5), 583–600. https://doi.org/10.1007/s11423-014-9347-4

Beavers, E., Orange, A., and Kirkwood, D. (2017) Fostering critical and reflective thinking in an authentic learning situation. *Journal of Early Childhood Teacher Education*, 38(1), 3–18. https://doi.org/10.1080/10901027.2016.1274693

Counseling, D. A. N., Parks, D. I., and Suranata, K. (n.d.). Problems experienced by young children and their implications for the implementation of guidance and counseling in kindergartens. 89–95.

Dennis, L. R. and Stockall, N. (2014) Using play to build the social competence of young children with language delays: Practical guidelines for teachers. *Early Childhood Education Journal*, 43(1), 1–7. https://doi.org/10.1007/s10643-014-0638-5

Jonassen, D. H. (2000) Toward a design theory of problem-solving. *Educational Technology Research and Development*. https://doi.org/10.1007/BF02300500

Yu, K. C., Fan, S. C., and Lin, K. Y. (2015) Enhancing students' problem-solving skills through context-based learning. *International Journal of Science and Mathematics Education*, 13(6), 1377–1401. https://doi.org/10.1007/s10763-014-9567-4

Emerging Perspectives and Trends in Innovative Technology
for Quality Education 4.0 – Kusmawan et al (eds)
© 2020 Taylor & Francis Group, London, ISBN 978-0-367-25803-0

Code-switching in Whatsapp group of Indonesian Participating Youth (IPY) 37

Ika Tri Yunianika
Universitas Terbuka, Tangerang Selatan, Indonesia

ABSTRACT: This research aimed to investigate the types, reasons, and dominant topics of code-switching in the chatroom of the WhatsApp group of Indonesian Participating Youth (IPYs) of the Ship for Southeast Asian Youth Program (SSEAYP) 37. The research was conducted using a qualitative descriptive approach. The participants of this study were 13 members of the SSEAYP 37 alumni WhatsApp group. The result of the study showed that the types of code-switching among the participants consists of 65 intra-sentential, 20 inter-sentential, and 3 tag switching. The internal factors that affected the group members to produce code-switching were natural occurrences, less formality, millennial identity, and a point of emphasis. The external factors were difficulty in communicating in Bahasa Indonesia or commonly used English, the same level of language proficiency among group members, and the difference in meaning. In addition, the most dominant topics in code-switching were related to technology, organizational terminology, and music.

1 INTRODUCTION

People's ability to use more than one language has been increasing in recent times. This ability is called bilingualism (Valdés and Figueroa, 1994). Bilingualism has brought up the production of code-switching in communication (Sutrismi, 2014). It is common for bilingual persons to switch the language in their dialogue (Kim, 2006). Research on the nature of bilingual speech is critical because of its significant influence on educational research and discourse of bilingual classrooms (Boztepe, 2003).

Communication takes place not only in the real situation but also in cyberspace. The use of online media as a tool to communicate has gradually improved because of technological development in the industrial revolution 4.0. Many different online media facilitate communication; WhatsApp is one of the most popular tools for communicating nowadays by Indonesians who have smartphones. It is mostly used because of the variety of its friendly features and multi-functions. One of those features is the WhatsApp group, which makes it easy for one particular organization to be in one group that is able to communicate with each other, share ideas, and exchange information. Code-switching has also occurred in the chat room of the alumni of the Ship for Southeast Asian Youth Program WhatsApp group (SSEAYP) 37's. Members tend to do code-switching between the Indonesian language and English when interacting in the group.

Consequently, the purpose of this study was to analyze (1) types of code-switching in the written conversation of the IPY 37 WhatsApp group; (2) the reasons for code-switching in the written communication; and (3) topics that triggered code-switching in the group.

2 LITERATURE REVIEW

Code-switching is generally defined as the practice of using different language variations in a single conversation (Wijanti, 2014). Cárdenas-Claros and Isharyanti (2009) stated that code-switching occurs when a bilingual speaker uses more than one language in a speech above the clause level to convince his or her audience. Hoffmann (2014) categorized code-switching into three types: inter-sentential switching, intra-sentential switching, and tag switching.

Inter-sentential switching occurs between the clause or sentence limit, where each of the clauses or sentences is in one language or another. Intra-sentential switching is code-switching that occurs in a phrase, clause, or sentence limit. The emblematic switching is a code-switching that is a tag, an exclamation, and a specific set of phrases in one language.

3 METHOD

This study used a qualitative descriptive research method. The data for this study were taken from the chat transcript of the WhatsApp group members from January to March 2017. Document analysis was used to process the data. To generate the data findings, the researchers analyzed the open questionnaire and interview of IPYs 37 WhatsApp group members. Participants in this study were 13 people consisting of 7 women and 6 men.

Data analysis was conducted by analyzing the types of code-switching in the chat conversation, which are classified into three categories: (1) intra-sentential switching, (2) inter-sentential switching, and (3) tag switching. Furthermore, the data were calculated and classified to determine subject domination in code-switching, while the questionnaires were analyzed for the internal and external factors of participants in code-switching and code-mixing.

4 RESULTS

4.1 *Types of switching*

According to the analysis, there were three types of code-switching: inter-sentential switching, intra-sentential switching, and tag switching in the written chat of the group

The most frequent type of code-switching that occurred in the group was intra-sentential switching. There were 65 intra-sentential switchings (74%) in the chat room. The following are examples of intra-sentential switching in the group chat:

A: *Innalillahi wainna ilaihi rojiun...turut berduka cita ...*

B: *Gw gk* recognize *sih yg mana orgnya, tapi bener kata bang oon...tetap jaga kesehatan, jaga pola hidup guys.*

Meanwhile, the researcher found 20 inter-sentential switchings of about 23% in the group. The examples of inter-sentential switching in the WhatsApp group of IPY 37 include the following:

A: *Ngahaha nggausah di respon gapapa kok Triaaa.* Just enjoy any birthday gift given by your husband.

B: Lol

On the other hand, tag switching was the least type of code-switching that was used in the group. Only three tags were switching in the chat of IPY's WhatsApp group from January to March 2017. Examples of tag switching in the group are the following:

A: *Aku tak bisa di Bekasi, Tria bisa kali tuh*

B: Sorry *baru liat hp lagi.* So, *gimana?*

4.2 *Reasons for code switching*

According to the data analysis from questionnaires, there were internal and external factors in utilizing code-switching. These factors are discussed in the following sections.

Internal Factors

4.2.1 *Natural occurrences*

Five respondents of this research stated that the daily use of English was one of the primary reasons for the occurrence of code-switching, which enabled its natural flow in WhatsApp groups. Another stated that she had been exposed to English since early childhood when she lived abroad; therefore, code-switching occurred naturally. Vogt (as cited in Tajudin, 2013) explained that the use of different languages was identified as a result of its contact phenomenon, which is indicated by the natural occurrence of code-switching.

4.2.2 *Expressing less formality*

Two respondents stated that they utilized code-switching while chatting in WhatsApp groups of IPY 37 to control the situation and to be informal. According to them, members of IPY's 37 WhatsApp group were alumni of SSEAYP, and they were very close as a team in 2010.

4.2.3 *Millennial identity*

One of the respondents stated that he switched to English because he wanted to be more millennial. The use of English is fast becoming a trend among many young people. These findings support the ideas of Piantari et al. (2011), in accordance with the number of codes in online media platforms, especially Facebook, which shows that respondents have bilingual ability.

4.2.4 *Emphasize a point*

The desire to emphasize a point is becoming a reason for code-mixing (Muthusamy, 2009; Hoffmann, 2014). This statement was expressed by a respondent, which is in line with the research results conducted by Piantari et al. in 2011, which stated that the respondents want to emphasize particular meanings to enable them to switch their languages from one into another.

External Factors

4.2.5 *Difficult to express in Bahasa Indonesia or commonly used in English*

Three respondents stated that they were unable to obtain the right word in Bahasa Indonesia to match their intended words or expressions. Among them, two were of the opinion that some words were commonly used in English, such as indeed, happy birthday, etc. One stated a great number of SSEAYP's terminologies were in English, which was difficult to translate into Bahasa Indonesia. This was in line with a study conducted by Wijanti (2014), who found that there were some words or expressions that cannot fit in Bahasa Indonesia but contain more profound meaning when written in English. Certain languages can only own some concepts. On the contrary, Nurhamidah (2017) assumed that the language barrier was not the dominant rationale of code-switching; instead, it was attributed to cultural and situational aspects.

4.2.6 *The same level of english ability*

Three respondents expressed that the awareness of having the same ability in English encourages them to switch their Indonesian into English in the WhatsApp group chats. In 2015, Kongkerd stated that code-switching was sometimes used to present local identity by group members.

4.2.7 *The difference in meaning*

Members of IPY's WhatsApp group came from different provinces, and they sometimes experienced a few different words or phrases used. Therefore, code-switching occurred in the chat room of the WhatsApp group to make conversations more understandable. One respondent chose this reason.

Table 1. Dominant topics in code switching.

No	Topics	Sample words
1.	Technology	speaker, transfer, subtitle, file, upload, print, share, link, back up
2.	SSEAYP terminology	attire, host, Post to Post edutainment, Pen A Friend
3.	Music	lyrics, standard recording, taking, mixing, mastering, midi, keyboard
4.	Health	medical check-up, diagnose, sick
5.	Working terminology	resign, employee, report, since, bright, well done, host

4.2.8 Dominant topics in code switching

Based on the results of an open questionnaire to 13 respondents, the most dominant topics in the use of code-switching in the WhatsApp group of IPY 37 were daily topics (9 respondents), technology (2 respondents), and organizational/SSEAYP (2 respondents). Pujiastuti (2007) stated that the topic of conversation had a strong influence on the use of code-switching. Moreover, according to the data analysis from the transcript of WhatsApp groups, the topics that mostly encourage respondents to use code-switching are presented in Table 1.

5 CONCLUSION

In conclusion, the type of code-switching that mostly occurred in WhatsApp groups of IPY 37 was intra-sentential, inter-sentential, and tag switching. The internal factors responsible for its occurrences were that it was natural, expressing less formality, millennial identity, and ability to emphasize a point. The external factors were difficulty in expressing in Bahasa Indonesia or commonly used in English, the same level of language proficiency among group members, and the difference in meaning. Furthermore, the most dominant topics in code-mixing were related to technology, organizational terminology, and music.

REFERENCES

Boztepe, E. (2003) Issues in code-switching: competing theories and models. *Issues in CS: Competing Theories and Models*, 3(2).

Cárdenas-Claros, M. S. and Isharyanti, N. (2009) Code-switching and code-mixing in internet chatting: between 'yes, 'ya,' and 'si' a case study. *The Jalt Call Journal*, 5(3),67–78.

Hoffmann, C. (2014) *An introduction to bilingualism.* London and New York: Routledge.

Kim, E. (2006) Reasons and motivations for code-mixing and code-switching. *Issues in EFL*, 4(1), pp. 43–61.

Kongkerd, W. (2015) Code switching and code mixing in Facebook conversations in English among Thai users. *Executive Journal*, 35(1), 126–132.

Muthusamy, P. (2009) Communicative functions and reasons for code switching: A Malaysian perspective. *Language & Society*, 5, 1–16.

Nurhamidah, I. Code-switching in WhatsApp-exchanges: Cultural or language barrier? In *Proceedings of education and language international conference.*

Piantari, L. L., Muhatta, Z., and Fitriani, D. A. (2011) Alih kode (code-switching) pada status jejaring sosial Facebook mahasiswa, *Jurnal al-azhar Indonesia seri humaniora*, 1(1).

Pujiastuti, A. (2007) Code-switching as a multilingual strategy in conversations among Indonesian graduate students in the US. *UII Journal of English and Education*, 1(2).

Sutrismi, S. (2014) The use of Indonesian English code mixing in social media networking (Facebook) by Indonesian youngsters. Universitas Muhammadiyah Surakarta.

Tajudin, T. (2013) The occurrence of code switching on personal message of Blackberry messenger. *Journal of English and Education*, 1(2),103–114.

Valdés, G. and Figueroa, R. A. (1994) *Bilingualism and testing: A special case of bias.* Ablex Publishing.

Wijanti, W. Bahasa Indonesia/English code switching. In *International conference on economics, education and humanities.*

Emerging Perspectives and Trends in Innovative Technology
for Quality Education 4.0 – Kusmawan et al (eds)
© 2020 Taylor & Francis Group, London, ISBN 978-0-367-25803-0

Development of interactive whiteboard media using Realistic Mathematics Education models in early childhood

Eem Kurniasih & Lusi Rachmiazasi Masduki
Universitas Terbuka, Tangerang Selatan, Indonesia

ABSTRACT: This study aimed to describe the development of interactive whiteboard media with the learning model Realistic Mathematic Education (RME) in improving mathematics learning outcomes in early childhood. This study used the Borg and Gall development method with step 6, which is a limited test. The results of this study found an average rating of 89.25% by media experts and 85.75% by material experts, which indicated that the media product interactive whiteboard with the learning model RME was very valid for use in kindergarten. An average teacher response of 88.50% and an average student response of 84.50% means that teachers and students, in general, assess this product as very practical to use in learning mathematics in class. The average posttest results in the experimental Kekancan Mukti Semarang kindergarten class were better than for the control class, that is, 76.75 > 72.75, and t arithmetic > t table is 2.36 > 1.70, so that this product is effectively used as a learning medium in PAUD children.

Keywords: kindergarten, media interactive whiteboard, Realistic Mathematics Education

1 INTRODUCTION

1.1 *Background*

Of the several subjects presented at PAUD, mathematics is one that the system needs for practicing one's reasoning. It is hoped that through mathematics instruction children can increase their abilities, be able to develop skills, and be able to apply the mathematics they have learned in their daily lives. Mathematics is also a means of thinking through the process of determining and developing various kinds of science and technology. As mathematics is a method of thinking logically, systematically, and consistently, all life problems that require careful solutions can refer to mathematics.

Learning media is needed to support the achievement of learning objectives. Therefore we need appropriate media to make it easier for students to understand the material, especially material motion systems in humans. The media used should be interactive because the teaching–learning process itself always involves interactive activities. The purpose of teaching–learning interactions is to help children in a particular development. Therefore the learning process must be conscious of the goal (Sardiman, 2014). One interactive learning media is the interactive whiteboard, which can foster children's creativity. An interactive whiteboard is designed to enable passive learning so that children become active.

In mathematics, there is a learning model that can make it easier for children to understand the material: the Realistic Mathematics Education (RME) model. The underlying principles for RME's primarily follow Freudenthal's views, two of which are that mathematics must be connected to reality and mathematics must be seen as a human activity. Accordingly, RME is based on the concepts that, among other things, in the learning process students should be allowed to reinvent mathematics through the guidance of teachers (Gravemeijer, 1994), and that the rediscovery (reinvention) of ideas and mathematical concepts must start from exploration of various situations and problems of the real world.

Mauro et al. (2014) concluded that interactive whiteboard media can improve understanding of mathematical concepts, and students become more active in learning. Furthermore, research by Kevser and Mehmet (2019) concluded that students could receive material more easily by using interactive whiteboard media. The engagement of students in learning increases more than when using conventional methods.

1.2 *The research objectives*

The purpose of this research is to know the results of learning mathematics using interactive whiteboard media with Realistic Mathematic Education (RME) models in early childhood.

2 RESEARCH METHODS

The research was conducted in Kekancan Mukti Kindergarten Semarang from January to August 2019. The subjects of this study were students of the Kekancan Mukti Kindergarten Semarang class in the 2018/2019 school year. The instruments used in this study were 10 multiple-choice test questions with a time allocation of 30 minutes. In addition, a questionnaire was given to material experts and media experts as validation of interactive whiteboard media with the RME model. The research design used in this study is the Borg and Gall model using six stages: (1) potential and problems, (2) data collection, (3) product design, (4) design validation, (5) design revision, (6) testing the product. The data collection technique consisted of obtaining documents, including the list of students' names. The data collected are the evaluation test values on real numbers, which are then analyzed using a sample normality test (Lilliefors test), a sample homogeneity test (two variance similarity test), and a right-sided t-test. The analyses were then used to determine the effectiveness of the results of the learning process as an indicator of the value of individual learning outcomes and classical learning completeness and to see the average results of the experimental class and the control class.

3 RESEARCH RESULTS AND DISCUSSION

From the results of the questionnaire assessment given to media experts and material experts, the average rating by media experts was 89.25% and by material experts, 85.75%, which means that the media product interactive whiteboard with the RME model is very valid for use in kindergarten.

Based on interviews and questionnaires completed by teachers and students of Kekancan Mukti Semarang, the average response by teachers was 88.50% and by students, 84.50%, which means that the media product interactive whiteboard with the RME model can properly be used in kindergarten.

Based on the results of the posttest in the Kekancan Mukti Semarang kindergarten, the average value of the experimental class was better than that of the control class, 76.75 > 72.75, so that this learning media product was classified as effective to use in kindergarten.

The results of this study are in line with those of Önal Nezih (2017), which concluded that the participants had positive perceptions of the use of the interactive whiteboard in the mathematics classroom. Specifically, they found it beneficial because it enabled students to understand the course better, enabled students to be engaged in meaningful learning and effective engagement in the classroom, increased students' concentration, and saved time.

Research by Kevser and Mehmet (2019) who concluded that students could receive material more efficiently by using interactive whiteboard media in the mathematics classroom. The engagement of students in learning increases more than when conventional methods are used.

4 CONCLUSION

Based on the formulation of the problem, research data analysis, and discussion of the problem, the following conclusions can be drawn:

1. Mathematical learning outcomes using interactive whiteboard media with the Realistic Mathematics Education (RME) model in early childhood shows students are more active. This is seen from better results in the experimental classes than in the control class.
2. Interactive whiteboard products with the RME model are very feasible for use in the learning process at the PAUD level, as can be seen from the results in terms of validity and effectiveness.

REFERENCES

Erdener, K. and Kandemir, M. A. (2019) Investigation of the reasons for students' attitudes towards the interactive whiteboard use in mathematics classrooms. *International Journal of Research in Education and Science*, 5(1), 331–345.

Gravemeijer, K. P. E. (1994) *Developing realistic mathematics education*. Utrecht, the Netherlands: Freudenthal Institute.

Hasan, M. (2011) *Early childhood education*. Yogjakarta: Diva Press.

Kevser and Mehmet (2019) Investigation of the reasons for students' attitudes towards the interactive whiteboard use in mathematics classrooms.

Mauro, D. V., Lieven, V., and Jan, E. (2014) Interactive whiteboards in mathematics teaching: A literature review. *Journal of Education Research International*, 5(1), 1–16.

Nezih, O. (2017) Use of interactive whiteboard in the mathematics classroom: Students' perceptions within the framework of the technology acceptance model. *International Journal of Instruction*, 10(4), 67–86.

Sardiman (2014) Interaction and teaching and learning motivation. Jakarta: Rajawali Press.

Emerging Perspectives and Trends in Innovative Technology
for Quality Education 4.0 – Kusmawan et al (eds)
© 2020 Taylor & Francis Group, London, ISBN 978-0-367-25803-0

Education from the urban marginal society's perspective

Novia Wahyu Wardhani, A.T. Sugeng Priyanto, Eko Handoyo & Martien Herna Susanti
Faculty of Social Sciences, Universitas Negeri Semarang, Indonesia

Sabar Narimo
Faculty of Teacher Training and Education, Universitas Muhammadiyah Surakarta, Indonesia

ABSTRACT: This was a descriptive study that aimed to portray education from the perspective of the marginal urban community, by studying two cities in order to achieve the country's goal of educating the entire nation. Marginal people in Indonesia are residents who need special attention in order to meet their education needs. Education has not had a big impact on changing their lives. Therefore, the researchers wished to study education from the perspective of the marginal urban community, which has been a topic of research from 2017 to 2019 in communities in the city of Surakarta and Demak district. The ethnographic approach was used with three data sources: participant observation, open and in-depth interviews, and documentation. The results show that education for marginal urban communities is an activity that takes up time, energy, and money, so it should be directly applied in order to improve people's lives.

1 INTRODUCTION

The marginal community refers to people with low incomes or poor people. Marginal society is a society whose level of awareness of rights and obligations is very low, even though one indicator is the creation of civil society, and to achieve the goals of the nation an awareness of their rights and obligations as citizens is necessary. The rights and obligations that were examined in this study is educational rights and obligations in marginal urban communities because this is an aspect that has caused very complex and multidimensional problems.

Marginal people are poor people who have an average per capita expenditure per month below the poverty line. The poverty data released by Badan Pusat Statistik stated that the percentage of poor people in urban areas in March 2018 was 7.02% (Badan Pusat Statistik Indonesia, 2018). Also, the percentage of the poor population in Central Java Province was 11.32%, followed by the city of Surakarta with 9.08%, while Demak Regency had a staggering 12.54% from data that was collated by the Central Statistics Agency (BPS, 2019). Although this number has decreased from year to year, it still shows a fairly high poverty rate.

Urban marginal people see education as something that is not so important in life because of the constraints of time, cost, and energy, whereas one of the visions of the State and Nation of Indonesia in the 1945 Constitution is to educate the nation's population. Furthermore, rights and obligations in the field of education are affirmed in article 31, paragraph (1) of the NRI Constitution Article, which states that every citizen has the right to education and paragraph (2), which states that every citizen is obliged to attend basic education and the government is obliged to finance it (Indonesia, 1945). The policy to accommodate the mandate of the 1945 Constitution of the Republic of Indonesia has to do with the 12 years of compulsory education with a great deal of assistance ranging from scholarships to free schools.

A common phenomenon in the city of Surakarta is that parents do not allow their children to study in both formal and nonformal educational institutions, which results in the free schools being eventually closed because there are no students to teach. In contrast to Surakarta, it was observed in Demak during the research that schools existed where there were only a small number of students, even though some were a bit aged for compulsory education. It was also discovered that the low interest of these children was due to the influence of their parents, who did not wish their children to go to school because of their belief that graduating from elementary school is the same as graduating from junior high school and from high school as laborers. Another cause is the factory environment, which requires a large number of workers with low education. This then leads researchers to become interested in exploring education from the perspective of marginal urban communities with the aim that the policies that will be issued by the government will answer the needs and expectations of the people involved.

2   RESEARCH METHODS

This study portrays education in marginal urban communities in Surakarta City and Demak Regency. The researchers used the ethnographic approach, which depict the characteristics of different people from other societies. Data were obtained from the results of observations of three participant families in each city from a pattern of community life. The second technique used was in-depth and open interviews obtained from six informants in each area consisting of scavengers, street children, and beggars. This interview was conducted with the researchers entering and becoming part of their daily lives to obtain valid data. Observations and interviews were conducted to explore and depict their daily lives as related to education. The third technique is documentation by collecting data in the form of documents of community participation in education and photos of their daily activities for seven days. Data analysis techniques were carried out with Nvivo 12 software, beginning with coding research data obtained in the field, then reducing the data, and finally drawing conclusions from the research data.

3   RESULTS AND DISCUSSION

From observations made on the characteristics of marginal communities in Surakarta City and Demak Regency, we noticed different characteristics that are distinct from what is seen in other marginal communities. The people of these two communities are not migrants, but they are indigenous people who can be classified into the cultural poverty level; on average, these are young families who experience economic shortages even though they live in cities.

Marginal people in the two cities are more concerned with activities that can make money because they believe that all they need to live is money. Issues like this are very worrisome, especially when we consider the living condition of the people who live in marginal areas and children who are accustomed to living on the streets. How to meet needs that seem as though they cannot be fulfilled is one of the obstacles preventing people from choosing life-enriching activities such as getting an education. When people are consumed with thoughts of how to put food on the table, then they are bound to only think of how to get money to meet their daily needs. Marginal people prioritize activities that they consider capable of giving them money so that their welfare and family burdens can decrease. This affects their children, who are often forced by their parents to work in order to make a living. When other children are in school, this group of children goes to work as laborers because their parents won't allow them to study.

Therefore, apart from the need for money, it turns out that environmental conditions also affect these children. The lack of support from the social environment to create a conducive learning atmosphere makes these children less motivated to learn. Many factories require low-income workers who do not have the will to acquire a high school education. The environment in which they live, which has on average only elementary and junior high school graduates, also does not motivate them to strive for greater heights. They have no desire for school. The environment and the need for money also affect their thinking about education. Their culture or mindset about education is different from the situation in more advanced societies because they are content as long as their needs are met. They don't see education as important; in fact, they view it as a waste of time, money, and energy. Apart from that, if they don't get an education, then they tend to attribute it to fate.

Educational obstacles in the marginal urban community include economic, social, and cultural, as shown in Figure 1.

This is in line with Ki Hajar Dewantoro's view, stating that three spaces influence the development of students: family, school, and community. The actual education process certainly involves the role of family, community, and the school environment. So if one of the three does not work well, it will affect the sustainability of education itself (Syafii, 2018).

Marginal or marginalized groups have different sociological constructions other than the social groups in general (Muttaqin, 2014). Usually, marginal communities are associated with poverty and naive living (Wayan Nitayadnya, 2016). Thus, marginal society is a symbol used to refer to people with a low economic status or those who live on the poverty line.

Development of education always leads to higher quality education but is accompanied by increasingly high costs (Husna, 2017). Many policies for education for marginalized communities have been tried by the government, especially in the city of Surakarta, such as cheap education programs and Surakarta City Community Education Assistance (BPMKS) (Harsasto and Firdaus, 2013) while the Demak government has also issued a follow-up policy from Central Java on the Family Hope Program (PKH) and provided scholarships for poor students.

The funds needed to obtain education and to support life make families unable to provide space for human expression. Parents tend always to force their children to make money by working. Market intimidation, governance regimes, and family pressure eventually give birth to alienated individuals/alienated groups. To be able to succeed in having a quality education, parental attention, financial support, a conducive social environment, and freedom for children to pursue education based on their interests and talents are issues that must be considered (Wayan nitayadnya, 2016). Education is an investment and an opportunity to compete to get a chance for a better life in the future and to be involved in the development process (Ustama, 2010). Not only that, but some studies also say that there is a significant relationship between education level and income level (Kurniawati et al., 2017).

According to Jerome S. Arcaro, improving the quality of education can be understood in two ways: (1) quality improvement is greatly associated with costs of education, and (2) quality improvement is associated with the ideals of hope to achieve a better life (Yusuf, 2014). When education is not in demand by the community, it must return to look at costs and expectations. Both cannot be abandoned in the quest to increase community participation in the administration of education.

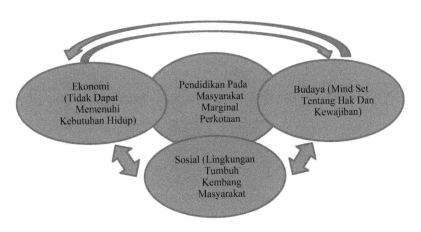

Figure 1. Factors affecting community views on education.

The functional, structural theory says that a system of society is likened to a body consisting of interrelated parts, united with one another, and each has a role (Hajar, 2017). Individual parts cannot function without being connected to the other parts. Changes that occur in one part will cause an imbalance and in turn will create changes in the other parts (Rasyid, 2015). Like the economic situation that is completely lacking but is accompanied by the fact that there is no increase in welfare, the impact of higher education causes people's mindset to change the function of education and their motivation toward education because the problem of education cannot be separated from economic problems (Syamsurijal and Widiansyah, 2017). Thus, prosperity has a role to play in the environment and culture of the community.

4 CONCLUSION

Education is less influential in the lives of the marginal urban communities in Indonesia because they feel that education only wastes energy, money, and time. The biggest influence on their views on education is their economy, social environment, and culture. Thus, the government should enforce the attainment of a good education on these communities so as to improve their standards of living because education for the marginal urban communities can be directly applied in order to improve their lives.

REFERENCES

Badan Pusat Statistik Indonesia (2018) Profil Kemiskinan di Indonesia Maret 2018. *Badan Pusat Statistik*, 29(5), 1–8.
BPS (2019) Badan pusat statistik Kabupaten Demak, (0714), p. 321023. Available at: http://sp2010.bps.go.id/index.php/site/tabel?tid=327&wid=3500000000
Hajar, K. (2017) Pemberdayaan Pendidikan Pada Masyarakat Kaum Miskin Kota : Studi Tentang Peran TAABAH dalam menghadapi Komunitas Ledhok Timoho, Kelurahan Muja Muju, Kecamatan Umbulharjo Yogyakarta.
Harsasto, P. and Firdaus, L. K. (2013) Citizen inclusion Dalam Praktik governance: Penelusuran Dalam Kebijakan Pendidikan Kota Surakarta, 2(2), 49–61. doi: 10.14710/politika,2,2,49-61.
Husna, F. (2017) Inovasi Pendidikan Pada Kaum Marginal, 11, 76–88.
Indonesia (1945) Undang-Undang Dasar Negara Republik Indonesia. (2), 1–19.
Kurniawati, L., Nurrochmah, S., and Katmawanti, S. (2017) Hubungan Antara Tingkat Pendidikan, Status Pekerjaan Dan Tingkat Pendapatan Dengan Usia Perkawinan Pertama Wanita Di Kelurahan Kotalama Kecamatan Kedungkandang Kota Malang. *Jurnal Preventia*, 2(1).
Muttaqin, A. (2014) Pola Keberagaman Masyarakat Marginal. Komunika, 8, pp. 129–156.
Rasyid, M. R. (2015) Pendidikan Dalam Perspektif Teori Sosiologi. *Aladuna*, 2(2), 274–286. doi: 10.1007/s00415-006-0359-9.
Syafii, A. (2018) Perluasan dan Pemerataan Akses Kependidikan Daerah 3T (Terdepan, Terluar, Tertinggal). *Jurnal Manajemen dan Pendidikan Islam*, 4(2), 153–171. doi: 10.12928/psikopedagogia.v1i2.4603.154.
Syamsurijal and Widiansyah, A. (2017) Peran Ekonomi dalam Pendidikan dan Pendidikan dalam Pembangunan Ekonomi. *Jurnal Ekonomi Pembangunan*, XVII(2), 1–9.
Ustama, D. D. (2010) Peranan Pendidikan Dalam Pengentasan Kemiskinan. *Dialogue* (Paris), 6(1), 1–12.
Wayan Nitayadnya, I. (2016) Perubahan Pola Pikir Kaum Marginal Terhadap. 28(2), 181–196.
Yusuf, M. (2014) Membangun Pendidikan yang Bermutu menuju Masyarakat Madani. *Jrr*, (1), 1–9.

Emerging Perspectives and Trends in Innovative Technology
for Quality Education 4.0 – Kusmawan et al (eds)
© 2020 Taylor & Francis Group, London, ISBN 978-0-367-25803-0

Reconstruction of creative movement to instill good character in kindergarten children

Siti Aisyah, Titi Chandrawati & Dian Novita
Universitas Terbuka, Tangerang Selatan, Indonesia

Putu Aditya Antara
Ganesha Educational University, Indonesia

ABSTRACT: This research proffers solution to development of values and manners in Undiksha Kindergarten Lab School's children by specifically discussing efforts made to foster children's character, which is developed through the creative movement model. The study used the development research method, which is conducted in various stages and carried out through nine main steps. The results of the pretest t-test were 0.425 > 0.05, with a probability of $p > 0.05$. This indicates that the control class and the experimental class have the same initial ability. The results of the posttest t-test showed differences between the values of children's character that were taught in the classroom with creative movement and without creative movement, as indicated by the significance value of 0.000 < 0.05, where if $p < 0.05$ then it shows that the use of creative movement in the material value of character can increase the development of children's character.

1   INTRODUCTION

The results of pre-research observations of kindergarten school children that were conducted by researchers in Buleleng Sub-district and specifically at Undiksha Kindergarten Lab School showed that when children play, some are not able to wait for their turns (stand in queue); they scramble, lack patience, and also like to disturb friends and ultimately to cause a fight. Based on this description, the teacher's role as a facilitator and mediator needs to be more creative and innovative. Schools should be able to prepare learning methods that facilitate children to develop their values and all their potential. A variety of play activities can be done in kindergarten, but teachers must choose the right game and identify behaviors that they want the children to develop. The right game is done to develop the values of character using a model of creative movement. This game is played simultaneously or in groups while following songs and music that have been previously taught. By adding songs and movements to each game, the children will be more cheerful and happy to participate in this play activity.

2   LITERATURE REVIEW

The creative movement model is used as a means of self-expression by a child according to his or her imagination. By expressing various imaginations, children can develop various other potentials. According to Dodge and Colker (2000: 247), creative motion contributes greatly to the social-emotional, cognitive, and physical development of children. When children move with freedom, quality stimulation is provided to their emotional and social development. Furthermore, Hawkins (2003: 12) reaffirms the activities that need to be carried out in the implementation of creative movements, which are associated with the creative process of feeling (absorbing), experiencing, imagining, manifesting, and giving shape.

The development of children's character is greatly influenced by the condition and social environment of the children, including parents, peers, and the surrounding community. Lynch and Simpson (2010: 10) explained that character could be introduced and developed from early childhood, by, among others, hanging out with friends, sharing and helping, working together, communicating, and being able to solve problems, as well as understanding the rules and expressing feelings. Pamela May (2011: 24) said that children learn to develop good character when they feel comfortable and feel free to explore their feelings. Through freedom in socializing, children will easily interact. But this freedom must still be responsible.

Based on the foregoing explanations, researchers can conclude that character is the ability to foster and establish relationships with others in terms of communication and interaction, which can create good relationships through attitudes, words, and behavior. Aspects of character that need to be developed in children from an early age include cooperation, communication, empathy, understanding the rules, and being responsible.

3 RESEARCH METHOD

The development research methods of Dick and Carrey were used in this study, starting from needs analysis to evaluation and effectiveness testing using the experimental method to see the differences between the experimental and control groups.

4 DISCUSSION

To test the creative movement, the developed sample independent t-test was used with the help of the computer program SPSS version 16.0. The results of the pretest t-test were 0.425 > 0.05, with the probability of $p > 0.05$. This indicates that the control class and the experimental class have the same initial ability. The results of the posttest t-test showed differences between the values of children's character in the classroom being taught by using creative movement and without creative movement, indicated by the significance value of 0.000 < 0.05, where if $p < 0.05$ then it shows that the use of creative movement in the material value of character can increase the development of children's character. This supports the validation results, which state that the creative movement, when developed, can increase the value of children's character.

Based on the results of the tests conducted, it can be concluded that the development of creative movement in stimulating children's character has been shown to be successful, and this is evidenced by the statement of a child that the child is interested in doing creative movement because it is more fun and it frees creativity in children. There is also the belief that with creative movement, children can become more interested in moving freely in order to express themselves. This is possible because the movements made certainly make children feel comfortable because the creative movement is a basic movement that is contemporary and flexible. According to Redfern (1982: 117), contemporary art movements do not pay attention to movements that must be studied specifically. Therefore, when children are being stimulated in kindergarten institutions, various art movements that are taught should be contemporary and should allow children to make movements that are usually done in the environment of each child.

5 CONCLUSION

The following conclusions are reached based on the results of the research and foregoing discussion. When carried out appropriately by children, creative movement will provide quality improvement and development in their physical balance and coordination ability, understanding of rhythm and tempo and give them the ability to predict the events that will occur next and also give them high body awareness. Movements developed to stimulate children's values are a collaboration with collaborative movements, communication with eclectic movements, empathy with facial expressions, understanding rules with appropriate movements, and responsibility with collaborative independent movements. The development of creative movement in stimulating children's character is successful because the validity test results show validity, and this is also proven by the statement of a child that the child is interested in doing creative movement because it is more fun and frees creativity in children.

The development of the concept of the creative movement model is a positive appreciation of many artists for the great work of Rudolf Laban (1976: 12), who succeeded in making a dance notation that is a recording of the basic movements of dance that can be used by people to re-study a dance movement. This dance's basic movement is the focus of the development of the creative movement in early childhood, as Laban emphasized that dance learning in public schools (nonvocational) should place more emphasis on creative dance learning that is able to develop children's personality. This creative movement learning is not geared toward the end result of producing a performance that has high artistic value as created by a choreographer. This statement is very reasonable because every child has a natural urge to display movements such as "dance" and unconsciously it is a good way to introduce dance early to the child, in addition to which the child will be able to recognize his or her personality and values and natural character and be provided opportunities for developing the ability of spontaneous expression through movements that are freed (free dance).

REFERENCES

Berk, L. E. and Winsler, A. (1995) *Scaffolding children's learning: Vygotsky and early childhood education.* NAEYC Research into Practice Series, Vol. 7. ERIC.

Dakhi, Emiria YZ, et al. (Budi Pekerti Team) (2001) *Flying with two wings: Success of Budi Pekerti training.* Jakarta: PT Grasindo.

Djaali and Muljono, P. (2008) *Measurement in the field of education.* Jakarta: Grasindo.

Eliason, C. (2016) *A practical guide to early childhood curriculum*, 10th ed. Weber State University: Pearson.

Hawkins, A. M. translated by I. Wayan Dibia (2003) *Move according to conscience: New method of creating dance.* Jakarta: Ford Foundation and the Indonesian Arts Society.

Hurlock, E. B. (1978) Meitasari Tjandrasa's translation of *Language, Child development*, Vol. 2. Jakarta: Erlangga.

Leseho, J. and Maxwell, L. R. (2010) Coming alive: Creative movement as a personal coping strategy on the path to healing and growth. *Canada: British Journal of Guidance & Counseling*, 38(1).

Makmun, Abin Syamsudin (2007) *Educational psychology: Systems tools capital teaching.* Bandung: PT. Teen Rosdakarya Offset.

Munandar, Utami (1999) *Developing gifted child creativity.* Jakarta: PT. RinekaCipta.

Papalia, Olds, Feldman (2002). *A child's world infancy through adolescence.* New York: McGraw-Hill.

Pease, A. and Pease, B. (2009) *Why men can only do one thing at one time and women can't stop talking.* Nayluvar translation. Jakarta: PT Cahaya Insan Suci.

Schunk, D. H. (2012) *Learning theories: An educational perspective.* Boston: Pearson Education.

Sedyawati, E., et al. (1997) *Guidelines for cultivating noble cultures.* Jakarta: Balai Pustaka.

Smith, J. (1976) Dance composition: A practical guide for teachers. Surrey: Unwin Brothers.

Emerging Perspectives and Trends in Innovative Technology
for Quality Education 4.0 – Kusmawan et al (eds)
© 2020 Taylor & Francis Group, London, ISBN 978-0-367-25803-0

The utilization of cow dung as a biogas fuel in Cibiru Wetan, Bandung

Maman Sudirman, Angga Sucitra Hendrayana, Nana Setiana & Rasdjo D. Suwardi
Universitas Terbuka, Tangerang Selatan, Indonesia

ABSTRACT: The continuous decrease in the availability of an energy supply encourages society to determine alternative energy sources, such as biogas, which is an anaerobic fermentation process consisting of organic waste and methane bacteria. This simple process produces energy in the form of methane with the ability to create heat. The raw organic material used in this process is waste products from dung. Biogas fuel as a substitute for kerosene was practically implemented in Cikoneng, Cileunyi district, Bandung. As a result, cows were no longer allowed to roam along the road, as they were caged with their dung processed as fertilizer against its initial indiscriminate displacement, which polluted the soil and environment.

Keywords: biogas, cow dung, society

1 INTRODUCTION

1.1 *Background of the activity*

Cibiru Wetan is one of the villages in the Cileunyi district, where a large number of people participate in field and livestock farming. According to Halimah (2007), most of the residents of Cibiru Wetan Village are livestock farmers (31.65%). The most common animals whose refuse is piled up twice a day are cows and sheep. However, they are not properly managed, thereby leading to the indiscriminate disposal of their waste products on soils. This research, therefore, offers a useful method to help eradicate this problem, in line with the writings of Muhammad et al. (2017) regarding the application of biogas making technology that at the same time generates compost.

1.2 *Purpose of the activity*

a. Provide information on the use of dung for new energy production, called biogas, and increase the utilization of cow manure waste as household energy fuel (Subeni et al., 2013).
b. Provide training to ensure a place for dung is available for the fermentation and processing of biogas as a substitute for liquefied petroleum gas (LPG).
c. Form a group that is able to determine other sources of new energy.
d. Increase knowledge of the utilization of livestock waste in creating sustainable energy and managing the environment.
e. Increase the skill for the production of easy and affordable biodigesters.
f. Encourage people to utilize livestock waste as a sustainable source of energy and fertilizer.

1.3 *Benefits of the activity*

a. This activity provides information on the importance of biogas to livestock farmers for adequate management of dung as a daily source of energy.
b. It provides information on the new energy source.

1.4 Dedication to society

a. A fermentation place for dungs and biogas is created.
b. Through this activity, people are expected to provide cadres willing to analyze new energy sources.
c. The society in the village becomes independent in fulfilling their energy needs and actively takes a role in managing the environment. A clean, healthy, and comfortable hamlet environment free from pollution from the smell of cow dung is created (Ibrahim et al., 2017).
d. More people make biodigesters, using free energy and not depending on chemical fertilizer or LPG.

2 RESULTS AND DISCUSSION

The volunteer activity started by carrying out a survey in Cikoneng, Cibiru Wetan village, which showed a decrease in the number of cattlemen capable of becoming IbM (Science and Technology for Society) partners in the production of biogas technology. Observation also showed various types of occupation by the village, from sellers, cultivators, farmers, civil servants, factory laborers, constructer workers, drivers, and cattlemen.

The information obtained during the training, village heads, and public figures stated that their subordinates had no knowledge regarding the process of making biogas from cow dung. The refuse was piled up in gardens, disposed of, and used as plant fertilizer.

The introduction of a simple biodigester technology, which acts as a fermentation place for the production of methane gas from cow dung, created a medium for an understanding and practice of the process (Figures 1 and 2).

2.1 Some aspects behind this activity

a. The implementation of science and technology toward society funding IbM in making biodigester from cow dung is still at the introductory stage.
b. There is a spirit of partnership.

2.2 Some factors hindering this activity

a. Changes in the schedule are needed to fit society and management.
b. It is in an area with inadequate water supply, thereby leading to thick fermentation, which is unable to produce gas.

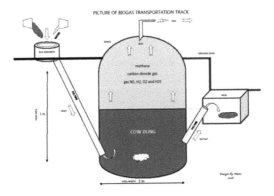

Figure 1. Biodigester scheme.

Figure 2.   Observation and technical explanation to partner at a biodigester construction place.

3   REVIEW OF THE FINDINGS

Generally, partners were satisfied with the Student Creativity Program IbM activity, which was evident from the attitude and behavior of the lecturer. The ratings showed four Very Satisfying answers and three Satisfying answers with 73%, 74%, and 24%, respectively.

3.1   *Result of dedication toward society*

The dedication program toward society is based on Science and Technology IbM, which helps to manage the waste potential and promote awareness of environmental health, understanding of basic technology, and profits associated with livestock dung. In detail, some benefits gained by society are listed in Table 1.

Table 1.   The outcome of the activity.

No.	Elements	Pre-IbM	Post-IbM
1.	The development of a fermentation place for livestock dung	Unable to develop a place for the fermentation of dung	Society is able to create a fermentation place for livestock and produce biogas.
2.	Budget savings	No plan on the economic side that could be controlled	The people in Cibiru Wetan village are able to save their gas usage.
3.	Alternative fuel	There is no teenager interested in the simple technology	This has the ability to create national cadres interested in determining new energy sources.
4.	Management energy	Depend on the LPG supply	Society gains in dependence in fulfilling its energy needs
5.	Awareness toward environment cleanliness	Unaware of environment sanitation	People take an active role in managing the environment.
6.	The potential of new energy source	Some people still made use of woods as fuel.	More people obtain free energy.
7.	Organic fertilizer	Dungs are indiscriminately disposed of without adequate treatment.	Society acquires a good fertilizer quality.
8.	Fertilizer and alternative fuel	It was enough to buy urea fertilizer and LPG gas from the agent.	People no longer depend on chemical fertilizer and LPG supply.

4 CONCLUSIONS AND SUGGESTIONS

4.1 Conclusions

The following conclusions were drawn from the utilization of cow dung as a source of biogas fuel for Cibiru Wetan Village, Bandung Regency:

1. There was an increase in knowledge within society, with cattlemen partnering in utilizing cow dung as a source of fuel.
2. Society gained additional skill (soft skill) in making and managing cow dung as biogas and plant fertilizer.

4.2 Suggestions

The inadequate technique and the emerging problems associated with the production of biodigester, excuses related to its implementation, and the program' success, led to the following suggestions:

1. The need to train people to ensure adequate gain benefit
2. The need to spread information to different areas

REFERENCES

Halimah, U. (2007) Desa-cibiru-wetan-2006. Available at http://uun-halimah.blogspot.co.id (accessed March 13, 2018).
Ibrahim, Idris, M. and Bunyamin, B. (2017) Peningkatan Kesejahteraan Masyarakat Desa Tertinggal Melalui Biogas Kotoran Sapi. *Jurnal Bakti Saintek*, 1(1), 33–45.
Muhammad, F., Hidayat, J. W., and Wiryani, E. (2017) Pembuatan Biogas sebagai Energi Alternatif dari Limbah Organik Berbasis Peternakan Terpadu dan Berkelanjutan di Ungaran, Kab. Semarang. *Jurnal Abdi Insani Unram*, 4(1).
Pemerintah Desa Cibiru Wetan (2015) Sejarah-desa-cibiru-wetan. Available at: http://wartacileunyi.blog spot.co.id (accessed March 13, 2018).
Subeni, Sukoco, Surono, U. B. (2013) Pembuatan Biogas dari Limbah Sapi dan Pemanfaatan Limbah Biogas Sebagai Pupuk Organik. *Agros*, 15(1), 207–213.
Sudarmawan, S. (2014) Biogas Kotoran Sapi Jadi Energi Alternatif Kenaikan Elpiji. Available at: https://www.tribunnews.com tribunnews.com.madiun
Wahyuni, S. (2017) Biogas Hemat Energi Pengganti Listrik, BBM, dan Gas Rumah Tangga. PT Agro Media Pustaka, Jakarta Selatan.

*Emerging Perspectives and Trends in Innovative Technology
for Quality Education 4.0 – Kusmawan et al (eds)*
© *2020 Taylor & Francis Group, London, ISBN 978-0-367-25803-0*

Implementation of School-Based Management in SMP Negeri 1 Bantul

Wartomo
Universitas Terbuka, Tangerang Selatan, Indonesia

ABSTRACT: School-Based Management is a management concept that offers autonomy to schools to make decisions in an effort to involve all school components effectively and efficiently to improve the quality of education. This study aims to determine the process of planning, implementation, and evaluation in improving the quality of education using a qualitative descriptive approach. Data was collected by interviews, observation, and study documentation. The research subjects were principal, vice-principal, teachers, and school committee of SMP Negeri 1 Kasihan Bantul. 1. The results revealed that program planning is based on the school's vision, mission, and goals. Work programs were compiled by the school's component through revising the previous year's work program, and they were verified by the principal. The substance leads to efforts to improve education quality, but the results are not clearly stated. Furthermore, it was also found that the program implementation is managed by the school's component by preparing written instructions, such as 2013 Curriculum documents, organizational structure, job description and division of teachers and educational staff, academic regulation, and school rules. Finally, based on the result, it was revealed that program evaluation focuses highly on academic programs rather than the effectiveness and efficiency of learning and teacher performance, implementing the school's self-evaluation, and school accreditation.

1 INTRODUCTION

Education is a shared responsibility among the government, parents, and community. School-Based Management (SBM) provides broader autonomy to schools, and it has greater authority and responsibility to make decisions according to the needs, abilities, and demands of schools, communities, or stakeholders to improve education quality at schools.

Application of SBM in school management is considered successful if it is able to elevate the quality of education and learning processes and products. According to Usman (2013: 543), quality in education includes the quality of inputs, processes, outputs, and outcomes." Educational input is declared as good quality if it is ready to be run, while the education process is considered a success if it is able to create an atmosphere of PAKEM (Active, Creative, Enjoyable, and Meaningful Learning). Furthermore, the output is considered to have a high quality of students' academic and non-academic learning outcomes are high. Quality is beneficial to the world of education because it could lead to (1) increase accountability (accountability) of schools to community and/or government who have provided all costs to schools, (2) guarantee the quality of its graduates, (3) work more professionally, and (4) grow healthy competition.

Ministry of National Education (2008: 112) establishes procedures for activities which can realize the prepared school programs as follows: (1) socializing school programs to teachers, school staff, students, and parents through formal and school meetings; (2) arranging a priority scale based on school's financial condition and human resources;(3) dividing tasks and responsibilities to person or division in charge based on their respective abilities; (4) conducting monitoring and evaluation for each activity to supervise school's targets. (5) making a report on the implementation of the school program, observing and noting challenges encountered in implementing the program, and determining efforts to overcome them; (6) compiling a follow-up program concerning future school programs.

The fact is various efforts to improve the quality of schools have been carried out; however, it was found that in practice, the education quality remains low. One contributing factor is the lack of community participation in educational institutions.

The purpose of this study is to investigate an increase or decrease in accountability (of schools to community and/or government who have provided all costs to schools, to guarantee the quality of its graduates, to work more professionally, and to grow healthy competition.

2 METHODS

This study employed a qualitative descriptive approach. It was carried out in SMP Negeri 1 Kasihan Bantul from 12 April to 12 October 2017. The subjects were the principal, vice-principal, teachers, and SMPN 1 committee in Kasihan Bantul District. Data were collected by conducting an interview, observation, documentation, and triangulation. Next, the data were analyzed by descriptive qualitative with the following steps, such as data reduction, data display, and conclusions or verification.

3 RESULTS AND DISCUSSION

Based on the results, it was found that the program plan in the school's vision, mission, and goals formulation has involved all stakeholders, particularly the board of teacher and school committee. In this case, the school committee has contributed to providing input. Likewise, all school development teams have actively participated in formulating education policies and programs, including the Annual Work Plan (RKT) and School Activity Plan and Budget (RKAS) Danim (2007: 108) states that: the first job of principals in implementing SBM is to convince as many teachers as possible that they have obligation, opportunity, and challenges to be involved in various forms of planning and problem solving which as of now have been carried out only by school's administrators. Thus, those involved, especially teachers, are expected to understand school policies and school goals written on the school's walls indicating the school's effort in socializing the policies to all of the components at the school and related parties (Mulyasa, 2009).

SMP Negeri 1 Kasihan Bantul has made good use of available resources through the job division in the education unit component. Furthermore, the school also made several guidelines governing various aspects of management, including academic regulations, organizational structure, and job descriptions. The evaluation program team has numerous tasks, such as improving the quality of education, supervising particularly regarding the attendance of students and teachers; thus, work programs and learning activities can run well. Evaluation of school work programs is carried out by the principal, school supervisors, and accreditation bodies. The evaluation has not become a comprehensive culture in this school. Internal school program evaluations are more focused on the achievement of learning programs or student learning outcomes. However, the evaluation has not examined the effectiveness and efficiency of the learning process and teachers' performance. The program's accountability and the implementation costs were reported to funders, while the teachers' evaluation was conveyed verbally in a meeting attended by the head of the school committee.

4 CONCLUSION

SMPN 1 Negeri 1 Kasihan Bantul has increased accountability to the government and their community as funders in implementing education. This school has implemented progress in the education quality, which can be examined in the RKS describing the school's objectives within a four-year period. Also, the objectives are related to the quality of graduates that are expected to be achieved and the improvement of the school's components, which support the progress of education quality. Furthermore, all school's components worked professionally indicated by teachers who have prepared learning tools such as syllabus and lesson plans. The preparation of learning tools refers to the Process Standards of the SNP. Although the teachers have applied the CTL learning approach, the school committee and parents have not been involved in implementing the learning process. To promote healthy competition and the program implementation in improving the education quality, the school curriculum development team has prepared written implementation instructions, such as KTSP documents, organizational structure, job division of teachers and education staff, academic regulations, and school rules which will be thoroughly and continuously evaluated and developed.

REFERENCES

Danim, S. 2007. New Vision for School Management. Jakarta: Earth Literacy.

Ministry of National Education. 2010. Technical Manual for Preparation of High School Work Plans. Jakarta; Directorate of High School Development.

Ministry of National Education. 2008. Education and Training for School Principals, Management of Development and Implementation of SBC. Jakarta: DirJen.PMPTK.

Harun, C.Z. 2009. Education Resource Management. Yogyakarta: Pena Persada.

Irianto, Y.B. 2011. Education Renewal Policy. Jakarta: Raja Grafindo Persada.

Komariah A. and T.C. 2010. Visionary Leadership Towards Effective Schools. Jakarta: Earth Literacy.

Murniati, A.R. 2008. Strategic Management the Role of School Principals in Empowerment. Bandung: Youth Citapustaka Media Pioneer.

Mulyasa, E. 2009. Implementation of KTSP, Independence of Teachers and Principals, Jakarta: Bumi Aksara.

Syafaruddin. 2012. Principal Management and Leadership. Jakarta: Earth Literacy.

Nanang F. 2004 The Concept of School Based Management (SBM) and School Board (Bandung: Paniaka Bani Quraisy, 2004), p. 16.

Rohiat. 2009. School Management - Basic Theory and Practice. Bandung: Refika.

Emerging Perspectives and Trends in Innovative Technology
for Quality Education 4.0 – Kusmawan et al (eds)
© 2020 Taylor & Francis Group, London, ISBN 978-0-367-25803-0

Exploratory Factor Analysis of the HEdPERF's scales on Balikpapan State Polytechnic

Saiful Ghozi, Aditya Achmad Rakim & Mahfud
Balikpapan State Polytechnic, Balikpapan, Indonesia

ABSTRACT: The purpose of this study is to analyze the construct validity of HEdPERF's scales using Exploratory Factor Analysis (EFA) technique in Balikpapan State Polytechnic context. The respondents are students from among seven departments in Balikpapan State Polytechnic. The 5-steps Exploratory Factor Analysis Protocol was followed for running EFA. The differently results from originally-six factors of HEdPERF by Abdullah Firdaus (2004) were obtained in this study, i.e., academic aspect, access, delivery service, campus reputation, staff attitude, program reputation, and employability graduate. Further factor analysis of Confirmatory Factor Analysis (CFA) can be carried out to evaluate the unidimensionality of the scales. This study is also attempting to find the right dimensions for service quality in the Indonesian polytechnic sector.

1 INTRODUCTION

The HEdPERF model was developed by Abdullah Firdaus (2004), which was specifically designed to assess the service quality of higher education institutions (HEI), and answer many researcher's doubts about the generalization of the use of the SERVQUAL model (Parasuraman et al. 1988). The scales consisted of 13 items instruments adapted from SERVPERF (Cronin & Taylor 1992), and 28 items resulted from the literature review and qualitative research, i.e., focus group discussion, piloting HEdPPERF scales on student samples in Malaysia and expert validation. The further study of HEdPERF utilization in Malaysian Polytechnic has confirmed that the HEdPERF was put forward as the most appropriate dimensions for the Malaysian polytechnic sector (Jalasi 2015). There is, in our view, still limited reputable literature of service quality measurement for higher education in Indonesia, particularly on vocational higher education. Some of that is presented in Table 1 below.

For this reason, an exploratory factor analysis (EFA) will be implemented on 41 HEdPERF attributes on Balikpapan State Polytechnic student respondents to find out the factors formed in the context of vocational higher education commonly named Polytechnic in Indonesia.

2 EXPLORATORY FACTOR ANALYSIS (EFA)

There are two types of factor analysis, i.e., Exploratory Factor Analysis (EFA) and Confirmatory Factor Analysis (CFA). The most difference is that EFA attempts to know the number of factors and does not identify which items load on their factors (Tabachnick & Fidell 2007). In *EFA,* the investigator has no expectations of the number or nature of the variables. Thus, it allows the researcher to explore the main dimensions to generate a theory or model from a relatively large set of latent constructs often represented by a set of items. The EFA is, in fact, sequential and linear, also involving many options. Therefore, to avoid potential

Table 1. Reputable pieces of literature related service quality measurement for higher education in Indonesia.

Literatures	Attributes
Ardi et al. (2012)	Item attributes were adapted from 5C TQM Model of Academic Excellence (Rajendran et al. 2005): (1) Commitment of top management; (2) Course delivery; (3) Campus facilities; (4) Courtesy; (5) Customer feedback and improvement
Sumaedi et al. (2012)	Item attributes were a combination between Athiyaman (1997), Soutar & McNeil (1996), and Hill (1995): (1) curriculum, (2) facilities, (3) contact personnel, (4) social activities, education counselors,(5) assessment, and (6)instruction medium
Napitupulu et al. (2018)	The 14-items attributes based on two dimensions, i.e., six-items of learning space and seven-items of the campus environment.

oversights in running EFA, the 5-steps EFA Protocol (Williams et al. 2010) can be used, i.e. (1) Is the data suitable for factors? → (2) How will the factors be extracted → (3) What criteria in determining factor extraction? → (4) Selection of rotational method → (5) Interpretation and labeling.

3 METHODOLOGY

The survey was taken place from June to July 2019 to among seven study program in Balikpapan State Polytechnic, Indonesia. Samples were taken using a non-probability random sampling technique. Only 235 from 369 administered questionnaires were valid due to many reasons, e.g., unreturned or uncompleted questionnaires. The adequacy of the sample size on the population size of 1,035 was met on the margin of error 5.6%, confidence level 95%, and variance in the population 50% (Krejcie & Morgan 1970),(Systems 2019). Seven-point of Likert scales had been used to measure the level of importance and performance. Scales were ranging from 1(strongly disagree) to 7 (strongly agree) for each importance and performance parts. EFA was applied on 41 items of HEdPERF questionnaires. The 5-step EFA protocol stated by Williams et al. (2010) was followed.

4 RESULT AND DISCUSSION

The KMO index was 0.924, which exceeded the recommended value of 0.6 (Kaiser 1974). While Bartlett's Test of Sphericity resulted in significant value at $\rho = 0.01$. Coefficients of correlation are also significant at $\rho = 0.01$ for most of the variables. It indicates a strong indication for the suitability of the factor analysis (Firdaus 2006). The principal component analysis (PCA) was used. Only factors of eigenvalue >1 were extracted (Kaiser's criteria), the communalities in variables are between 0.461 and 0.819, and the seven factors can explain 68.374 percent of the total variance. The rotation technique used was orthogonal with the varimax method. Factor loadings of 0.4 and above are considered significant at p-value = 0.05 with a sample size of 200 respondents (Hair et al., 2014), while n = 235 in our study. The investigation of Cronbach's alpha coefficient shows that all items were demonstrating internal consistency and established scales, respectively because Cronbach's alpha values were above 0.70. The results of seven factors, n items/factors, eigenvalue, percentage of variation, Cronbach α, factor loadings, and commonalities, are presented in Table 2.

From EFA, deferent number of factors in HEdPERF's is obtained. There are seven named factors i.e., academic aspect, access, delivery service, campus reputation, staff attitude, program reputation, and employability graduate. To perform the unidimensionality of these

Table 2. Result of factor analysis.

	Eigen Value	Percentage of Variation	Cronbach's alpha	Factor Loadings	Communalities
F1 Academic Aspect	16.70	13.598	0.906		
(Q2) Caring and courteous (by academic staff/lecture)				0.818	0.71
(Q3) Responding to request for assistance (by academic staff/lecture)				0.712	0.60
(Q1) Knowledgeable in the course content (academic staff/lecture)				0.710	0.65
(Q5) Positive attitude (by academic staff/lecture)				0.702	0.71
(Q8) Sufficient and convenient consultation (by academic staff/lecture)				0.698	0.69
(Q7) Feedback on progress(by academic staff/lecture)				0.698	0.69
(Q6) Good communication (by academic staff/lecture)				0.670	0.68
(Q4) Sincere interest in solving problem (by academic staff/lecture)				0.658	0.63
F2 Access	3.17	11.903	0.925		
(Q34) A fair amount of freedom				0.732	0.75
(Q39) Student's union				0.715	0.69
(Q36) Easily contacted by telephone				0.638	0.69
(Q37) Counseling services				0.631	0.68
(Q35) Confidentiality of information				0.614	0.69
(Q32) Service within a reasonable time frame				0.580	0.74
(Q33) Equal treatment and respect				0.556	0.58
(Q40) Feedback for improvement				0.538	0.71
F3 Delivery service	2.33	10.314	0.906		
(Q22) Caring and individualized attention (by administration staff)				0.820	0.81
(Q21) Sincere interest in solving problem (by administration staff)				0.742	0.69
(Q23) Efficient/prompt dealing with complaints(by administration staff)				0.732	0.82
(Q24) Responding to request for assistance(by administration staff)				0.710	0.74
(Q25) Accurate and retrievable records				0.574	0.71
F4 Campus Reputation	1.82	9.920	0.869		
(Q17) Ideal campus location/layout				0.739	0.66
(Q13)Recreational facilities				0.724	0.65
(Q10) Hostel facilities and equipment				0.713	0.73
(Q11) Academic facilities				0.683	0.69
(Q38) Health services				0.670	0.74
(Q41) Service delivery procedures				0.616	0.59
(Q9) Professional appearance/image				0.460	0.60
F5 Staff Attitude	1.56	9.233	0.882		
(Q28) Positive attitude				0.743	0.78
(Q29) Good communication (administration staff)				0.706	0.75

(Continued)

Table 2. (*Continued*)

	Eigen Value	Percentage of Variation	Cronbach's alpha	Factor Loadings	Communalities
(Q30) Knowledgeable of systems/ procedures				0.667	0.76
(Q27) Convenient opening hours				0.636	0.66
(Q26) Promises kept				0.503	0.70
F6 Programme Reputation	1.40	8.586	0.802		
(Q15) Variety of programmes/ specializations				0.776	0.69
(Q16) Flexible syllabus and structure				0.768	0.70
(Q12) Internal quality programmes				0.529	0.60
(Q14) Minimal class sizes				0.503	0.46
(Q19) Educated and experience academicians(academic staff/lecture)				0.478	0.64
F7 Employability graduate	1.07	4.820	0.747		
(Q18) Reputable academic programs				0.561	0.74
(Q20) Easily employable graduates				0.552	0.62
(Q31) Feeling secured and confident				0.518	0.59

scales, the CFA can be run using Structural Equation Modeling (SEM). The CFA is useful to analyze the underlying seven-factor model where individual items in the model are examined to see how closely they represent the same construct (Firdaus 2006). The adequacy of samples on the margin of error 5.6% is also being a limitation of this study. The improvement of adequacy validly samples to exceed the margin of error 5% can be conducted in the next factor analysis study.

5 CONCLUSIONS

Seven factors had been produced from EFA in the context of Balikpapan State Polytechnic, i.e., academic aspect, access, delivery service, campus reputation, staff attitude, program reputation, and employability graduate. Further factor analysis of Confirmatory Factor Analysis (CFA) can be carried out to evaluate the unidimensionality of the scales.

REFERENCES

Abdullah Firdaus, 2004. Managing service quality in the higher education sector: a new perspective through the development of a comprehensive measuring scale. In Proceedings of the Global Conference on Excellence in Education and Training: Educational Excellence through Creativity, Innovation & Enterprise.

Ardi, R., Hidayatno, A. & Zagloel, T.Y.M., 2012. Investigating relationships among quality dimensions in higher education. Quality Assurance in Education, 20(4), pp.408–428.

Athiyaman, A., 1997. Linking student satisfaction and service quality perceptions: the case of university education. European Journal of Marketing, 31(7), pp.528–540.

Cronin, J.J. & Taylor, S.A., 1992. Measuring Service Quality: A Reexamination and Extension. Journal of Marketing, 56(3), pp.55–68.

Firdaus, A., 2006. The development of HEdPERF: a new measuring instrument of service quality for the higher education sector. International Journal of Consumer Studies, 30(6), pp.569–581.

Hill, F.M., 1995. Managing service quality in higher education: the role of the student as primary consumer. Quality Assurance in Education, 3(3), pp.10–21.

Jalasi, A. Bin, 2015. Measuring Service Quality in Malaysian Polytechnic : Applying HEdPERF Model as New Measurement Scales for Higher Education Sector. Business Research Colloquium (BRC), pp.1–12.

Kaiser, H.F., 1974. An index of factorial simplicity. Psychometrika, 39(1), pp.31–36.

Krejcie, R. V & Morgan, D.W., 1970. Determining Sample Size for Research Activities. Educational and Psychological Measurement, 30(3), pp.607–610.

Napitupulu, D. et al., 2018. Analysis of Student Satisfaction Toward Quality of Service Facility. Journal of Physics: Conference Series, 954, p.12019.

Parasuraman, A., Zeithaml, V.A. & Berry, L.L., 1988. Servqual : A Multiple-Item Scale For Measuring Consumer Perceptions of Service Quality. Jornal of Retailing, 64(1), pp.12–40.

Rajendran, G., Sakthivel, P.B. & Raju, R., 2005. TQM implementation and students' satisfaction of academic performance. The TQM Magazine, 17(6), pp.573–589.

Soutar, G. & McNeil, M., 1996. Measuring service quality in a tertiary institution. Journal of Educational Administration, 34(1), pp.72–82.

Sumaedi, S., Bakti, G.M.Y. & Metasari, N., 2012. An empirical study of state university students' perceived service quality. Quality Assurance in Education, 20(2), pp.164–183.

Systems, C.R., 2019. Sample Size Calculator., p.1. Available at: https://www.surveysystem.com/sscalc.htm [Accessed March 21, 2019].

Tabachnick, B.G. & Fidell, L.S., 2007. Using Multivariate Statistics SEVENTH ED., NewYork: Pearson.

Williams, B., Onsman, A. & Brown, T., 2010. Exploratory factor analysis: A five-step guide for novices. Journal of Emergency Primary Health Care (JEPHC), 8(3), pp.1–13.

*Emerging Perspectives and Trends in Innovative Technology
for Quality Education 4.0 – Kusmawan et al (eds)
© 2020 Taylor & Francis Group, London, ISBN 978-0-367-25803-0*

Is financial education needed in elementary school?

D.S. Puspitarona, I. Abdulhak & Rusman
Indonesia University of Education, Bandung, Indonesia

ABSTRACT: The importance of acquiring the six literacies of basic life skills has grown in attention, resulting in an increase in programs and policies across countries worldwide. The Indonesian government recognizes literacy as a national movement; as such, it welcomes any number of diverse programs with the common goal of establishing literacy across multiple life skills. One of the six literacies considered to be a fundamental life skill is financial literacy. Thus, it has become a priority for program development. Developing financial literacy through formal education or more non-formal financial training programs has become a trending issue among teens, adults, the working sector, young couples, and retirees. But, does financial education need to begin as early as elementary school? This case study will examine the current financial literacy of elementary school students via survey, as well as identify potential areas of need across several elementary schools in Bandung, Indonesia.

Keywords: Elemetary School, Financial Education, Financial Literacy

1 INTRODUCTION

The importance of financial literacy programs was brought to attention in 2008 at the G-20 Countries Meeting put on by the Organization for Economic Co-operation and Development (OECD). This awareness resulted in the formation of the International Networking on Financial Education (INFE); specific areas of interest named include a promotion, research, and education of international corporations on the issues surrounding financial (OECD, 2012) (Lusardi, 2015) The Asia Pacific Economic Cooperation (APEC) designated financial literacy as an essential life skill that every child should learn. Furthermore, in 2016, the World Economic Forum issued an agreement that the six literacies—language, numeracy, scientific, digital, financial, culture, and citizenship-be required life skills learned in the 21st century.

Students' understanding of basic financial knowledge is the foundation for pursuing more in-depth financial education programs later on and presents an urgent need to be studied. The existence of a financial education curriculum aims to improve overall financial literacy-an important component needed to develop positive coping skills in the face of uncertain life situations (Aflatoun, 2012). Children have a particular need for financial literacy. First, it is prudent for children to be individually prepared to be tough, wise, reasonable, principled, responsible, and capable of decision-making for their own interests (CUNA, 2005) (Habschick, 2007). Second, the optimal window of time for the introduction of various life skills (including dealing with a range of daily life problems) is best introduced to children and pre-teens. (Triling and Fadel, 2009). Third, financial education is imperative to improving financial literacy because it increases the knowledge and skills of adolescents as they deal with the real-life problems that occur later in age (Blue, 2016). Finally, from an educational standpoint, the preparation and development of such initiatives should be carried out on an ongoing basis. In other words, the assessment of 15-year-old students—at the end of junior high school and junior high school—should reflect knowledge of adequate financial education long before the children enter their junior high school.

Therefore, there is a need for a specific investigation to assess if financial education is appropriate in elementary school students. Empirically conducted research should be targeted on current conditions, in places where surveys reveal that both financial literacy and education needs in elementary schools have not yet reached the standard level. This research furthermore is intended to fill this gap by providing an overview of the empirical conditions of elementary school students through an assessment of their needs for financial education.

2 LITERATURE REVIEW

Financial literacy competence is not solely defined by monetary knowledge, but rather, it is a concept ranging from the introduction of financial management to the ability to control financial expenditure by distinguishing between needs and wants (Cole, Sampson and Zia, 2009), (Aflatoun, 2012) The need for acquired financial literacy in children is well documented (CUNA, 2005) (Habschick, 2007), (Triling and Fadel, 2009), (Blue, 2016)The general understanding of financial literacy is defined as the knowledge and skills to process financial information, which includes financial concepts and motivation. In addition, the confidence to make effective financial decisions in various contexts such as financial awareness, financial knowledge, and understanding financial concepts is important (Cole, Sampson and Zia, 2009), (OECD, 2012), (Holden, 2009), (Lee, 2010), (Xu and Zia, 2012). Reflection on the impact of individual financial decision-making is another critical component that must be learned (Huston, 2010) (Hidajat, 2015), (Fernandes, Lynch Jr and Netemeyer, 2014), (Aflatoun, 2012).

By the time a preteen has entered the upper elementary school level, they have interacted socially and gained various influences from peers. Consumptive activities are familiar to most, such as buying food, toys, funny and unique hobby items, and attending activities with their friends that require funds, and so forth. This is certainly influenced not only by the financial status of and influence from their parents but by their own financial education as provided by their experiences and education received at school (Cochran, 2010; Senjiati, 2018). Some children believe that their social status and interaction with friends is heavily influenced by having certain clothes, accessories, and items. Moreover, feelings of self-acceptance in social situations come from pride when buying new goods (Courhanea, Gaileya and Zomb, 2008)

literacy capabilities needing to be mastered as a) being knowledgeable and informed about the issues of money and asset management, banking, investment, credit, insurance and taxes; b) the comprehension of basic concepts of money and asset management (eg investment value fluctuation and insurance risk); and c) the application of such knowledge to plan, implement, and evaluate financial decision-making. Financial education provides opportunities for children and adolescents to have increased social and economic power. Indeed, financial literacy competencies help them to be more "financially intelligent & resource smarter" in their lives, along with strengthened creativity, productivity, and overall competence towards the achievement of their own goals (Fernandes, Lynch Jr and Netemeyer, 2014; Drever, A I, 2015). The results of Aflatoun & UNICEF's research (2012) determined that children completing a financial education program showed positive changes in several areas: saving behavior (children changed how and where they saved money); saving preferences (strong increase in the percentage of children who save at school and the amount of money saved); risk preferences (children became more aware and attuned to potential risks in financial transactions); and locus of control (children felt that they had more control over their actions and decisions).

3 RESEARCH METHODOLOGY

A population of elementary school students in Bandung-Indonesia comprised of fifth-grade students were selected for the survey, which was conducted from March to May 2019 at four public and private elementary schools, and numbered 139 pupils' total.

Questionnaires were used in the process of collecting data, where students self-reported as a means to explore their need for financial education, in the following categories:

a. Student experience with money interaction
b. Student knowledge of financial management
c. Student experience with financial education at home
d. Student experience with financial education at school

4 RESULTS AND DISCUSSIONS

Elementary school student surveys revealed that one third had received an allowance from their parents since kindergarten (35%), another third since 1st-2nd grade (39%), one fifth from 3rd-4th grade (19%), and the smallest percentage from 5th grade (6%). Only 1% of students did not yet receive an allowance until after a 5th-grade level.

Most students reported that they acquired financial education experiences at home. Members of their family (father, mother, older brothers, and sisters, grandfather, grandmother) often provided simple and practical financial management advice concerning allowance management. Most students reported a lack of financial education from their teachers at school on ways to manage their allowance.

As paraphrased from elementary school student survey answers, they define competent financial management as the ability to: 1) save money by avoiding wasteful spending; 2) allow their allowance to accumulate over time; 3) save a portion of their allowance; 4) convert money into venture capital; 5) anticipate bankruptcy from over-spending and inadequate saving; 6) understand the purpose of using money; 7) categorize money to be used, saved and/or donated; 8) consider shopping only when necessary; 9) shop frugally; 10) save money safely; 11) utilize a shopping plan; 12) count money properly and quickly; and 13) delay desires and prioritize needs.

Students obtain the concepts and definitions of good financial management by paying attention to their daily interactions with parents, family, and friends. Examples of sound money management, such as observing the responsible salary and business practices of their parents, as well as attention to allowance management of their friends and relatives, inspired students how to approach their own financial management in a better way.

5 CONCLUSION

Students attending Bandung City Elementary School reported their involvement with money interactions as managing their allowance as well as making simple economic choices from a young age. However, it was revealed that they did not yet have relevant financial experiences—neither at school nor at home—in the use of their allowance in an appropriate way. Even though the role of adults (teachers and parents) in their financial education was very significant in that it provided role models for financial management, this research found the comprehensive financial education in elementary schools to be lacking.

This empirical finding, therefore, supports the conclusion that in order to increase financial literacy in students, formal financial education (both in separate and integrated content) in subjects at the elementary school level is necessary. Despite student exposure to simple economical and financial systems in their daily lives, formal financial education in schools will prepare them to better face future problems in their lives.

REFERENCES

Aflatoun, U. (2012) *Child Social and Financial Education*. New York: Division of Communication UNICEF.

Blue, L. E. (2016) "Financial Literacy Education with Aboriginal People: the Importance of Culture and Context," *Financial Planning Research Journal*, pp. 91–105.

Cochran, C. A. (2010) *Financial literacy in teens*. Caldwell College.

Cole, S., Sampson, T. and Zia, B. (2009) *Financial Literacy, Financial Decisions, and the Demand for Financial Services: Evidence from India and Indonesia*. Ameri.

Courhanea, M., Gaileya, A. and Zomb, P. (2008) "Consumer credit literacy: What price perception?," *Journal of Economics and Business*, 60, pp. 125–138.

CUNA (2005) *Credit Union National Association (CUNA) Thrive by FiveTM: Teaching Your Preschooler About Spending and Saving*.

Drever, A I, et al (2015) "Foundations of Financial Well-Being: Insights into the Role of Executive Function, Financial Socialization, and Experience-Based Learning in Childhood and Youth," *The Journal of Consumer Affairs*, 49(1), pp. 13–38.

Fernandes, D., Lynch Jr, J. G. and Netemeyer, R. G. (2014) "Financial Literacy, Financial Education, and Downstream Financial Behaviors," *Management Science*, 60(8), pp. 1861–1883.

Habschick, et al (2007) *Survey of Financial Literacy Schemes in the EU27*. Hamburg: Evers-Jung Financial Services Research and Consulting.

Hidajat, T. (2015) "An Analysis of Financial Literacy and Household Saving among Fishermen in Indonesia," *Mediterranean Journal of Social Sciences*, 6(5), pp. 216–222.

Holden, K. et al (2009) "Financial Literacy Programs Targeted on Pre-School Children: Development and Evaluation," *La Follette School Working Paper*, pp. 1–25.

Huston, S. J. (2010) "Measuring financial literacy," *The Journal of Consumer Affairs*, 44(2), pp. 296–316.

Lee, N. (2010) *Financial literacy and financial literacy education: What might be the components of an effective financial literacy curriculum?* University of London.

Lusardi, A. (2015) "Financial literacy skills for the 21st century: evidence from PISA," *Journal of Consumer Affairs*, 49, pp. 639–659.

OECD (2012) *Organisation for Economic and Cooperative Development International Network on Financial Education (OECD - INFE). OECD/INFE High-Level Principles on National Strategies for Financial Education*. Paris: OECD Publishing.

Senjiati, I. H. (2018) "Literasi Keuangan Syariah Bagi Anak School Age (Studi Kasus Pada Siswa Kelas 2 SD Darul Hikam Bandung)," *Amwaluna*, 2(2), pp. 33–55.

Triling, B. and Fadel, C. (2009) *21st Century Skills – Learning for Life in Our Times*. San Fransisco: Jossey-Bass A Wiley Imprint.

Xu, L. and Zia, B. (2012) "Financial Literacy around the World an Overview of the Evidence with Practical Suggestions for the Way Forward. Washington, D. C: The World Bank Development Research Group Finance and Private Sector Development Team."

Emerging Perspectives and Trends in Innovative Technology
for Quality Education 4.0 – Kusmawan et al (eds)
© *2020 Taylor & Francis Group, London, ISBN 978-0-367-25803-0*

Management of practical courses in the early childhood study program: Challenges for Open University of Indonesia

Marisa, Titi Chandrawati & Dian Novita
Universitas Terbuka, Tangerang Selatan, Indonesia

ABSTRACT: This study aims to determine the management and implementation of courses practicing in the Early Childhood Education Study Program. The evaluation was done using the "The Hierarchy of Policy Process" framework of Bromley (1989). The analysis is conducted on three levels of policy-making hierarchy, 1) Level of policy-making itself, 2) Level of strategy formulation to implement the policy, and 3) the operational Level in implementing the policy. The approach used in this research is "Context, Input, Process, Product/CIPP" by looking at these four components in the implementation of tutorials in practicing subjects. The results of this study are 1). Policies related to the assignment of courses with practical assignments have been made at UT, contained in the UT Catalog as reference for all related units in the implementation of face-to-face tutorials. However, UT Catalog has not been used as the main reference in implementing tutorials, either by tutors or students. 2). Formulation of strategies for conducting practicing course tutorials has been done by preparing tutorial support components in the Distance Learning Program Unit in the region. However, there is less synergy between UT policy and implementation strategy of practicing subjects in the field, 3). Tutor's perception of the course tutorial process is very good, seen from the learning strategy used in the tutorial, 4). Students' perceptions of the course tutorial process are inadequate because the information contained in the UT Catalog is not fully understood 5). Implementation of tutorials for practice courses takes place, especially when tutors have a high commitment to helping students learn.

Keywords: Practice Courses, Face-To-Face Tutorials, Practical Assignments, Teachers Early Childhood Education

1   INTRODUCTION

One of the goals of the Undergraduate Early Childhood (EC) Program, the Indonesia Open University (Universitas Terbuka/UT), is to equip EC educators with the ability to manage educational development activities. These include planning and implementing development activities, assessing learning processes and outcomes, promoting student development, and improving development activities based on assessment results (Universitas Terbuka Catalog, 2016).

In the EC Study Program, there are various courses with competencies in cognitive level 3, namely "application," meaning that students must apply and experience the competencies themselves. The course has been designed to allow students to practice the application of concepts. For this reason, as many as 22 individual courses are designed, which all include practical assignments in them. These include nine practicing courses involving students and 13 practice subjects that do not involve students.

Results of the observation period of the Face-to-Face Tutorial in 2016.1 and 2016.2 revealed the ineffective implementation of activities designed for EC Program practical courses in many UT Regional Offices (UT Learning Center, 2016). The results of Marisa's study (2016) show that students were asking for tutorials of higher quality. This could be achieved with the provision of instructional media, specific case studies, and the opportunity to practice concepts learned from the material. Face-to-face tutors have also suggested for real-life practice to be conducted in the EC courses. In the results of the discussion of UT lecturers with tutors at Regional Offices, it was found that tutors also did not practice the subjects they were responsible for.

The data above shows that there is a possibility that students have not received teaching that aligns with the standard outlined in the course objectives, and thus have not been able to achieve competency through the courses that they have enrolled in. In light of the problems mentioned above, the research problems can be formulated as follows: "How to manage and administer practical courses in the EC Teacher Education Study Program." The following outline the issues to be resolved in this study:

a. What policies are related to the management and implementation of practical courses?
b. What are the perceptions of tutors on the course tutorial process in practice?
c. What are the perceptions and expectations of students about the course tutorial process in practice?
d. What is the implementation of a tutorial for practice courses?

Therefore, this study aims to determine the management and implementation of practical courses in the EC Education Study Program. In more detail, this study aims to conduct an analysis of a) Policies related to the management and implementation of practical courses, b). Tutor's perceptions of the effectiveness of the practical courses, c). Students' perceptions and expectations of the tutorials for the practical courses, and d). Management and implementation of tutorials for practical courses.

2 METHODS

Conducting research into current policies will allow us to evaluate the management and implementation of courses in practice. Research is based on the approach Context - Input - Process - Product (CIPP) (Scriven et al., 1983). This evaluation will investigate the four levels in the management and implementation of practical courses in the EC Teacher Education Study Program. The objects that are the target of each type of evaluation in this study are as follows:

1. Evaluation of Context (context), in this context, the implementation of course tutorials that require students to practice their learning. The importance of tutorials consisting of real-life teaching practice is one of the efforts made by the Universitas Terbuka to improve the quality of its graduates.
2. Evaluation of Inputs (Input) to evaluate the readiness of the UT Center, Study Program, and UPBJJ in managing and organizing course tutorials with practical assignments, which include strategies for implementing policies, procedures, fees, and schedules that have been established for the course tutorials with practical assignments.
3. The evaluation of the process (Process), to analyze the potential for inconsistency in the implementation of tutorial courses with practical assignments to be anticipated before the program is implemented. The evaluation process will also evaluate the implementation of activities in the tutorial with practical courses.
4. Product Evaluation (Product), to evaluate the achievement of student learning outcomes in course tutorials with practical assignments, including the results of learning in face-to-face tutorials. The research samples are UT leaders, students, tutors, and supporting documents related to conducting face-to-face tutorials in practicing courses.

Data was collected through analysis of various policy documents related to practice subjects and data from the Academic and Student Administration Bureau and UT Regional Offices. In addition, data collection was also conducted through face-to-face meetings with all netted respondents (students, tutors, UT lecturers, and UT Coordinator for Learning Support and Materials Services at Regional Offices). The instrument used was a questionnaire for tutors and students, as well as interview guidelines.

The research was conducted from April to November 2017. The research locations were at 4 Universitas Terbuka Regional Offices, namely Purwokerto, Bandung, Bogor, Samarinda, Jakarta.

The research data was analyzed qualitatively and quantitatively through reviewing the number of practicing subjects, the number of students participating in the tutorial and the number of tutors, as well as the data on student learning outcomes in the 2016.1 and 2016 semester.2. Data from various documents (guidelines/technical guidelines) will be analyzed qualitatively descriptive (Morgan, 1998).

3   CONCLUSION AND RECOMMENDATIONS

This study can be concluded in the following points:

1. Official policies of the tutorial course were created at the UT Center and are documented in the UT Catalog to act as a reference for all relevant work units in conducting face-to-face tutorials. However, such policies have not been strictly implemented in the tutorials. Other findings indicate that UT Head Office's readiness in supporting the implementation of learning for practicing courses is supported by the availability of Face-to-Face Tutorial Guidelines along with technical instructions. Nevertheless, the guideline does not contain tangible instructions on the implementation of practicum/practicum courses.
2. The strategy of holding tutorials on practicing courses was formulated by preparing the supporting components of the tutorial in the Distance Learning Program Unit in the area. However, there is no synergy between UT's policies and DLPU's strategies, which has resulted in a difficult implementation of practice courses in the field.
3. Tutors believe that hosting course tutorials is a very good practice and has strong learning objectives to achieve great improvement for students.
4. Students perceive the course tutorials to be ineffective in practice, as the information documented in the UT Catalog has not communicated its objectives effectively.
5. Implementation of tutorials for practical courses is most effective if the assigned tutor has a strong commitment to helping students learn.

The recommendations given for the implementation of practicing courses are as follows:

1. Policies on practicing courses documented in the UT Catalog should be more focused on academic aspects, namely the provision of information related to the curriculum of study programs and practice courses at an operational level, in the Study Orientation New Students (OSMB) and in the process of training and debriefing of tutors.
2. For this reason, guidance to students in OSMB needs to be supplemented with material about the curriculum of the study program and practice courses, and strategies for applying them in the tutorial.
3. Debriefing to tutors should be supplemented with material about the study program curriculum and practice courses, and strategies for applying them in the tutorial.
4. A complete study needs to be conducted relating to the specific elements of the learning system for practicing courses that do not correspond to the implementation of face-to-face teaching tutorials and its materials. This will ensure that the learning is in accordance with the teaching strategy for the courses.

REFERENCES

Bromley, Daniel W., (1989) Economic Interests and Institutions, Cambridge, Basil Blackwell.
Creswell, John W., (1994) Research Design: Qualitative, Quantitative, and Mixed Methods Approaches, California, Sage Publication.
Universitas Terbuka Catalog, 2016.

Madaus, George, Michael S Scriven, and Daniel L. Stufflebeam., (1983) Evaluation Models: Viewpoints on Educational and Human Service Evaluation. Boston, Kluwer-Nijhoff Publishing.

Richey, RC, (2001) Instructional Design Comptencies: Standards. Syracuse, New York: Clearinghouse on Instructional and Thnology.

Simonson, M; Smaldino, S. E & Zvacek, S. Teaching and Learning at A Distance (6years), Information Age Publishing, North Carolina (2015).

Smith, PL & Ragan, TL, Instructional Design. Upper Saddle River, NJ Merril Prentice Hall. Inc. 2003.

Stupak, Nataliya, Towards Improvement of Performance Policies: Addressing Policy Implementation, http://www.esee2011.org/registration/fullpapers/esee2011_6e5025_1_1305697321_4924_2163.pdf

Emerging Perspectives and Trends in Innovative Technology
for Quality Education 4.0 – Kusmawan et al (eds)

Use of interactive video program base on a problem-based learning approach in Universitas Terbuka

S. Prabowo & Andayani
Universitas Terbuka, Indonesia

ABSTRACT: The primary aim of the Universitas Terbuka (UT) is to assist students willing to attend higher education through an open and distance learning system. It provides a learning medium capable of improving the quality of distance learning using an interactive video program based on problem-based learning (PBL). The research aimed at theoretically reviewing the interactive video based on PBL. Half of the respondents in a field study stated that it had not been optimally utilized because of the materials package, and this motivated students to determine solutions to the problems, thereby easily understanding the learning material in the modules. The implication of this article was the need to develop and evaluate interactive video programs based on PBL for UT students.

1 INTRODUCTION

The rapid development of science and technology requires continuous improvement in the quality of human resources. However, one of the main limitations faced by people in developing themselves and their careers is time and place. Therefore, Universitas Terbuka (UT) provides an opportunity for higher education through an open and distance learning system, with the aim of providing quality tertiary education for Indonesian people. UT provides various printed materials such as (BMP or module) and nonprinted learning materials such as interactive video programs, audio graphics, and computer-assisted instruction (CAI) programs.

This article was written in accordance with the results obtained from the use of BMP and nonprinted materials in 2018. It showed that the use of nonprinted teaching materials in the form of interactive video programs and CAI by students was still below average because they were viewed as less attractive (Krisnadi and Prabowo, 2018). Furthermore, it was unable to ensure that students used them when needed. Therefore, the BMP and nonprinted materials such as interactive video programs need to be arranged in interrelated and complementary ways to encourage students to use both forms simultaneously.

1.1 Significance

This article discusses the concept of integrating nonprinted teaching materials into the printed form using a problem-based learning (PBL) approach. Presently, UT is improving and enriching the form of printed teaching materials by integrating nonprinted materials and the PBL approach.

2 LITERATURE REVIEW

The implementation of UT provides opportunities for individuals with difficulties in assessing education because of several factors, such as distance, time, age, physical challenges, and learning resources. The limitations of these learning resources are accommodated by the distance learning process using various teaching materials. Printed materials, also known as modular, contain the scientific meaning of the prepared materials, which enable students to learn independently, thereby minimizing the need for tutors. UT also develops nonprinted teaching materials into BMP, which aims to enrich its concepts such as audio, interactive video, and CAI provided in one package for students (Suparman, 2014). The combination of printed and nonprinted teaching materials is required to carry out open and distance education (Moore and Kearsley, 2012). The effectiveness of learning is supported by the diversity of the materials provided following students' needs. De Simone (2014) stated that using a package of quality teaching materials tailored to the type of media that has been prepared is the key to active learning. Various electronic media are used to improve the quality of learning services for UT students, such as interactive video programs into BMP.

The application of interactive video programs is related to the efficiency and effectiveness of learning, student comfort, and a more dynamic classroom atmosphere. The program is also easier to use and equipped with various other interactivity features (Papadopoulou and Palaigeorgiou, 2016). Multimedia learning material, such as interactive videos, are used to create dynamic reading comprising of sentences, sound, video, and animation (Munir, 2013). Smaldino et al. (2008) stated that video programs accommodate almost all topics for all learning styles and dimensions.

Problem-based learning (PBL) is a learning strategy that presents problems that are closely related to real and significant cases for students to carry out further exploration and observation (Arends, 2015). First, it provides an alternative way of teaching to help provide an atmosphere of learning in response to rapid changes (Hussain et al., 2007) and increases motivation and a high level of thinking (Baden and Major, 2004; Apriliadewi, 2017). In the context of learning, there are several PBL indicators that answer questions or problems stimulated by students' curiosity to study further. Second, it studies the problem from various scientific angles such as biology, sociology, and economics. Third, PBL is a strategy that directs students to determine real solutions to problems (not imaginary). Fourth, it encourages students to produce products after investigating and determining solutions to the problem under study, as well as building cooperation between them (Arends, 2015; Apriliadewi, 2017).

According to Hung (2013), PBL is a complicated strategy and requires high-level thinking skills. The first step in its implementation is a tool that tends to help students internalize the conceptualization process and encourages them to explore (Barret et al., 2011). Besides, it is essential to design those that are appropriate to the learning needs of students and are divided into three experiences: real, stimulating, and digital forms while the goals include conceptual, collaborative, and lifelong learning to build knowledge, problems, learning, and facilitators (De Simone, 2014).

In practice, UT consists of a Smart Online Teacher Portal (Guru Pintar Online) and an Open Educational Resource (OER) aimed at triggering critical and creative thinking skills. An illustration used as a problem in video streaming between 5 and 10 minutes was developed based on learning problems experienced by kindergarten and elementary school teachers. Furthermore, the solution presented in streaming video is the experience of teachers who have successfully overcome learning problems (Rahayu and Andayani, 2015). The same pattern used to develop video streaming as an open-source learning method is used to create additional teaching materials that are not printed using interactive video programs.

3 METHODOLOGY

This concept study produced a framework for integrating nonprinted teaching materials (interactive videos) into printed (BMP or module) using a PBL approach, which is followed by research and development. This is carried out via the following procedure: (1) analyzing student needs through focus group discussions; (2) developing learning objectives to be achieved; (3) analyzing the topic of the material to be packaged in interactive video programs and BMP forms; (4) developing interactive video scripts and writing BMP in an integrated manner with a PBL approach; (5) producing teaching materials; (6) conducting small-scale trials; (7) improving the materials; and (8) conducting larger-scale trials.

4 RESULTS AND DISCUSSION

The interactive video program is a combination that is provided as a complementary teaching material to BMP. According to Arends (2015), PBL syntax is divided into five stages that are applied as follows: (1) guide students in raising problems relevant to the concept being discussed and presentation of various events described through BMP and also interactive videos; (2) explain the stages in arranging students in written form in BMP with clear and detailed procedures; (3) provide assistance, support, and guidance in investigating issues written in BMP, without the need to include other sources, such as contact and email addresses for the opportunity to communicate and interact; (4) produce products that are presented visually through interactive video programs; and (5) analyze and evaluate problem-solving processes in two formats through visualized BMP in the form of videos that are presented in fragments and last for 10 minutes. The effectiveness and duration are related to the convenience of students, with a shorter duration provided if necessary (Smaldino et al., 2008). UT is diversified across large cities, small towns, and rural areas, as well as ages, with a simple and large PBL-based navigation component. This interactive video program is applied relative to the benefits of the PBL approach, which has an impact on learning outcomes (Tambunan et al., 2018)

Figure 1 shows the framework of an interactive video integrated into printed material using a PBL approach.

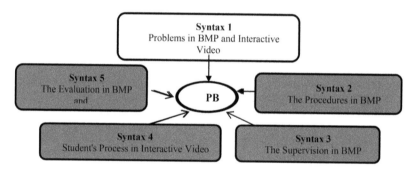

Figure 1. The framework of integration of the interactive video program into BMP based on the PBL approach

5 CONCLUSIONS AND DIRECTIONS FOR FUTURE STUDIES

The use of interactive video programs based on the PBL approach is expected to increase the effectiveness of learning using the right strategy. Therefore, the development of interactive video programs using BMP requires further study through an in-depth evaluation of various concepts related to the application of PBL. Also, studies are needed to develop and assess the importance of the program, which acts as a supplement to BMP among UT students.

REFERENCES

Apriliadewi, P. A. (2017) An analysis of the implementation of problem based learning in learning English at the XI grade science class of SMA negeri 1 Singaraja in the academic year 2015/106. *International Journal of Language and Literature*, 1(1),11–18.
Arends, R. I. (2015) *Learning to teach*, 10th ed. New York: McGraw-Hill Education.
Baden, M. S., and Major, C. H. (2004) *Foundations of problem-based learning*. New York: Society for Research into Higher Education & Open University Press, McGraw-Hill Education.

Barret, T., Cashman, D., and Moore, S. (2011) Designing problems and triggers in different media: Challenging all students. In *New approaches to problem-based learning: Revitalizing your practice in higher education*, p. 18. New York: Routledge.

Hung, W. (2013) Conceptualizing problems in problem-based learning. In *Learning, problem-solving, and mind tools*. New York: Routledge.

Hussain, R. M. R., et al. (2007) Problem-based learning in Asian universities. *Studies in Higher Education*, 32(6),761–772. doi: 10.1080/03075070701685171.

Krisnadi, E. and Prabowo, S. (2018) Kajian Pemanfaatan Bahan-bahan Pembelajaran Non Cetak Oleh Mahasiswa, Jakarta.

Moore, M. G. and Kearsley, G. (2012) *Distance education: A systems view of online learning*. Belmont, CA: Wadsworth, Cengage Learning.

Munir (2013) *Multimedia: Konsep & Aplikasi dalam Pendidikan*. Bandung: Alfabeta.

Papadopoulou, A. and Palaigeorgiou, G. (2016) Interactive video, tablets and self-paced learning in the classroom: Preservice teachers perceptions. In *13th International conference on cognition and exploratory learning in the digital age (CELDA)*, Mannheim, Germany, p. 8.

Rahayu, U. and Andayani (2015) Portal Guru Pintar Online (pintu interaksi antarguru secara online). In Belawati, T., Damayanti, N. S., and Puspitasari, K. A. (eds.), *Universitas Terbuka di Era Informasi*. Jakarta: Universitas Terbuka.

De Simone, C. (2014) Problem-based learning in teacher education: Trajectories of change. *International Journal of Humanities and Social Science*, 4(12),17–29.

Smaldino, S. E., Lowthre, D. L., and Russel, J. D. (2008) Instructional technology and media for learning. Upper Saddle River, NJ: Pearson Education.

Suparman, A. (2014) *Teknologi Pendidikan dalam Pendidikan Jarak jauh: Teori dan praktek*. Jakarta: Universitas Terbuka.

Tambunan, L., Rusdi, R., and Miarsyah, M. (2018) Effectiveness of problem based learning models by using e-learning and learning motivation toward students' learning outcomes on subject circulation systems. *Indonesian Journal of Science and Education*, 2(1),96–104. doi: 10.31002/ijose.v2il.598.

Emerging Perspectives and Trends in Innovative Technology for Quality Education 4.0 – Kusmawan et al (eds)
© 2020 Taylor & Francis Group, London, ISBN 978-0-367-25803-0

Target situation analysis of writing for academic purposes in Indonesiantertiary education: Users' insight

Audi Yundayani
STKIP Kusuma Negara, Jakarta, Indonesia
Universitas Terbuka, Tangerang Selatan, Indonesia

Dian Kardijan
Universitas Siliwangi, Tasikmalaya, Indonesia

ABSTRACT: In higher education, students' academic writing competence is an important aspect that contributes to their academic performance. Previous research showed that writing skills had not been taught based on specific academic purposes. At the college level, this skill is generally regarded as the heart of academic life. Need analysis has not been applied, leading to students' inability to achieve the competence needed. This paper describes the users' insight related to target situation analysis in an academic writing course at non-English education study programs. A descriptive survey was used by involving students, instructors, and heads of study programs. The data were collected using a questionnaire and interviews. The findings are presented in terms of (1) the learning objective, and (2) required knowledge and skills. These findings can be used as a basis for developing a better writing course for academic purposes.

1 INTRODUCTION

Generally, the English for Academic Purposes (EAP) course is designed as part of English for Specific Purposes (ESP) to meet certain requirements that all college students have to meet. The development of EAP can be viewed as the result of the low generalizability of ESP courses (Sabariah and Rafik-Galea, 2005). EAP aims to enable students to create and use abilities that can enhance the learning of students and the advancement of students in an educational learning setting (Spirovska, 2015). The majority of undergraduates not only require language support at university but also that the supports should be oriented towards academics rather than general English (Evans and Green, 2007). This suggests that the mastery of study skills within the EAP domain will allow students to perform in academic context because the focus of the EAP is the development and use of students' English skills that will facilitate them in the learning process. The significance of writing as the core of academic life will be the most demanding in academic literacy practices. The reason is that writing is a convenient way to judge the performance of students. The academic achievement of a student can be readily assessed by his writing skills (Valli and Priya, 2016). Students' writing product represents their previous knowledge, experience, and ability. It is important for academic institutions to develop students' academic writing habits.

The most significant phase in ESP is the needs analysis, focused on defining the requirements and attitudes of students, including their shortcomings, desires, and interests. Needs analysis is a tool to get a picture of the students' needs as a basis of course design in accordance with learning objectives. It is an umbrella term that covers many elements, incorporating the objectives and backgrounds of students, their language skills, their reasons for taking the course, their preferences for teaching and learning, and the circumstances they need to interact in (Hyland, 2006). Needs are understood as a gap between what is expected and what is happening at the moment, and it can be seen as necessities, lacks, and wants (Hutchinson and Waters, 1987). The result of need analysis can be an element to design or review the whole learning components to establish key learning outcomes based on the requirements.

Analysis of the target situation focuses on the requirements of students by the end of a learning course. Essentially, it is a matter of asking questions about the target situation and the attitude of the different respondents in the learning process towards that situation (Hutchinson and Waters, 1987). The results of the analysis of the target situation will respond to questions: (1) Why should students learn the language? (2) How is the language to be used? (3) What are the content areas that should be covered? (4) Who will students use the language with?; (5) Where is the language used by the students?; and (6) When is the language used by the students? These frameworks are described in the course's learning goal. The objective of the analysis of the target situation is to determine what students need to be able to do in English as a result of the course and to obtain this result through activities that reflect those of the target situation (Munby, 1981). The upshot of the target situation analysis will provide data on the objectives and needs of the student, including the abilities and language needed for the context in which the language will be used.

The findings from the preliminary research show that the learning program in writing for academic purposes has not been designed based on needs analysis in the ESP field. The program performed was only based on the teachers' intuition without mentioning the objective of the program clearly. It impacted the inability of students to achieve students' competencies in writing for academic purposes. In addition, the learning components were not suitable for the students' requirements, including instructional materials. The current learning program did not reflect the students' future roles covering required linguistic skills and knowledge that they should perform based on their subject. Students tended to face some difficulties in achieving writing skills for academic purposes communicatively. They cannot present their ideas clearly in writing based on academic contexts. Accordingly, this research is aimed at portraying the result of target situation analysis in writing for academic purposes at non-English education study programs. The upshot was described in terms of (1) the learning objective; and (2) required knowledge and skills of writing for academic purposes. There are some relevant studies (Taghizadeh, 2019); (Sari, Kuncoro and Erlangga, 2019), but the target situation analysis of writing for academic purposes in the Indonesia context has not been explored widely. Therefore, this study will contribute to the language education field.

2 METHOD

The research involved thirty students of non-English education study programs. They were chosen through purposive sampling technique. To support the data, three English lecturers as instructors and five heads of non-English education departments were involved in identifying target situation analysis data. The data were collected through distributed questionnaires. To broaden the findings, semi-structured interviews were conducted with the instructors and heads of non-English education departments. The data collected through the questionnaire was submitted and administered. The interviews aimed to identify the problem more openly by asking respondents' opinions and ideas. In this study, the reviewed documents included the syllabus, lesson plans, instructional materials, and students' writing test data that were provided to get the description of the current situation. The data analysis employed a qualitative approach.

3 FINDINGS AND DISCUSSION

The research findings reveal the learning objective as well as the required knowledge and skills from the users' perspectives. The research findings describe the learning goal from the insight of the user's required knowledge and skills. The results of the research showed that (1) the purpose of the academic writing course is to obtain the ability of the students to arrange and develop good sentences in a discourse related to each discipline; (2) the students must be given the knowledge and skills of (a) rhetorical - functional in the form of description, narrative, definition, and classification; (b) academic genres in the form of essays and reports; (c) process of writing; (d) summary; and (e) paraphrasing. These findings relate to the nature of the EAP. As the branch of English for specific purposes, the EAP sphere, writing course, is aimed at meeting the needs of students in order to be able to write in the academic field. One of the goals is to acquire new knowledge. Specifically, writing is seen at the university level not only as a standardized communication scheme but also as an important instrument for learning (Weigle, 2002).

Moreover, the findings emphasize that the main function of English writing skills in higher education is to develop the students' knowledge, not just to convey information. Writing for academic purposes is essentially an ability to incorporate data from various sources and past studies into appropriate research fields. Even the most initial academic paper in-tegrates facts, thoughts, concepts, and theories with quotations, paraphrases, summaries, and short references from other sources (Jordan, 1997). In combining read-ing and writing, there are two main underlying purposes, such as using a source text. The first is knowledge sharing, where students demonstrate their ability to identify and present suita-ble source text materials in the course displaying a broader ability to understand sources and address the subject at hand. The second is the transformation of understanding, where students use source text material more substantially to create a broader theme (Bereiter and Scardamalia, 2013). Writing English in the EAP field with a process approach is the most appropriate way to develop academic texts. What it says shows that the aspect of writing English for academic purposes requires skills to inte-grate ideas, facts, concepts, and theo-ries from various sources, either by quoting, inter-preting, summarizing, and using refer-ences.

The result of students' perception can be seen as corresponding to an assessment of the target situation, an assessment of the learning situation, and an analysis of the pre-sent situ-ation. It should determine the communication requirements arising from an assessment of the communication in the target situation (Basturkmen, 2010). Further-more, it is a basis for determining the direction of the learning program objectives pre-cisely so that the feasi-bility, usefulness, and effectiveness of the learning program will be in accordance with the expected needs. Hopefully, the research findings would be beneficial in redesigning and developing a better academic writing syllabus for English lecturers to achieve the learning goal.

4 CONCLUSION

The upshot of the target situation analysis enables the course designer to define students' requirements for academic reasons over the duration of their writing abilities as well as their area shortcomings. This study was conducted to describe the outcome of target situation analysis in academic writing skills in non-English education study programs in terms of learning goals and knowledge required. Hopefully, the research findings would be beneficial to English lecturers in redesigning and developing a better academic writing syllabus to achieve the learning goal. English writing skill for academic purposes indicates that the required cognitive abilities are available to students in higher education. These skills are related to reading comprehension skills. Reading skills are needed to understand learning materials dominated by the use of English subtitles. Thus, writing skills are absolutely neces-sary to update the reading understanding, the content, so that students can write well. This skill is not easy because students are not only supposed to convey the message, the written essence of the original source, but they must have the ability to express ideas using their own words, structure, and style.

REFERENCES

Basturkmen, H. (2010) "Developing courses in English for specific purposes. Springer."

Bereiter, C. and Scardamalia, M. (2013) "The psychology of written composition. Routledge."

Evans, S. and Green, C. (2007) "Why EAP is necessary: A survey of Hong Kong tertiary students," *Journal of English for Academic Purposes*, 6(1), pp. 3–17.

Hutchinson, T. and Waters, A. (1987) "English for specific purposes. Cambridge university press."

Hyland, K. (2006) "English for academic purposes: An advanced resource book. Routledge."

Jordan, R. R. (1997) *English for academic purposes: A guide and resource book for teachers.* Cambridge University Press.

Munby, J. (1981) *Communicative syllabus design: A sociolinguistic model for designing the content of purpose-specific language programmes.* Cambridge University Press.

Sabariah, M. and Rafik-Galea, S. (2005) "Designing test specifications for assessing ESL writing skills in English for academic purposes," in *ELT concerns in assessment*, pp. 149–167.

Sari, R. K., Kuncoro, A. and Erlangga, F. (2019) "NEED ANALYSIS OF ENGLISH FOR SPECIFIC PURPOSES (ESP) TO INFORMATIC STUDENTS.," *JEELL (Journal of English Education, Linguistics and Literature)*, 5(2), pp. 26–37.

Spirovska, E. (2015) "SELECTING AND ADAPTING MATERIALS IN THE CONTEXT OF ENGLISH FOR ACADEMIC PURPOSES-IS ONE TEXTBOOK ENOUGH?," *Journal of Teaching English for Specific and Academic Purposes*, 3(1), pp. 115–120.

Taghizadeh, M. (2019) "The Effect of Needs Analysis on Designing ESP Courses in Iranian Context 'A Case Study,'" *English Language Teaching Letters*, 1(1).

Valli, K. S. and Priya, N. V. (2016) "A task-based approach to develop the writing skills in English of students at college level," *International Journal of Applied Engineering Research*, 11(3), pp. 2145–2148.

Weigle, S. C. (2002) "Assessing writing. Ernst Klett Sprachen."

Emerging Perspectives and Trends in Innovative Technology
for Quality Education 4.0 – Kusmawan et al (eds)
© 2020 Taylor & Francis Group, London, ISBN 978-0-367-25803-0

Pushing toward internationalization of higher education: Challenges for a newcomer

Siti Hadianti
Universitas Terbuka, Tangerang Selatan, Indonesia

Bobi Arisandi
Universitas Pamulang, Indonesia

ABSTRACT: Internationalization of Higher Education (IHE) in Indonesia has been developing in recent years. The Indonesian government has been bringing internationalization into higher institutions, as clearly shown by the implementation of a new accreditation instrument. However, many challenges need to be faced by newcomers. The study reported in this article aimed to investigate the challenges for a newcomer to the IHE using qualitative research with a case study design. The research was conducted in one of the private universities in North Lampung, Indonesia, which has implemented one international cooperation program. The participants were the head of the Office of International Affairs, while the six students were in a student–teacher exchange program with Thailand and the Philippines. The research showed that the newcomers experienced human resource, managerial, and financial challenges. It was suggested that the institution should train its lecturers and students to improve their knowledge and skill to enable them to join the international community and make an international collaboration to share the funding.

1 INTRODUCTION

The Internationalization of Higher Education (IHE) in Indonesia has been developing in recent years. According to Knight (2003), IHE is "the process of integrating an international, an intercultural, or global dimension into the purpose, functions or delivery of postsecondary education." Practically, it is "the process of commercializing research and postsecondary education, and international competition for the recruitment of foreign students from wealthy and privileged countries to generate revenue, secure national profile and build an international reputation" (Khorsandi Taskoh, 2014).

Moeliodihardjo (2015) stated that many universities had established an international office to offer support to international students, since they are likely to generate significant revenue to the universities. The international office is responsible for providing information on the programs, facilities, and learning environment. The IHE was developed to ensure globalization in the field of economy, social, and technology. Globalization is different from internationalization. According to Currie and Thobani (2003), "...globalization represents neo-liberal, market-oriented forces enabling a borderless world, and internationalization represents arrangement between nation-states primarily cultivating greater tolerance and exchange of idea."

The action of the Indonesian government in bringing internationalization into higher institutions is shown by the implementation of a new accreditation instrument. The instrument has been socialized recently by the Board of National Accreditation for Higher Education (BAN-PT). In the report guide of Akreditasi Perguruan Tinggi (APT) version 3.0, the new instrument focused on output and outcome. Therefore, the assessment would be done only on the implemented cooperation. The upcoming higher education accreditation will make use of the instrument, consequently leading to a new challenge for universities in Indonesia since they need to fulfill the demand to achieve good accreditation and improve Education Quality Continuously (EQC).

Therefore, for newcomers, this comes as a challenge. A newcomer is a novice in a particular activity or situation. In this case, a newcomer is an institution that had implemented the international agreement. The creation and implementation of international cooperation are two different entities. The newcomer universities in internationalization must sign a Memorandum of Understanding (MoU) and Memorandum of Agreement (MoA). However, its implementation is a major problem for universities, and it can be eliminated through the construction of research related to internationalization. The research was predicted to answer a question about what types of challenges newcomers faced in implementing internationalization. This research took place in one of the private universities in Lampung with hidden names and participants for ethical reasons.

2 METHOD

This is qualitative research where a case study design is used. According to Creswell and Poth (2017), "qualitative research is a means for exploring and understanding the meaning individuals or groups ascribe to a social or human problem." For data collection, the research considered documentation, interview, and questionnaire. The collected data will be organized, categorized, identified, and concluded in a scheme relevant to Merriam's qualitative research and the case study applications in education (Malik and Hamied, 2014).

The research used a purposive sampling technique in which the participants were the head of the office of international affairs and six students attached to an exchange program with Thailand and the Philippines for one month. These students were called ST (student) while the lecturers became the head of the Office of International Affairs or Kantor Urusan Internasional (KUI) and were called HK (head of KUI).

3 RESULTS AND DISCUSSION

On conclusion of the research, the results were compiled based on the data from documentation, interview, and questionnaire in three main areas: human resources, financial, and managerial sector as represented in Figure 1.

3.1 *Human resources*

Lack of human resource qualifications was the first problem the institution experienced when it conducted internationalization. Qualified international staff was responsible for the smooth operations of internationalization within the institution. Unfortunately, the findings indicated a lack of qualified staff in the international office, which led to hibernation from 2011 to 2016. The first cooperation conducted and implemented was a student–teacher exchange program as an initiative of one organization in Southeast Asia.

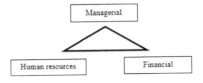

Figure 1. The three areas of challenge in internationalization.

Based on the interview with the head of KUI, the major problem was ensuring international cooperation in the presence of a lack of understanding, since creating international cooperation is not a trivial matter. The person in charge must have a sufficient background of expertise concerning the international project. In this case, the HK was not included in the criteria. The only reason for the appointment is the award from a European scholarship program, which revealed plenty of internationalization programs. However, the willing participants did not qualify for the program for various reasons, such as insufficient English proficiency. The following is the transcript of the interview.

R   :   *What are the main obstacles to international cooperation in your institution?*
HK :   *I think the biggest challenge for lecturers and students who are not interested in joining the international event because of the language problem. Moreover, it is difficult to have new cooperation that can be implemented.*

The head of KUI had already persuaded the lecturer and faculty staff to join the international program in the form of joint research or visiting lecturer of international conferences abroad, but they were not interested. The case was similar for students because they had poor English proficiency besides having originated from the English Education department. In addition, some of the students were from other departments such as Mathematics Education, which was confirmed from the participants' questionnaire responses.

3.2   *Managerial*

Based on the questionnaire and interview, a number of students suffer from problems such as lack of competence in the international community, especially in the language and pedagogic areas. Another problem is culture shock, since some of the students had never gone abroad. The answer is shown in the following questionnaire:

R   :   *What are the challenges as a participant in the international program?*
HK :   *These challenges include the English language and class management in teaching. Besides, adopting the new culture results in a difficult life.*

The interview also confirmed two participants, ST 6 and HK, citing that there were no preparations or pre-departure training offered by the institution. Such cases of poor management result in disadvantages for participants in the program. The following are the result of the interview:

R   :   *What is the preparation before students do the international event? If any.*
HK :   *There are no preparations since we were not taken through; the briefing from the organization was enough.*
S6 :   *No, there is no pre-departure training. I think that is very important, considering none of us had gone abroad.*

Unfortunately, that area was neglected by the university, as testified by the participants in the international program, where none was given pre-departure training from the institution. Lack of management from the university is to blame for this matter. The stakeholders do not understand the management of the international program. They consider having MoU or MoA between two institutions at the start and the end of the program without giving students some management training.

3.3 *Financial*

Financial matters were crucial since the researcher found the institution lacks the commitment to funding. Based on the questionnaire, interview, and documentation, it was revealed that in the international program, the students were charged for the ticket and accommodation. This was contrary to the MoU applied in the agreement. The reality was further confirmed by the questionnaire below:

R : *What are the challenges when you become a participant in the international program?*
S : *The main challenge is the financial issue to cover travel expenses to and from Indonesia. Even though the university gives some help, there is a need for a lot of money for expenditures.*

In the MoU, it is stated that the receiving and sending university needs to provide the funds for accommodation. A strong commitment to financial issues solves human resources and managerial problems. In the area of human resources, the issue mainly relates to English proficiency. It was found that the head of KUI already suggested a language program. However, the financial issue was brought to the surface. The policymaker did not confirm his agreement related to this matter, and as a result, the program never became a reality. Furthermore, several proposals for self-improvement workshops for increasing international office competence were also being ignored because of financial disagreement.

4 CONCLUSIONS AND SUGGESTIONS

The research concludes that there are three main challenges for a newcomer in higher education internationalization: human resources, financial, and managerial. These challenges are related to each other, so it is important to find an amicable solution.

Based on these conclusions, the following suggestions were outlined. The human resource problem can be solved by the actualization of the language training center for the problem in the area of English proficiency. For the poor management issue, the KUI staff can be trained and asked to follow international or national training abroad that is commonly conducted by the well-established international office to boost their confidence and broaden their relations. The financial issue can be solved by creating a mutually beneficial partnership, which helps ease the financial burden caused by internationalization. The program can be a joint seminar by inviting guest lecturers from universities abroad, where the financial obligation can be shared among the institutions holding the event.

REFERENCES

Creswell, J. W. and Poth, C. N. (2017) Qualitative inquiry and research design: Choosing among five approaches.
Currie, D. H. and Thobani, S. (2003) From modernization to globalization: Challenges and opportunities. *Gender, Technology and Development*, 7(2), 149–170.
Khorsandi Taskoh, A. (2014) A critical policy analysis of internationalization in postsecondary education: An Ontario case study.
Knight, J. (2003) Updated definition of internationalization. *International Higher Education*, 33.
Malik, R. S. and Hamied, F. A. (2014) *Research methods: A guide for first time researchers*. Bandung: UPI Press.
Moeliodihardjo, B. Y. (2015) Higher education in Indonesia. In *International Seminar on Massification of Higher Education in Large Academic Systems*.

Emerging Perspectives and Trends in Innovative Technology
for Quality Education 4.0 – Kusmawan et al (eds)
© 2020 Taylor & Francis Group, London, ISBN 978-0-367-25803-0

Utilization of digital teaching materials from internet networks to increase student motivation in learning science

Husnul Khotimah
Jakarta Islamic Spiritual Development Institute, Jakarta, Indonesia

Nandang Hidayat
Pakuan University, Bogor, Indonesia

Yetty Supriyati & Asep Supena
State University of Jakarta, Indonesia

ABSTRACT: This study aims to increase students' motivation to learn science through the use of digital teaching materials from internet networks integrated in 5E learning models (Engage, Explore, Explain, Elaborate, and Evaluate). The study used the Classroom Action Research method. The study was conducted at SDN 3 Tapos, Depok, in the 2017/2018 school year. This study concludes that the use of digital teaching materials from internet networking as a media in science learning activities can increase students' learning motivation in science. The 5E learning model integrates digital learning media from internet networks so that it effectively increases students' motivation to learn science. The right integration of digital learning media comes under the Explore phase of learning.

Keywords: digital teaching materials, internet networking, learning media, motivation to learn science

1   INTRODUCTION

1.1   *Background*

The quality of interaction between teachers and students in learning activities is determined by many factors, including the selection of approaches, methods, techniques, and learning media used. The success of teachers in choosing factors that are, first, in accordance with the characteristics of teaching materials and students and, second, supported by learning media that can encourage and maintain curiosity, interest, and student learning motivation is thought to be able to support optimal learning outcomes (Bates, 2015; Hidayat and Rostikawati, 2018a). However, many elementary school teachers are still reluctant in using sophisticated digital-based teaching materials and media. To uncover field facts, a preliminary survey and observation were conducted in five elementary schools, both public and private, in Tapos, Depok sub-district. The results of the preliminary survey indicate that the teachers there used conventional media. The involvement of students was almost even. This study tries to find a solution to the problem by improving learning processes through the use of the abundant digital teaching materials on the internet as a media for learning science. The learning model used is the 5Es – engage, explore, explain, elaborate, and evaluate.

1.2 Research problem

The research problem formulation is: 1) Does the use of digital teaching materials from internet networks as learning media increase students' motivation to learn science? 2) How can digital learning media from internet networks be integrated into the learning phase of 5Es so that it effectively increases students' motivation to learn science?

2 LITERATURE REVIEW

2.1 The nature of motivation to learn science

Motivation can be defined as several processes, internal or external to an individual, causing an attitude of enthusiasm and persistence in carrying out certain activities (Anderman and Gray, 2015). Students will be more motivated and successful in the academic environment when they believe that they are included and accepted (Yeager, Walton, and Cohen, 2013; Acun-Kapikiran et al., 2014). The ability to achieve various educational goals is important in enhancing true scientific literacy (Impedovo, Ginestié, and Williams, 2017) or multidimensional scientific literacy (Bybee et al., 2006; Hassan and Rajab, 2014). Moreover, learning science can help students develop creativity and imagination and become better able and more willing to continue learning science.

2.2 Digital media and the 5Es learning model

Digital literacy can be definedas the awareness, attitude, and ability of individuals to properly use digital tools and facilities to identify, access, manage, integrate, evaluate, analyze, and synthesize digital resources; to build new knowledge; to express themselves through media; and to communicate with others in a constructive social context (Hidayat, Nandang, Khotimah, 2019). Digital competence and digital literacy are related yet not identical (Ibrahim, 2018). Digital competence relates to the technical use of ICT (Gallardo-Echenique et al., 2015). Digital literacy and digital competence are needed by teachers in the current era. In this study, the use of digital teaching materials as learning media was integrated into learning activities using the discovery-inquiry approach. The learning model with the discovery-inquiry approach consists of five steps, called the 5Es (Bybee et al., 2006; Gejda and LaRocco, 2006).

3 RESEARCH METHODS

This study applied Participatory Action Research (PAR) (Baum, MacDougall, and Smith, 2006); Brydon-Miller, Greenwood, and Maguire, 2003). In relation to the field of teacher assignments, this method is often called Classroom Action Research or CAR (Stringer, 2010). The CAR procedure in this study employed the Kemmis and Taggart (1988) model. The study was conducted for one semester at SDN 3 Tapos, Depok City, in the year 2017/2018, with 35 Grade 5 students and two teachers acting as data sources. The research data included quantitative and qualitative data regarding students' learning motivation observed from their activities and involvement (on-task and off-task) during the learning process taking place. Data collection applied observation techniques.

4 RESULTS AND DISCUSSION

4.1 Results

Student behavior was grouped into on-task and off-task behaviors. Observational data recapitulation during treatment or action in the first cycle is presented in Table 1.

Table 1. Recapitulation of off-task student data in the first cycle.

| Behavior during learning activities* | Students off-task | | | |
	Total	%	Average duration (minutes)	%
Chat	9	25.71	29.10	32.33
Sleepy	13	37.14	21.18	23.53
Annoying friends	18	51.43	24.40	27.11
Running in the classroom	15	42.86	17.18	19.09
Not paying attention to the teacher	13	37.14	32.32	35.91
Not working on worksheets	7	20.00	19.26	21.40
Not involved in discussion	9	25.71	39.21	43.57
Not involved in group work	17	48.57	43.35	48.17
Average	-	36.07	28.25	31.39
N	35	-	90.00	-

In Table 1, the average percentage of students behaving off-task during the first cycle is 36.07%, with the average duration of time measured at 31.69%. In the first cycle, the percentage of students behaving was 63.93%, with an average time percentage of 68.61%. Meanwhile, the success indicators set for both are 80%. With the success indicator not being achieved, the process must enter the second cycle by first revising the learning steps and digital learning media used. Observations regarding the implementation and accuracy of each step of learning as well as the use and presentation of teaching materials or digital media show that almost all the learning steps that have been designed can be implemented and are relatively precise. However, there are three steps of learning that seem not to havenot been implemented properly: 1) the engaging phase; 2) the exploration phase; and 3) the explaining phase.

Based on the findings of weaknesses as presented, a discussion was then conducted to identify a number of alternative learning actions and the selection of more appropriate digital instructional materials. Then, a redesign of the learning implementation plan was carried out. Observations data recapitulation in the second cycle is presented in Table 2.

Table 2. Recapitulation of Off-task Student Data in the Second Cycle.

| Behavior during learning activities* | Students off-task | | | |
	Total	%	Average Duration (minutes)	%
Chat	4	11.43	9.14	10.16
Sleepy	9	25.71	13.26	14.73
Annoying friends	7	20.00	12.28	13.64
Running in the classroom	3	8.57	8.14	9.04
Not paying attention to the teacher	5	14.29	14.17	15.74
Not working on worksheets	7	20.00	3.24	3.60
Not involved in discussion	3	8.57	7.12	7.91
Not involved in group work	6	17.14	17.22	19.13
Average	-	15.71	10.57	11.75
N	35	-	90.00	-

Table 2 shows that 15.71% of students behaved off-task during the second cycle, with an average time duration of 11.75%. Thus, at the end of the second cycle, the success indicators of 80% on-task activity were successfully achieved; hence no further cycle action was needed.

4.2 *Discussion*

The findings of this study indicate that the use of digital teaching materials from internet networking during science learning activities can increase student motivation. The use of digital technology as learning media allows students to experience real learning experiences, and they can interact with other individuals without the need to meet face to face (Kassim, Razaq, and Ahmad, 2010). The use of digital technology as a new technology in learning will radically change what students learn, how they learn, and where they learn (Warschauer, 2007). Thus, the use of digital teaching materials as learning media motivates students to learn independently. This study also found that media and digital teaching materials can effectively increase students' motivation to learn science if integrated into the discovery-inquiry learning approach of the 5Es.

5 CONCLUSION

The use of online digital teaching materials can increase students' motivation to learn science. The integration of digital teaching materials from the internet is most effective in increasing students' motivation to learn science as part of the explore phase in the inquiry 5E learning model.

ACKNOWLEDGMENTS

We would like to thank the Ministry of Religion of the Republic of Indonesia, the Islamic Spiritual Development Institute Jakarta, and the University of Pakuan for their support, both moral and material, so that this research could be carried out well.

REFERENCES

Acun-Kapikiran, N., Kapikiran, S., Adel, T., Zainal Ariffin, A., Ahmed, A. S. P., Sonn, T. P. ... Rakes, C. R. (2014). School effectiveness and school improvement: alternative perspectives. *School Effectiveness and School Improvement*. https://doi.org/10.1080/09243453.2012.680892.

Anderman, E. M., and Gray, D. L. (2015). Motivation, learning, and instruction. In *International Encyclopedia of the Social & Behavioral Sciences: Second Edition*. https://doi.org/10.1016/B978-0-08-097086-8.26041-8.

Bates, A. W. (Tony). (2015). Guideline for designing, teaching and learning. In *Teaching in a Digital Age*. https://doi.org/10.1017/CBO9781107415324.004.

Baum, F., MacDougall, C., and Smith, D. (2006). Participatory action research. *Journal of Epidemiology and Community Health*, 60, 854–857. https://doi.org/10.1136/jech.2004.028662.

Brydon-Miller, M., Greenwood, D., & Maguire, P. (2003). Why action research? *Action Research*. https://doi.org/10.1177/14767503030011002.

Bybee, R. W., Taylor, J. A, Gardner, A, Scotter, P. V, Powell, J. C., Westbrook, A, and Landes, N. (2006). The BSCS 5E Instructional Model: Origins, Effectiveness, and Applications. Bscs, 1–19. https://doi.org/10.1017/CBO9781107415324.004.

Gallardo-Echenique, E. E., de Oliveira, J. M., Marqués, L., and Esteve-Mon, F. (2015). Digital competence in the knowledge society. *Journal of Online Learning and Teaching*, 11(1).

Gejda, L., and LaRocco, D. (2006). *Inquiry-Based Instruction In Secondary Science Classrooms: A Survey of Teacher Practice*. 37th Annual Northeast Educational Research Association Conference.

Hassan, S. S. S., and Rajab, M. (2014). Science teaching styles and student intrinsic motivation: Validating a structural model. *Pertanika Journal of Social Science and Humanities*.

Hidayat, Nandang dan Khotimah, H. (2019). Pemanfaatan Teknologi Digital dalam Kegiatan Pembelajaran. *Jurnal Pendidikan & Pengajaran Guru Sekolah Dasar*, 2(1), 10–15.

Hidayat, N., and Rostikawati, R. T. (2018a). Energize learners to use scientific approach. Jurnal Pendidikan Ilmiah.

Hidayat, N., and Rostikawati, T. (2018b). The effect of the scientific approach with comic intelligent media support on students' science competencies. Journal of Educational Review and Research, 1(1), 38–50.

Ibrahim, W. (2018). Cloud computing implementation in libraries: A synergy for library services optimization. *International Journal of Library and Information Science*, 10(2), 17–27. https://doi.org/ 10.5897/ijlis2016.0748.

Impedovo, M. A., Ginestié, J., and Williams, J. (2017). Technological education challenge: A European perspective. *Australasian Journal of Technology Education*. https://doi.org/10.15663/ajte.v4i1.50.

Kassim, Z. B., Razaq, A., and Ahmad, B. (2010). E-Pembelajaran: Evolusi Internet Dalam Pembelajaran Sepanjang Hayat. Globalisasi, Teknologi dan Pembelajaran Sepanjang Hayat.

Mayer, R. E. (2014). Incorporating motivation into multimedia learning. *Learning and Instruction*. https://doi.org/10.1016/j.learninstruc.2013.04.003.

Stringer, E. (2010). Action research in education. In *International Encyclopedia of Education*. https://doi. org/10.1016/B978-0-08-044894-7.01531-1.

Warschauer, M. (2007). The paradoxical future of digital learning. *Learning Inquiry*, 1, 45–49. https://doi. org/10.1007/s11519-007-0001-5.

Yeager, D., Walton, G., and Cohen, G. L. (2013). Addressing achievement gaps with psychological interventions. *Phi Delta Kappan*, 94(5),62–65. https://doi.org/10.1177/003172171309400514.

Emerging Perspectives and Trends in Innovative Technology
for Quality Education 4.0 – Kusmawan et al (eds)
© 2020 Taylor & Francis Group, London, ISBN 978-0-367-25803-0

Parenting in the digital era: Dealing with the impact of technological developments on children's moral development

Arini Noor Izzati & Lidwina Sri Ardiasih
Universitas Terbuka, Tangerang Selatan, Indonesia

Audi Yundayani
STKIP Kusuma Negara, Jakarta, Indonesia

ABSTRACT: Technological development has a significant influence on all members of society, including children. This study aims at: 1) describing the current growth of Information and Communication Technology (ICT); 2) identifying the effect of ICT growth on the nation's younger generation; and 3) analyzing the role of parents as children's primary partner in addressing ICT growth in the digital age. This descriptive qualitative study employed observation and document study for data collection. The results indicate: 1) that the number of internet users in Asia is the biggest in the world; 2) the benefits, as well as the negative impact of ICT growth on the younger generation; 3) five significant things to consider when discussing the effect of ICT growth on the moral development of children. These findings can assist parents and educators in preparing their children to become strong and competitive individuals in the digital age.

1 INTRODUCTION

In the digital era, the field of education is closely intertwined with current technological developments. Various educational institutions have begun to use Information and Communication Technology (ICT) to help students in their learning process. Many educators have used online learning, or e-learning, using the internet as a primary tool for communication with students, including children and adolescents.

Taking these conditions into account, it is possible for children to access the internet and, whether by accident or on purpose, find content that they should not see, such as violence or pornography. Therefore, educators and parents need to be vigilant and pay more attention to the activities of their children, especially when accessing the internet. The negative impact of ICT will pose dangers to children, especially on their brain development.

From the explanation above, this paper aims to: 1) describe the current condition of the development of Information and Communication Technology (ICT); 2) identify the impact of ICT development on Indonesia's younger generation; and 3) analyze the role of parents as the main companion of children in addressing the development of ICT in the digital age.

2 METHODOLOGY

This descriptive qualitative study employed observation and document study in the data collection. All the collected data were selected and presented in accordance with the objective of the research.

3 FINDINGS AND DISCUSSION

3.1 *Development of information and communication technology globally and its impact on child growth*

Based on data from world statistics in 2019 regarding internet users in the world, the number of internet users in Asia has reached 49.8% of total users globally. Customers or users of mobile equipment in Asia make up 55% of mobile equipment users globally. In addition, the data also show that the development of broadband internet in Asia also leads the world. The total number of users in Asia reached 2.9 billion at the end of 2012 and reached 4.4 billion users in 2019. This shows the rapid development of the use of mobile equipment at the Asian level (Jones, 2008).

The development of the use of the internet and mobile equipment in Indonesia is also fast. As a country with a population of around 265 million, the population of *netter* (those with online access) in the country reached 143 million in 2018 (Kristiansen, 2003). Mobile device users also exceeded Indonesia's population (Graham, 2002). This means that one person has more than one mobile device and that mobile devices are the primary media to connect to the outside world via the internet. With such widespread access, various kinds of sites can be easily accessed by anyone, anywhere, at any time.

The development of ICT has provided various benefits. We can easily get the latest information or news just by accessing various online sources. This provides a positive experiences for children, especially as they get used to positive things such as reading texts or digital books. This can also have a positive impact on their learning to socialize or interact in cyberspace. However, there are many negative impacts that arise and affect the development of children. Many sites have adult content and are related to drugs (legal or illegal), violent scenes, and very dangerous pornographic scenes. The head of the Ministry of Health's Intelligence Center, Jofizal Jannis, stated that, in a study of 4,500 adolescents in 12 major cities in Indonesia in 2007, it was found that the majority (97%) of the respondents claimed to have watched pornographic films (Kanuga and Rosenfeld, 2004). This is considered worrying because, if children continue to consume pornography, they may experience brain damage and behavioral disorders, including sexual disorder.

In addition to addiction to accessing pornography sites, many children and adolescents are addicted to playing online games. If they play games excessively, the children may experience a negative impact on their psychological and physical health (Robinson and Martin, 2009).

3.2 *Technological vs. development of moral development of children*

From a biological perspective, the children's developing brain is the center that coordinates the movements of all other body parts. The impact of the development of ICT on children has been discussed in the previous section, where the brain becomes one of body parts directly affected.

The brain's prefrontal cortex serves to plan, regulate emotions, control oneself, consider consequences, and make decisions (Robinson and Martin, 2009). When children access sites that attract their attention, one part of the brain, called the hypothalamus, will produce dopamine, a hormone that produces a sense of happiness or fulfillment of pleasure. When these hormones are still at normal levels, there will be no problems. If the child has started to become addicted to accessing the site excessively, it will have a negative impact (Anderson, 2000).

In preparing children and adolescents to be the nation's next generation of adults, educators and parents need to be equipped with various kinds of knowledge related to appropriate parenting to optimize the potential of their children in terms of aspects such as motor skills, language, cognition, emotion, and socialization (Boulianne, 2009). This is supported by Akhadiah and Listyasari, who state that a brain empowerment plan originating from the human learning process to actualize itself optimally in a learning system not only accelerates but also exalts mental life and brings human beings to a spirit of innovation development (Akhadiah, 2015).

Children must be formed into moral individuals by instilling positive values that support their development in facing a challenging future and intense global competition (Anderson, 1999). Based on this fact, parents face challenges in accompanying their children in this digital age. There are three patterns of parenting offered by Cevher-Kalburan and Ivrendi (2016), namely 1) overprotective, 2) permissive, and 3) democrative. The first pattern, *overprotective*, means protecting someone excessively, which can make someone uncomfortable. In this case, parents will monitor the child's every activity and not give the child an opportunity to have fun without parents' supervision. Permissive parenting is the opposite, where parents allow the child to do everything. Often, children are free to go beyond parental control. The third pattern is *democratic*, a more flexible parenting style in which parents continue to protect and supervise children but also provide freedom as long as the children can be responsible for what they do. Each of the three parenting patterns can be considered by parents as a way to form children into moral individuals, those who have positive values in their lives to help them face their future.

3.3 *Assistance to the moral development of children in the digital age*

In principle, all parents have great expectations for their children. They want their children to grow into healthy, smart, diligent, and accomplished individuals. Will the parents' dream come true? That cannot be predicted. Parents face a big challenge, and it will be not easy for them. Parenting requires a long time and a difficult process.

Therefore, parents must determine for themselves which parenting model to take in order to accompany their children. Children experience development in three different situations, namely at home, at school, and in the community. This should be taken into account by parents, who cannot continue to participate in children's activities when at school or when involved in social activities in the community. As a consequence, parents should create a pleasant home atmosphere where children feel safe and comfortable with their families. Therefore, parents need to pay attention to the following things.

First, parents need to instill spiritual values into their children as early as possible. It is the responsibility of parents to introduce the values inherited by their ancestors. These values are also important when dealing with the development of ICT, helping children be selective in accessing information and in maintaining the values or norms that apply in the family and community environments.

Second, parents should provide appropriate parenting patterns. The three parenting methods – overprotective, permissive, and democratic – each have their own advantages and disadvantages. An overprotective protective may be more appropriate for children under the age of three (toddlers) where the children are still vulnerable to their environment. The application of permissive parenting will have more negative effects at this age, because parents will be overwhelmed or unable to control the children's will and activity due to the children's freedom. When a child reaches adolescence, democratic parenting can be applied, once children can already determine for themselves what to do and can be invited to negotiate about the things that can and may not be done.

Third, parents must provide clear rules for the facilities provided. Many families provide devices to facilitate obtaining information. However, parents must be strict with children using these facilities when it is time to do schoolwork and meet their obligations as students.

Fourth, children need to be given the right model. That is, parents must give examples to children about what should be done. If parents forbid children to do something but parents do it themselves, this will have a bad impact on children, who may lose trust in their parents.

Fifth, parents must be aware of what exactly the games are that children love. There are some games that are packaged for children but with content for adults. It should be a concern for parents to remain vigilant and not technologically illiterate by continuing to monitor and learn more about the games that interest their children.

4 CONCLUSION

In this digital era, technology is developing rapidly. ICT has been used in all aspects of life to facilitate routine activities. By using the internet, everyone can easily get information even from other parts of the world. In addition, individuals can communicate with each other without limitations of time and space. This communication has an outsized impact on children and adolescents who are in a period of growth and of searching for their identity. It is not impossible that they will access sites with content such as scenes of violence and even pornography. In addition, online games offered to children may feel so interesting as to motivate the children to 'mingle' with the game for hours.

This needs to be a concern of parents. The negative impact of the development of ICT not only relates to the moral cultivation of children but also affects their psychological and biological development. {arents need to apply appropriate parenting so that children grow up with a sense of security and comfort with family and will be able to socialize naturally in school and community environments.

REFERENCES

Akhadiah, S. (2015) Filsafat Ilmu Lanjutan, Kencana.

Anderson, S. W., et al. (1999) Impairment of social and moral behavior related to early damage in human prefrontal cortex. *Nature Neuroscience*, 2(11), 1032.

Anderson, S. W., et al. (2000) Long-term sequelae of prefrontal cortex damage acquired in early childhood. *DevelopmentalNneuropsychology*, 18(3), 281–296.

Boulianne, S. (2009) Does internet use affect engagement? A meta-analysis of research, *PoliticalCcommunication*, 26(2), 193–211.

Cevher-Kalburan, N. and Ivrendi, A. (2016) Risky play and parenting styles. *Journal of Child and Family Studies*, 25(2), 355–366.

Graham, S. (2002) Bridging urban digital divides? Urban polarisation and information and communications technologies (ICTs). *Urban Studies*, 39(1), 33–56.

Jones, S. (2008) *Internet goes to college: How students are living in the future with today's technology*. Collingdale, PA: Diane Publishing.

Kanuga, M. and Rosenfeld, W. D. (2004) Adolescent sexuality and the internet: The good, the bad, and the URL. *Journal of Pediatric and Adolescent Gynecology*, 17(2), 117–124.

Kristiansen, S., et al. (2003) Internet cafe entrepreneurs: Pioneers in information dissemination in Indonesia. *The International Journal of Entrepreneurship and Innovation*, 4(4), 251–263.

Robinson, J. P. and Martin, S. P. (2009) Social attitude differences between internet users and non-users: Evidence from the general social survey. *Information, Communication & Society*, 12(4), 508–524.

Emerging Perspectives and Trends in Innovative Technology
for Quality Education 4.0 – Kusmawan et al (eds)
© 2020 Taylor & Francis Group, London, ISBN 978-0-367-25803-0

Audio visual media versus games media on mastery of English vocabulary for elementary school student

D.S. Bimo
Universitas Terbuka, Tangerang Selatan, Indonesia

M.Y. Rensi
Universitas PGRI Semarang, Indonesia

ABSTRACT: Audiovisual and games media are effective educational learning tools for improving information retention among students. The purpose of this study is to determine the best tool for improving the mastery of English vocabulary between audiovisual and game media. The research was quasi-experimental and utilized a randomized pretest and post-test comparison group design. The population involved two fifth grade classes at Manahan Surakarta Elementary School. The Fifth Grade A was made Experimental Group 1 (EG1) with 33 students, while Fifth Grade B became Experimental Group 2 (EG2) with 28 students. Also, cluster random sampling technique was utilized in this experiment. Data from the post-test results show that EG1 used with audiovisual performed better than EG2 using games. The study concludes that audiovisual media is a valuable learning tool for students' vocabulary retention in foreign language classes.

1 INTRODUCTION

Vocabulary is one of the most critical components for mastering a second language. An increased vocabulary awareness enables students to hold detailed conversations on several topics, allowing them to practice effectively the skills needed to learn a second language.

It is difficult for students in elementary schools to use English as a Second Language (ESL) properly. The lack of vocabulary slows the rate at which learners master ESL. Additionally, the mispronunciation of English words is very common. From a student's perspective, it is challenging to retain knowledge about vocabulary and correct pronunciation in case the words are not used in everyday learning. Also, students are bored and less motivated to take ESL while using print media, such as books, as an intermediary information delivery system.

Smaldino, Lowther, & Russell (2008) suggested the use of non-print media for an information delivery system. This model places learning media as an intermediary information source between teachers and students and facilitates interaction during lessons. The approach makes it possible for students to obtain knowledge from both information sources simultaneously. According to Scarrat & Davison (2012), the media is divided into three, including audial-visual, print, and e-media. Audial visual includes cinema/ film, television, radio, music video, and animation, while print includes newspapers, magazines, and comics. E-media comprises of internet, HP, computer games, and video games.

Heinich etal., (1985) defines audiovisual media as a way of facilitating learning using tools that demonstrate learning material for students to witness and carefully observe the lesson objectives directly. It allows a student to interact with a learning prop physically. Audio-visual learning uses several learning media such as video, radio, TV, LPs, tape recorders, and map images, among many other learning options. Audiovisual media is also a tool that helps to transmit knowledge or attitudes by turning abstract ideas into concrete imagery for the students to process (http://alaksamana.blogspot.com/2018/04/pengertian-dan-jenis-media-audio-visual.html.).

According to Yusantika (2018), both audio and visual media affect the listening ability of students. There are differences in cognitive skills, such as learning styles, which affect learning ability through audio and visual media. This means it is not enough for a teacher to play videos in every lesson. A variety of learning props should be used even within the audio and visual media group.

Brewster (2002) defines learning games as any fun activity which provides opportunities for students to practice foreign languages in a relaxed and pleasant atmosphere. Games might be played competitively, resulting in winners and losers, but this is usually conducted while seeking greater entertainment. For this reason, games have enormous potential to motivate students to learn.

Although students are often capable of learning a second language, they still need careful guidance because they have particular characteristics different from adults. Elementary students who are beginning language studies need an appropriate learning environment that allows for fun, games, motor movements, and challenges that provoke their curiosity. Students need creative and innovative teachers to meet these needs in a learning environment. Elementary school teachers help meet the needs of their ESL students using media in each lesson appropriately to demonstrate the learning objectives. Teachers should carefully select the learning media, which is most suitable for students' needs. However, foreign language teachers who use this strategy are successful in helping their students retain vocabulary and other needed language skills.

The purpose of this study is to determine the effects of the audiovisual and games media on improving vocabulary mastery for ESL elementary school students.

2 RESEARCH METHODS

The subjects in this research include fifth-grade students at Manahan Surakarta Elementary School. These students were divided into two test groups, Class Fifth Grade A comprising of 33 students formed experimental group 1 (EG1) and Class Fifth Grade B consisting of 28 students made experimental group 2 (EG2). A cluster random sampling technique was used in this study. This research method uses quantitative methods. Two tests were used as a data collection method.

EG1 and EG2 were both given randomized pretests to measure preexisting English vocabulary mastery. One group was taught using an audiovisual treatment while games were used in the other one. Students then took a posttest to quantify the growth of English vocabulary mastery.

The randomized pretest and posttest comparison groups are presented in Table 1.

This research focuses on measuring the effectiveness of audiovisual and games media for assisting ESL learners to master new vocabulary. The aim is to find out whether there is a significant measurable difference between the learning outcomes of students taught with these media.

Table 1. Randomized pretest-posttest comparison group design.

Group	Pretest	Treatment	Postest
EG1	R_1	Audio Visual (X_1)	R_3
EG2	R_2	Games (X_2)	R_4

3   DATA AND DISCUSSION

Differences in data collected from the EG1 pre-test and EG2 pre-test are presented in Table 2:

Table 2.   Pre-test data results from EG1 and EG2.

Source of variation	Experimental Group 1	Experimental Group 2
Sum	2092	1776
$-n$	33	28
x	63,39	63,43
Variance(s^2)	52,4962	42,6984
standard deviation (S)	7,25	6,53

Data shows that the mean of EG1 scores is almost the same as the EG2 scores. The difference is not significant, and the two groups are homogenized. The difference in the test results of EG1 and EG2 shows that t count = - 0,02, t table = 1.67, and α = 0,05 which is presented in Figure 1 below:

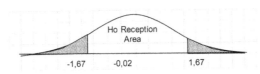

Figure 1.   Differences in the pre-test averages collected from EG1 and EG2.

Data shows that t count = - 0.02 is between the Ho reception area. The difference between EG1 and EG2 was not significant in this experiment. Data Results Differences in Post-test Experimental Group 1 and Experimental Group 2 are shown in Table 3:

Table 3.   Data on differences in Post-Test experimental group 1 and experimental group 2.

Source of variation	Experimental Group 1	Experimental Group 2
Sum	2743,3	2090,0
n	33	28
$\bar{X}$	83,13	74,64
Variance(s^2)	49,9865	70,2381
standard deviation (S)	7,07	8,38

The experimental group 1 is higher than group 2. Therefore, learning through audiovisual media is better than using games. The difference test results of two post-test meanings of the experimental group 1 and the experimental group 2 stated that t count = 4.29 > t table = 1.67, α = 0,05 which is presented in Figure 2 below:

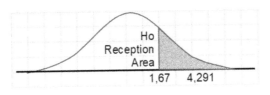

Figure 2.   Data on differences in the two posttest average of the experimental group 1 and group 2.

This shows that the experimental group 1 with audiovisual is better than the experimental group 2 with games. While learning English using audiovisual media, students see the material directly, and therefore it is easier to remember the information delivered. In the video media, there are two elements that are used simultaneously, audio and visual. In general, the audio allows students to receive information through hearing, while visual elements present the material through visualization. Learning in the experimental class shows the enthusiasm of students during the learning process, and they show seriousness in participating in learning and carrying out assignments. Also, students pay greater attention to material due to the use of audiovisual media. According to Karami, Amirreza (2019), audiovisual materials help learners to improve their knowledge of the target language. The vocabulary knowledge of the second language learner might be enhanced incidentally through watching videos and listening to words in meaningful contexts and communication. A summary of each research study and some guidelines for future research were provided.

4 CONCLUSION

Audiovisual is more effective than games media in achieving the desired learning outcomes. The use of audiovisual media needs to be adjusted to suit students' and environmental conditions. This is a useful alternative that ensures students are not saturated. It has the potential to increase their attention, activity, and motivation in vocabulary learning.

5 SUGGESTION

The use of audiovisual media should be considered in increasing students' English vocabulary. In practice, it should be developed in other ways and strategies, such as listening, reading, and writing. The approach has the potential to provide maximum results in case it is appropriately used.

REFERENCES

Brewster, J, Ellis, G, Girard, D. 2002.The Primary English Teacher's Guide. England: Penguin English.
Heinich, Robert, Russel, James D, Molenda, M. 1985. Instructional Media and the New Technologies Instruction. New York: Macmillan Publishing Company, a division of Macmillan, Inc.
Karami, Amirreza. 2019. Implementing Audio-Visual Materials (Videos), as an Incidental Vocabulary Learning Strategy, in Second/Foreign Language Learners' Vocabulary Development: A Current Review of the Most Recent Research. Journal on English Language Teaching, v9 n2 p60–70 Apr-Jun 2019, retrieved on August 23th, 2019.
Scarratt, E & Davison, J. 2012. The media teacher's handbook. New York: Routledge.
Smaldino, S.E, Lowther, D, & Russell, J.D. 2008. Instruction technology and media for learning. (9th ed). Upper Saddle River: Merril Prentice Hall.
Yusantika, F.D, Suyitno, I, Furaidah, 2018. The Effect of Audio and Audio Visual Media on Listening Ability of the Forth grade Students, vol. 3, no. 2, retrieved on April 24th, 2019, <http://journal.um.ac.id/index.php/jptpp/article/view/10544>.
_____.2018. Definition and Types of Audio-Visual Media.http://alaksamana.blogspot.com/2018/04/pengertian-dan-jenis-media-audio-visual.html.retrieved on August 23th, 2019.

Students' instructional success: A foundation of psychology perspective

W.A. Surasmi
Universitas Terbuka, Tangerang Selatan, Indonesia

Dwikoranto
Fisika, Universitas Negeri Surabaya, Indonesia

A.F. Ali
Universitas Terbuka, Tangerang Selatan, Indonesia

A. Rachman
Informatics Engineering, Institut Teknologi Adhi Tama Surabaya, Indonesia

ABSTRACT: Psychology can be defined as the study of a person's character. In the communication field, education is crucial in achieving learning goals: whether by conveying information conventionally or technologically, teachers need to manage their learning resources adequately. This study aims to determine the importance of psychological foundations on the learning success of elementary school students. This is qualitative research, with data collected from 80 students through literature, observation, and field research on several aspects related to the psychological foundation of the teacher and its impact on students. The tools used to obtain data were observation sheets, interview guidelines, and archival documents and photographs from the research site. The data collected were processed and arranged according to category, nature, and type through reduction, presentation, and verification/conclusion. The results of the study indicate that the teachers failed to pay adequate attention to the crucial points of the students' psychological foundation. Teachers can therefore do better at understanding learning styles, characteristics, potential, and interests.

1 INTRODUCTION

The success of educators is influenced by their ability to understand the educational and psychological foundation of classroom learning. Individual differences occur due to the different psychologies of students, which related not only to intelligence and talent but also differences in experience and levels of development, aspirations, ideals, and personality. Therefore, educators need to understand the development of individual students.

In elementary school, it is important to apply diverse psychological foundations to students with diverse learning challenges. These students possess unique features that need to be channeled to their potential. Teachers are facilitators that apply various teaching techniques to resolve students' learning problems. Education cannot be separated from problems, especially those related to individuals as actors. In the learning process, all children tend to have the desire to be more active.

The word "psychology" comes from "psyche," which means soul or breath of life, and "logos," which means science. Judging from its meaning, psychology is, therefore, defined as the science or study of the soul. One of the requirements of science, however, is the existence of objects studied, so it becomes improper to define psychology as the study of the soul, because the soul is an abstract notion. Still, related to the object of psychology, the soul in the form of individual behavior interacting with the environment can be studied and analyzed. Psychology can be therefore interpreted as a study of individual behavior interacting with the environment. It is divided into two parts: general psychology, which generally examines behavior; and particular psychology, which examines individual behavior in certain situations. This study focuses on psychology related to successful student learning, or educational psychology. This is categorized as a science because it possesses the requirements, namely ontology, epistemology, and axiology. Educational psychology is part of the science that examines individual behavior to determine facts, generalizations, and psychological theories in order to achieve the effectiveness of the educational process using scientific methods.

Formal educational activities, such as curriculum development, teaching and learning processes, evaluation systems, and guidance and counseling services, are some of the main techniques required according to educational psychology. Education as an activity that involves many people, including students, administrators, the community, and parents. Therefore, students' learning is most successful when everyone involved can understand individual behavior effectively.

Mastery of educational psychology is one pedagogic competency to be mastered. Muhibbin Shah (2003) stated that "in addition to the knowledge of teachers, candidates need knowledge of applied psychology that is closely related to the students' learning process."

2 METHODS

Qualitative research was carried out in Perdana Jombang Primary School. The data sources of the study were classes 1, 2, and 3, with a total of 80 students. Research techniques were obtained through literature study and field research with data collected through observations on several aspects related to the psychological foundation applied to students. Tools used to obtain data included observation sheets, interview guidelines, and archival documents or photographs from the research site. Findings were analyzed through reduction, presentation, and verification/conclusion.

3 DISCUSSION

Based on observations on elementary school education, the psychological foundation is one of the basic principles and concepts of education with implementation importance as a reference. According to Gerungan (Khodijah, 2006_, psychology comes from the Greek word "psyche," which means "soul," and "logos, "which means "knowledge." Therefore, etymologically, psychology means "the study of the soul, its symptoms, process, and background." However, the "science of soul" and psychology are different, because the study of the soul is more broadly based, involving imagination and speculation. In contrast, psychology is the study of the soul systematically obtained through scientific methods. Essentially, psychology is the study of psychiatric symptoms that are manifested in the form of behaviors for the benefit of humans or individual activities consciously and unconsciously. It is obtained through a particular scientific process, through the application of the basics, principles, or psychological methods and approaches in order to solve educational problems.

Psychological foundations include "learners diversity," characteristics, potential, uniqueness, kinds of intelligence, etc. In elementary school, students have different learning styles. According to Pidarta (2007: 206), learning is a behavior change that is relatively permanent, as a result of experience, and tends be implemented on other knowledge, along with the ability to communicate that knowledge to others. Psychologically, learning is defined as "an effort made by someone to obtain a conscious change in behavior from the results of their interaction with the environment" (Slameto, 1991: 12). According to experts, the learning style of each student is diverse and tends to determine patterns of behavior in gaining knowledge.

"Diverse learners" shows the uniqueness of student potential, with each requiring motivation to develop their potential and adjust to their uniqueness.

According to Nana Syaodih (1998), the development of every aspect is not always the same. There are three theories to development, namely the Stages, Differential, and Inclusive approaches. Darmodjo (1992) stated that elementary school students have unique and diverse characters and experience different intellectual, emotional, and physical growth, which leads to individual differences.

Diverse learners need to have multiple intelligences considered, with elementary school teachers needing to direct, guide, and develop these intellectual abilities. The theory based on the idea that intellectual abilities are measured through IQ tests is minimally useful, because such tests only emphasize logic (mathematics) and language skills (Gardner, 2003). According to Gardner, there are several categories of human intelligence, namely linguistic, mathematical-logical, visual-spatial, physical-kinesthetic, musical, interpersonal, intrapersonal, naturalistic, and existential.

According to De Porter and Hernacki (2002), learning styles are a combination of absorbing, regulating, and processing information. However, there are learning styles based on the modalities that individuals use in processing information (perceptual modality), namely visual, auditory, and kinesthetic. Such learning models provide the recipient with primary information easily absorbed without reading the explanation. From the above description, the psychological foundation is underlined as important and plays a role in the success of students' learning. Therefore, every educator, especially teachers in elementary school, need to understand how to teach based on educational psychology.

4 CONCLUSION

The results showed that, during the learning process, teachers failed to carry out psychology analysis on students. Therefore, the impact of student learning, situations, and outcomes did not match the learning objectives.

REFERENCES

Darmojo, H. dan. (1992) *Educational psychology*. Jakarta: Balai Pustaka, Jakarta.
De Porter, B., and Hernacki, M. (2002) *Quantum learning models*. Yogyakarta: Faculty of Psychology, Gajah Mada University.
Gardner, H. (2003) *Multiple intelligence*. (Translation of Dr. Alexander Sindoro). Batam Center: Interaksara.
Khodijah, N. 2006. *Educational psychology*. Palembang: Grafika Telindo Press.
Slameto. 2003. *Learning and the affecting factors*. Jakarta: Rineka Cipta.
Sukmadinata, N. S. (1998) *Platform for psychology of educational processes*. Bandung: Youth Rosdakarya, 1998.
Syah, M. (2003) *Educational psychology with a new approach*. Bandung: Teenager Rosdakarya.

Emerging Perspectives and Trends in Innovative Technology
for Quality Education 4.0 – Kusmawan et al (eds)
© 2020 Taylor & Francis Group, London, ISBN 978-0-367-25803-0

Intensive communication with parents key to success in handling autism in children: A case study in parental affection for autistic patients in Bogor

Stefani Nawati

Universitas Terbuka, Tangerang Selatan, Indonesia

ABSTRACT: Degenerative diseases, in general, have always come with political challenges related to how to overcome them, in connection with policies and procedures. When parents receive expert diagnoses that their children suffer from an illness, life is changed for all involved. Autism, a behavioral disorder that affects growth and development of children, affects family resources, especially for school-age children. Current government policies to accommodate children with special needs – inclusive schools and special curricula – are not yet available. This study aims to explore how caring for a child with autism affects parents. This research is qualitative research, a type of case study research, a social paradigm of universal definition, data collection through observation, interviews, and documentation. Data analysis uses Interactive Miles and Hubermann models.

Keywords: Autism, Government policy, Inclusive schools, Social actions, Communicative, parents

1 INTRODUCTION

Government policies in the field of education for children who have special needs are inadequate. Local government – in this case, the Department of Education – should program schools for children with special needs, including children with autism, children who are victims of drugs, and children with mental disorders. Special education is regulated in the Minister of Education Regulation No. 01 of 2008 concerning the Special Operational Education (SPOK) standard regarding student groupings. In addition, the Minister of Education Regulation No. 70 of 2009 states that inclusive education means the national education system to provide opportunities for all students who have abnormalities or who have the potential for special intelligence or talent to participate in education and learning in an educational environment together with other students.

According to Suyanto and Mudjito (2012: 5), there are three models for combining children with disabilities or special needs with non-needs children in a learning environment: the mainstream, integration, and inclusion. The inclusion model provides an opportunity for all students who have abnormalities and have intelligence potential, or special talents, to participate in education or learning in an educational environment together with other students in general (Permendiknas No.70 of 2009). The integration model means placing children with special needs in standard classes, where they follow lessons regularly from the teacher but, at certain times for particular subjects, can get special teachers in different classes, with separate teaching. Facilities and infrastructure for children who have special needs are inadequate. Parental complaints about learning facilities for children with special needs are very urgent. At present, such children and their families are treated improperly due to local government policies that have not allowed children the right to learn well. To overcome these problems,

Based on the above matters, the formulation of this problem is:

1. What is the experience of parents in handling children with special needs such as autism?
2. How do parents realize that their children have unique needs, such as people with autism?
3. How should parents communicate with children with special needs such as autistic patients?

2 RESEARCH METHODS

This research takes a phenomenological approach, meaning it explores: 1) the subjective experience or phenomenological experience of a person; and 2) a person's consciousness and essential perspective. This type of research is qualitative, as it intends to understand the phenomenon of what is experienced by the subject – for example, research on behavior, perception, motivation, action, is conducted holistically by means of descriptions in words and languages, in a natural context and utilizing various natural methods (Sukardi, 2003: 157).

The research was carried out in the City of Bogor. Bogor City has a large number of children under five, and children with special needs have not received maximum government attention, especially in education. By having special school facilities for autistic children, academic quality can be improved according to their abilities. To explain the program to improve the quality of education for children with special needs, researchers conducted interviews, made observations, and reviewed existing documentation, alongside parents, teachers, school committees, Puskesmas Regional Health Centers, and NGOs in Bogor City.

3 DISCUSSION

Normative or non-normative child development can be assess based on four principles:

1. Abnormalities occur in individuals who experience development issues and behavioral abnormalities.
2. Developmental disorders must be associated with typical development, major developmental tasks, and changes that occur throughout the normal life span. The critical thing is to differentiate from developmental "disorders" that can be tolerated.
3. The initial signs of non-normative/abnormal behavior must be studied seriously because they can lead to severe problems in the future.
4. Declaring child development normative or non-normative is based on medical models of average deviation and deviation from the ideal. Understanding of developmental psychopathology includes the study of the origins of changes and continuity of abnormal behavior throughout the life span.

The response that arises is related to biological and environmental factors, including permanent developmental irregularities, delayed sleep effects that will appear later in life, resilience, and even adaptability related to stress.

3.1 *Communication in Habermas's view*

For Jurgen Habermas, language is a medium of intersubjective relations. Language is a practical activity that involves speech-act cogniti tion and is based on grammatical rules. Grammatical rules in language form and develop social interactions because two subjects, present together, perform speech acts by obeying the grammatical rules. On this basis, Habermas said that interaction actors are those who have communication competencies. Actions in the world of life/lifeworld, according to Habermas, refer to the daily context of social interactions where subjects participate in sharing experiences, examining other people's arguments, and justifying their actions. Action will be socially meaningful when they gives access to others.

Intentional actions are meaningful because an action is aimed at another person and expects "reciprocity"; action, in this case, is called strategic action or communicative action. Strategic actions are instrumental because they treat others to achieve goals, while communicative action seeks to find a common understanding. Habermas considers communicative actions as human actions when people approach those with special needs using three characteristics of conscious and deliberate communication: those for the purposes of 1) reaching intersubjective agreement as a foundation; 2) reaching mutual understanding; 3) reaching a peaceful consensus about the steps to be taken in practical situations.

Communication with parents with autistic children involves the disclosure of a lifeworld. In the lifeworld, they are faced with a system that might or might not be appropriate. An action is taken to implement interventions that is deemed appropriate within their lifeworld. The purpose of communicative action is the understanding of various parties involved in the intervention system for handling children with autism.

4   CONCLUSIONS AND SUGGESTIONS

4.1   Conclusion

Special-needs children need to be treated well. One technique or method used to deal with the behavior of children with special needs is giving clearer instructions, namely by making simple drawings and using more customized communication methods for children.

4.2   Suggestion

Suggestions for parents need to have great patience in dealing with children with special needs by being objective and not blaming the child for mistakes made. The government education office in Bogor needs to pay serious attention to establishing appropriate facilities to help children with special needs.

REFERENCES

Act. No: 23/2014 concerning Tasks, Authorities, Obligations, Regional Heads. Published by the State Secretariat. RI.

Assauri, S. 2000. *Marketing Management, Basics, Concepts and Strategies*. Jakarta: Rajawali Press Publishers.

Bogdan, R. R., and Taylor, S. J. (1993) *Basic Qualitative Research*. Surabaya: Aneka Usaha.

Bogdan, R.C. and Biklen S.H. (1998) *Qualititative Research for Education: An Introduction to Theory and Methods*. Boston, MA: Allyn And Bacon.

Collemen, J. (2008) *Basic Theory of Social Sciences*. Bandung: Nasional.

Creswel, J. (1988) *Qualitative and Research Page: Choosing among Five Traditions*. New York: Sage.

Fandy, T. (2004) *Marketing Services*. Malang: Bayu Media.

Fisher, and Strauss A: Thir Succes, Symbolice. Chicago Tradition: Thom. Park and The inc.

Haris, P. N. and Kussusanti, I. (2008) *In House Marketing Training UPBJJ-UT*. Jakarta: C & G Training Network.

Kotler, P. (2005) *Marketing Management, Millennium Edition*. Upper Saddle River, NJ: Prentice Hall.

Kotler, P. and Armstrong, G. (2001) *Principle of Marketing, Issue 9*. Upper Saddle River, NJ: Prentice Hall.

Mainnes, D. K. (1977) Social Organization Interactive Tough in Symbolic Structure, Annual.

Ministry of Education and Culture of the University. (2016). *Catalog of Open Universities*. South Tangerang City: Publishers of Open University.

Ministry of Education and Culture. 1995. *1st Large Dictionary of the Indonesian Language*. Jakarta: Balai Pustaka.

Ministry of National Education. 2006. Catalog of Open University Education Programs. Jakarta: Open University.

Open University (2015) *Profile of Open Higher Making Education University Open to All*.

Simamora, H. (2002) *International Marketing Management, Volume 2*. Jakarta: Salemba Empat.

Swastha, B. and Sukotjo, I (2002) *Introduction to Modern Business: Introduction to Modern Corporate Economics, Issue 3*. Yogyakarta: Liberty.

Distance learners' readiness for online final examinations

Yeti Sukarsih & Sunu Dwi Antoro
Universitas Terbuka, Tangerang Selatan, Indonesia

ABSTRACT: This research aimed to determine distance learners' readiness to take the online remedial examination. Data was collected through questionnaires, with the regression analysis used to determine the factors affecting the independent ability of learners to operate computers, as well as the active effect and outcome of the tutorial. The results show that Tutorial Learning Support (X3) has no positive and significant influence on the ability of self-study towards online examination scores (Y) for remedial distance learners. These results prove that self-education, computer operating skills, and tutorial learning support for remedial distance learners have no influence on online exam scores. Therefore, other factors may influence the competence of distance learners.

1 INTRODUCTION

Distance education is a system that organizes the learning process for students by teaching and testing independent skills in their subjects. The ability to learn independently is a key requirement and comes from interacting with the environment and having set goals. An independent learner needs to have his or her own way of learning from mistakes, deciding what to do on their own, adapting to an environment and task, and assimilating suggestions (Shen, 1994).

Over the years, many studies have focused on the level of competency of self-directed learning for distance learners. However, information relating to the influence of this ability is usually related to computer operating skills, with the active participation of learners during the Online Examination System tutorial not widely conducted. In addition, there has not been any study of remedial students. The result of this research contributes to the proper management of the distance learning process regarding the improvement of students with low levels of understanding participating in a remedial course.

Universitas Terbuka (UT) is one of Indonesia's state universities that utilizes the distance learning system. This system requires students to possess independent learning skills, acquired under their own initiative. Distance learning systems are designed to bridge the gap between students and lecturers (Pandiangan et al., 2012). However, it is impossible to control the student learning process in a distance system.

Learning facilities are highly important. UT, as a distance institution, provides learning supports to students. It is important to note how significantly it supports and influences students' learning outcomes and the management system. The aim of this research, therefore, is to determine the effect of self-directed learning competence, computer operating skills, and student activeness in online tutorials on final examination scores.

2 METHODOLOGY

The subject of this study were remedial students with elementary junior and senior school teaching backgrounds looking to update their knowledge at Universitas Terbuka Yogyakarta.

A total of 154 students were randomly selected from 307 remedial students. Data were collected using questioners with a focus on multiple variables, namely self-directed learning, computer skill, and participation in online support tutorials. The data related to academic achievement were obtained from online examination scores conducted in 2018; it was also obtained from remedial students that registered in 2017.4. The data for the self-directed learning readiness scale, computer ability, and enrolment in the tutorial were collected using questionnaires.

The reliability level of the questionnaires was analyzed using the SPSS version 17.0 programs with content validity used as a measuring instrument. The reliability level of the questionnaire trial results was 0.955; therefore, it was interpreted that the questionnaire designed has the ability to be used as a data collection tool. To determine the influence of self-regulated learning ability, computer operating skills, and activeness in the tutorial to the scores of online examination, data were analyzed using ANOVA regression. It was used to test the hypothesis and coefficients between the independent and dependent variables.

3 RESULT AND DISCUSSION

3.1 Self-directed learning

Table 1. The level of self-directed learning competence.

	Frequency	Percent	Valid Percent	Cumulative Percent
Valid too low	14	9.1	9.1	9.1
Low	20	13.0	13.0	22.1
Moderate	50	32.5	32.5	54.5
High	40	26.0	26.0	80.5
Very high	30	19.5	19.5	100.0
Total	154	100.0	100.0	

Data is calculated by simple regression analysis, variance, and statistical -tests. The results of data analysis were used to determine the proof of hypothesis 1, as shown in Table 2.

Table 2. The correlation regression coefficient between self-directed learning and scores from the online examination.

Model	Unstandardized Coefficients		Standardized Coefficients		
	B	Std. Error	Beta	t	Sig.
1 (Constant) Self-directed learning (X1)	.837	.611	.099	1.369	.173
	.005	.004		1.222	.224

a. Dependent variable: scores from the online examination (Y)

This shows that self-directed learning (X1) does not affect the Online Exam Value (Y). In this case, the relationship between independent direct learning and student academic scores is studied using bivariate linear regression. As a result, the linear regression model is significant, which means that there is no relationship between independent learning and the scores achieved by students in online examinations.

Table 3. The correlation regression coefficient between computer skill and scores from the online examination.

Model	Unstandardized Coefficients		Standardized Coefficients	t	Sig.
	B	Std. Error	Beta		
1 (Constant) Computer Skills (X2)	.630 .017	.728 .013	.106	.865 1.308	.388 .193

a. Dependent variable: scores from the online examination (Y)

Table 3 displayed that computer operating skills (X2) do not affect the online exam value (Y).

3.3 *Active involvement in tutorial*

Table 4. The correlation regression coefficient between PGSD and PGPAUD students' activeness in online test scores Coefficients[a].

Model	Unstandardized Coefficients		Standardized Coefficients	t	Sig.
	B	Std. Error	Beta		
1 (Constant) Tutorial Involvement (X3)	1.331	.633	.032	2.103	.037
	.004	.011		.390	.697

a. Dependent variable: scores from the online examination (Y)

This showed that active enrolments in the tutorial (X3) do not affect the Online Exam Scores-(Y). The research conducted by Haryanto (2014) proved that there was no relationship between online tutorial enrolment and final examination scores.

The tutorial is a learning assistance service provided by UT with the aim of establishing an independent learning college for students. During tutorials, learning activities are carried out under the guidance of a tutor as a facilitator, with lessons considered difficult and student mastery important.

4 DISCUSSION

The findings of this study indicate that, in distance learning, there is no significant relationship between the ability of independent students of Yogyakarta Open University to study in remedial courses and those that repeat courses. The absence of a significant relationship between self-learning ability and achievement is in line with the results of a study conducted by Sitepu (2017) (Universitas Terbuka, 2018), which states that there is no significant relationship at the Open University of Yogyakarta Unit. A similar result was obtained from the research carried out on the Anatomical Pathology (PA) practice test at the Faculty of Medicine, University of Lampung.

Research analysis on the level of self-learning ability of participants in this study shows 45% of students had high self-learning abilities. Also, research conducted by Chou (2012) shows that students' independent learning abilities do not affect their outcomes.

The variable ability to operate a computer has no effect on the scores of online exam results. Furthermore, the ability to operate a computer does not correlate with the final semester assessment in the field. Therefore, this study analyzes the influence of students' proficiency in operating computers while considering the era of the so-called industrial revolution 4.0 and technological capabilities. The Open University, as a distance education institution, used information and communication technology (ICT) to gradually develop its distance education services, both for academic and administrative services (Farida and Yulia, 2014).

The active variable in following the tutorial is thought to have a positive contribution to the score of online exam results. The verification results using regression analysis show no relationship to the remedial students. Hurip Pratomo (1990) (Haryanto, 2014) carried out research to prove that there is no correlation between the frequency of student attendance in the tutorial and the semester exam scores. The research conducted by Wahyuningsih, Rusli, and Bintari (2015) showed that online access to tutorials does not relate to achievement in the final examination.

5 CONCLUSION

The ultimate goal of distance instructional delivery is to facilitate self-directed learning. Students with a high level of self-directedness are expected to perform better in this instructional activity. However, for the remedial students, there is no positive and significant influence on the ability of the online examination scores. Another factor that is supposed to influence final examination scores is participants' computer skill. However, it was proved that computer skill had no effect on these scores. The distance learning system also provides a face-to-face or online tutorial to support distance learners. The effect of tutorial learning support to online exam scores was no significance, and this shows that it does not affect remedial distance learners.

REFERENCES

Chou, P.-N. (2012) Effect of students' self-directed learning abilities on online learning outcomes: Two exploratory experiments in electronic engineering, *International Journal of Humanities and Social Science*, 2(6).

Haryanto (2014) Laporan Penelitian Korelasi Antara Nilai Tutorial Online (Tuton) Dengan Nilai Ujian Akhir Semester (Uas) Mahasiswa Jurusan Sosiologi Fisip-Ut Masa Ujian 2013, *Fakultas Ilmu Sosial Dan Ilmu Politik Universitas Terbuka*.

Pandiangan, P., et al. (2012) Aktivitas Mahasiswa Dalam Tutorial Online Mata Kuliah Manajemen Strategi Dan Kontribusinya Terhadap Hasil Belajar, *Jurnal Pendidikan Terbuka dan Jarak Jauh*, 13(1), 43–51.

Shen, W. M. (1994) *Autonomous Learning from the Environment*. Computer Science Press and W. H. Freeman and Company.

Wahyuningsih, S. S., Rusli. Y. and Bintari, A. (2015) Aksesibilitas Mahasiswa Pada Tutorial Online Program Studi Perpustakaan, *Pada Jurnal Pendidikan Terbuka dan Jarak Jauh*, 16(1), 29–38.

Universitas Terbuka (2018) *Katalog Sistem Penyelenggaraan Universitas Terbuka 2018/2019, Universitas Terbuka*. Available at: https://www.ut.ac.id/sites/all/files/images/2018/Mei/Katalog-Sistem-Penyelenggaraan-Universitas-Terbuka-2018-2019.pdf.

Emerging Perspectives and Trends in Innovative Technology
for Quality Education 4.0 – Kusmawan et al (eds)
© 2020 Taylor & Francis Group, London, ISBN 978-0-367-25803-0

Challenges in implementing a curriculum: A case study of SD Negeri Pudjokusuman 1 teachers

Monika Handayani & Alpin Herman Saputra
Universitas Terbuka, Tangerang Selatan, Indonesia

ABSTRACT: This research aims to describe the difficulties experienced by primary school teachers in the process of implementing the curriculum in 2013, using qualitative research with a case study method. The examined subjects include the fifth-grade teachers of SD Negeri Pudjo-kusuman 1, while the instruments used consisted of interview guidelines and observation sheets. Data evaluation techniques involved the use of Miles and Huberman analysis, encompassing data reduction and presentation, as well as conclusion drawing. Validity was examined using a credibility test through a triangulation of techniques and reference materials. The results showed that teachers had difficulty in developing supporting teaching materials for the 2013 curriculum, as they emphasized the completion of delivered learning materials and not thematic integration. Also, the teachers lacked creativity in learning to enhance study time, making it ineffective and inefficient, leading to a deficiency in understanding in authentic assessment implementation.

Keywords: Difficulties, Curriculum 2013

1 INTRODUCTION

Changes in the curriculum of Indonesia serve to improve on previous versions and also answer contemporary challenges. Hence, modification in the KTSP curriculum in 2013 aimed at improving educational patterns that previously prioritized cognitive aspects and lacked focus on character education. The purpose of this advancement, according to Permendikbud Nomor 67 (No), was to better prepare Indonesian people to live as individuals and citizens who are faithful, productive, creative, innovative, effective, and also able to contribute to societal life as well as to the nation, the state, and world civilization. In addition, it aimed to enhance people's productive and innovative characteristics, in an attempt to respond to global challenges.

The 2013 curriculum puts emphasis on learning based on a scientific approach, with the use of (1) constructivist models, encompassing discovery, project-based, problem-based and inquiry learnings (Copies of Appendix (No, 2014), and (2) the application of learners' character, encouraging students to obtain knowledge through the proper constructing of the understanding obtained, promoting personal learning experiences (Nn, 2014).

The implementation of the 2013 curriculum in primary school education occurs in the form of integrative thematic (Nomor) learning, which is based on themes that accommodate a combination of materials in several fields of science. In addition, it is also grounded on themes chosen, developed, and related to the content of the study material (Prastowo, 2013), designed to enable the containment of some materials in numerous areas of learning, thus providing students gain meaningful learning experiences.

This implementation was not conducted simultaneously in some areas, although it was made available to elementary schools, including the designation of SD Negeri Pudjokusuman 1, in the Yogyakarta city area, as pilot. Based on results obtained from the interviews with the school's principal, the teachers, in general, were reported to have followed the training and already understood the 2013 curriculum; and also the school had performed the implementation at the start of the academic session. SD Negeri Pudjokusuman 1 was chosen because of the infrastructure and human resources (teachers) present, perceived as supporting this implementation, and also because it is included in the top 10 excellent primary schools in the city of Yogyakarta (Nn, 2013).

Based on the outcomes from interviews with the fifth-grade elementary school teachers, their understanding of the curriculum is good enough from the explanation given. Moreover, observation results showed that they used student and also teacher handbooks as guides in the implementation process, although overdependence was perceived, and it was assumed that its constituent materials were not studied in depth. Therefore, the teachers explained, most of the material was not studied in one lesson; thus, it needed to be learned outside of school hours. Based on these problems, there is a rising interest in examining the cause of the difficulties experienced by fifth-grade teachers in SD Negeri Pudjokusuman 1 during the process of implementing the 2013 curriculum (Kristiantari, 2015).

2 RESEARCH METHODS

2.1 *Type of research*

This is qualitative research involving the use of a case study method, conducted between September and October 2015, in the fifth grade of SD Negeri Pudjokusuman 1 in Yogyakarta.

2.2 *Data*

The data sources in this study consist of all fifth-grade teachers at SD Negeri Pudjokusuman 1, and other additional sources include documents relating to the research subject. The data collection techniques encompassed non-participatory observation, structured interviews, and documentation, utilizing the instruments of observation sheets and interview guidelines.

2.3 *Data analysis techniques*

This study employed the Milles and Huberman analysis model, encompassing data reduction and presentation, as well as conclusion drawing. Therefore, validity assessment required the use of a credibility test conducted through the triangulation of techniques and reference materials.

3 RESULTS AND DISCUSSION

Based on the results of the study, the difficulty observed in applying the 2013 curriculum at SD Negeri Pudjokusuman 1 was as follows.

3.1 *Lack of supporting teaching materials for the 2013 curriculum*

Teachers utilize handbooks from the 2013 curriculum, which contain minimal material and are less helpful for students' deep understanding. Other books used include KTSP and worksheets, where learning activities lack scientific nuance, as they tend to prioritize the mastery of materials rather than the study process. Research by Harizaj and Hajrulla (2018) explained that teaching materials ought to be carefully chosen and centered on student characteristics. Thus, there is a possibility of overcoming the problem of their limitation, related to the 2013 curriculum, via an approach in which teachers develop the material obtained or create appropriate constituents.

However, the fact remains that teachers are too dependent on existing books and have never really attempted conducting developmental practices. Prastowo (2015) explained the propensity for reducing the quality of learning among educators who are too fixated on conventional teaching materials, without any encouragement to develop them. Therefore, when teaching materials from the 2013 curriculum are deemed to be lacking requirements, the teacher ought to be more selective in providing additional resources, because the study process is not only a place to transfer knowledge. Other resources can help instill and transfer behavior, knowledge, and skills into competence (Sehe et al., 2016).

3.2 The learning provided is not fully integrative thematically

The teachers focused more on completing and providing in-depth materials to equip students for the final exam; hence, they tend to use less supportive materials. Kristiantari (2015) conducted a study regarding teacher readiness in implementing integrative thematic learning, which reported that difficulties were due to the limitation of a schools' ability to provide the appropriate resources.

This was in line with the research conducted by Retnawati et al. (2017), where the difficulty was explained as being due to the lack of supporting facilities. Therefore, the limited materials present in the handbook require supplementary efforts from the teacher, through the use of KTSP and LKS books. This causes a deviation from the thematic approach, based on the assumption that teachers tend to prefer completing rather than integrating the material. However, thematic learning is expected to encourage students to understand the concept learned through direct and real experiences, subsequently establishing a connection in and between subjects (Majid, 2014). Assistance is, therefore, needed from the schools in order to prepare supporting materials that follow the 2013 curriculum.

3.3 Time spent on learning activities in the 2013 curriculum is less effective and efficient

Time used in learning the 2013 curriculum is assumed to be ineffective and inefficient; thus, the materials being studied are not completed, leading to the requisition of additional time outside of class hours because of the bulky nature of the activities and the insufficient facilitation of the delivery system. The results obtained in the research by Khan et al. (2016) showed teachers' learning plan techniques to be very effective for controlling performance and time management in the classroom, indicating the need to make appropriate and appropriate study plans. Moreover, Kaya et al. (2015) show efficient use of teaching time occurs when utilizing the student-centered approach, in contrast with a teacher-centered approach. Thus, through the 2013 curriculum, it is assumed that teachers are more efficient in managing study time because of the utility of student-centered learning models, although their inherent creativity is also required, especially in terms of preparing appropriate study plans.

3.4 Teacher difficulties in authentic assessment

In the 2013 curriculum, a change was observed in the method of the assessment conducted, including the evaluation of knowledge competencies, based on results centered on authenticity, measuring attitude, skills, and also knowledge, based on processes and results (Kunandar, 2014). In addition, the instruments and the number of portfolios of students in learning collected tend to be the main reason for difficulties in recapitulating all results. The research conducted by (Enggarwati, 2015) explained that the difficulty experienced was influenced by several factors, encompassing low understanding and creativity of teachers, student characteristics, lack of training in authentic assessment and insufficiency in time.

Based on the data obtained from the interview, it was reported that the limitation of a teacher's understanding of authentic assessment is due to the uncommonly large number of instruments used to develop the required instruments. Therefore, during this time, they tend to rely on several others provided in the teacher's handbook, an approach that was similar to the report in other studies conducted by Suwandi et al. (2019). This means that the incorrect application of appraisals was due to the fact that teachers do not use rubrics or assessment instructions as a result of inadequate understanding; thus, special training is recommended.

4 CONCLUSIONS AND SUGGESTIONS

4.1 *Conclusions*

Based on the results obtained, the following conclusions were made.

1. Teachers have difficulty in developing teaching materials that follow the 2013 curriculum, as seen from the use of conventional resources, by relying on supporting study materials obtained from KTSP books, which tend not to possess an integrative thematic approach. Meanwhile, there was also a deficiency in teacher creativity towards the development of appropriate materials.
2. The teachers emphasize the completion of learning materials; thus, the delivery is not thematically integrative. In addition, they also tend to prioritize the mastery of student material, with emphasis on deepening it for the final exam; thus, administration is attained through the use of KTSP and LKS books.
3. Lack of teacher creativity in managing learning reduces the effectivity and efficiency of study time. This is seen from the limited time used in learning, as teachers have to include additional time outside the school hours.
4. The deficiency of teacher understanding in implementing authentic assessment is seen from the difficulty observed in the process of recapitulating results of student learning portfolios, as well as the limitation in the relevant instruments.

4.2 *Suggestions*

Suggestions that can be given to improve the implementation of 2013 curriculum learning are as follows.

1. There is a need for further training to enable the teachers to dissect the 2013 curriculum, especially in the area of creating materials and in authentic assessment. This is conducted to overcome the associated difficulties of identifying supporting teaching materials and assessments.
2. Schools are expected to provide more supportive facilities and infrastructures, to promote the delivery of study materials, and to enhance appropriate teaching and learning activities.
3. Principals ought to routinely monitor the implementation process and make efforts to hold discussions with teachers concerning the difficulties experienced.

REFERENCES

Enggarwati, N. S. (2015) Kesulitan guru SD Negeri Glagah dalam mengimplementasikan penilaian autentik pada Kurikulum 2013. *Basic Education*, 5(12).
Harizaj, M. and Hajrulla, V. (2018) Selecting and Developing Teaching/Learning Materials in EFL classes.
Kaya, S. et al. (2015) Teachers' awareness and perceived effectiveness of instructional activities in relation to the allocation of time in the classroom. *Science Education International*, 26(3), 344–357.
Khan, H. M. A. et al. (2016) Exploring relationship of time management with teachers' performance. *Bulletin of Education and Research*, 38(2), pp. 249–263.
Kristiantari, M. R. (2015) Analisis kesiapan guru sekolah dasar dalam mengimplementasikan pembelajaran tematik integratif menyongsong kurikulum 2013. *JPI (Jurnal Pendidikan Indonesia)*, 3(2).
Kunandar, D. (2014) Penilaian Autentik, Suatu Pendekatan Praktis.
Majid, A. (2014) *Pembelajaran tematik terpadu*. Bandung: PT Remaja Rosdakarya.
Nn (2013) *Tahun 2013 tentang Kerangka Dasar dan Struktur Kurikulum SD', MI Pembelajaran Tematik Terpadu kelas I dan kelas IV*. Jakarta: Departemen Pendidikan dan Kebudayaan.
Nn (2014) *Tahun 2014 tentang Kurikulum 2013 Sekolah Dasar/Madrasah Ibtidaiyah*. Jakarta: Departemen Pendidikan dan Kebudayaan.
Prastowo, A. (2015) *Panduan Kreatif membuat Bahan Ajar Inovatif*. VIII. Jogjakarta: Diva Press.
Retnawati, H. et al. (2017) Teachers' difficulties in implementing thematic teaching and learning in elementary schools. *The New Educational Review*, 48, pp. 201–212.
Sehe, S. et al. (2016) The development of Indonesian language learning materials based on local wisdom of the first grade students in Sma Negeri 3 Palopo. *Journal of Language Teaching and Research*, 7(5), pp. 913–922.

Emerging Perspectives and Trends in Innovative Technology
for Quality Education 4.0 – Kusmawan et al (eds)
© *2020 Taylor & Francis Group, London, ISBN 978-0-367-25803-0*

Student perceptions of face-to-face tutorials in Handling Children with Special Needs (HCSN) classes in Universitas Terbuka (UT)

Mukti Amini
Universitas Terbuka, Tangerang Selatan, Indonesia

ABSTRACT: HCSN (Handling Children with Special Needs) is one of the courses provided in the Early Childhood Education (ECE) department in UT. This subject is difficult because it requires students to provide education services to various special needs children . This course includes tutorial assistance and both face-to-face and online tutorials. This study aims at determining student perceptions on the implementation of face-to-face tutorials for HCSN courses. The study was conducted at UT branch Serang in the semester 2018.2. The method used was descriptive research. The data collection was a questionnaire. The study concludes that the implementation of the tutorial in terms of tutor performance, learning resources, and media used is useful, but it needs to be improved in several areas – namely, a systematic tutor explanation and media diversification. This study suggests that tutors should get ongoing refreshment about the latest learning methods.

1 INTRODUCTION

UT, as one of the Open and Distance Universities in Indonesia, currently has several departments, including the ECE (early childhood education) department. The learning process at UT relies on independent learning, meaning that students are required to learn on their own initiative, both individually and in groups, without relying on meetings with lecturers. To facilitate students in independent learning, UT has provided various types of teaching materials, both print and non-print. UT also provides several forms of study assistance services, including tutorials.

The course for HCSN is one of those provided for ECE-UT students and involving tutorial assistance. The face-to-face tutorial (FFT) for the HCSN was held for eight meetings between tutee (students) and tutors (lecturers) in one semester. The meeting will discuss essential concepts that are quite difficult for students to understand.

The course in HCSN includes subjects that are challenging, because they introduce various forms of special needs for early childhood and how to handle them. Therefore, this study aims at exploring information from students about the implementation of face-to-face tutorials on HCSN.

1.1 *Tutorial*

Tutorials involve teaching others or providing learning assistance to someone. Learning assistance can be given by older people or those of the same age (Wardani, 2005). Tutorial as a learning aid in distance education can be given in various forms, for example face-to-face tutorials, over-the-air tutorials via radio broadcasts, written tutorials by correspondence, and online tutorials through the internet (Wardani, 2005). In face-to-face tutorials, communication that occurs between tutors and tutees naturally occurs directly. Tutorials are different from ordinary lecture activities. In the tutorial activity, the party expected to be more active is the tutee, while the tutor is only a facilitator (Hazard (1967), in Wardani (2005)); in lectures, lecturers usually dominate more activities.

Tutorial kits are a set of tools used to support, smooth, and increase the efficiency of the tutorial implementation. In conducting face-to-face tutorials, the form of tutorial kits can be used tutorial activity plans (RAT), tutorial activity units (SAT), tapes, pictures, CDs, VCDs, presentation outlines, and others. Before implementing the tutorial, tutors need to plan activities. RAT is a planning tutorial prepared for one course in one semester, while the SAT is a translation taken from the RAT for each tutorial meeting (Universitas Terbuka, 2005: 63).

1.2 Handling Children with Special Needs (HCSN)

HCSN is one of the courses in the ECE-UT program, weighing four credits and providing face-to-face tutorial assistance. This HCSN course is included as a practical assignment. This means that students are required to practice learning to handle children with special needs in one of their assignments. The competency that is expected to be possessed by students after studying this course is the ability to handle children with special needs by providing educational services tailored to the children with special needs to the extent of their function as ECE teachers (Hildayani, 2006). In addition to reviewing modules and following tutorials, students are also required to practice the material they have mastered. Students' abilities are evaluated with tutorial grades and a final examination.

2 METHODOLOGY

This research uses descriptive methods through surveys. This research was conducted at Serang regional service, Banten, for semester 3 ECE-UT undergraduate students in 2018.2. The selection of semester 3 is due to the HCSN tutorial given in the previous semester, namely semester 2, so it is expected that they still have a strong memory about the implementation of the face-to-face tutorial in the previous semester. The survey was conducted through a questionnaire in the form of a closed questionnaire. The questionnaires were collected from the field and totalled 177 sets. This instrument was then tabulated and graphed, and then the percentage was calculated.

3 RESULT AND DISCUSSION

Based on the results of the questionnaires to students, their opinions about the tutors' performance in the HCSN course when giving a face-to-face tutorial can be presented in the following table.

Table 1. Student Opinions about Performance Tutor.

No	Aspect	Statement	Answer (%)					
			SA	A	NAg	SNA	NA	Average
1.	Tutorial model	Tutors use certain models that can be scientifically justified	29	69	1	1	0	3.3
2.	Explanation of material	Tutors provide essential concepts from courses	39	57	3	1	1	3.3
3.		Tutors master the material of the course that is documented	40	56	4	0	0	3.4
4.		Tutors provide enrichment of material and examples that are easily understood	41	55	2	1	1	3.3

(Continued)

Table 1. (*Continued*)

No	Aspect	Statement	SA	A	NAg	SNA	NA	Average
					Answer (%)			
5.		Tutors use language that is easily understood	42	52	5	1	1	3.4
6.		Tutors give a conclusion after each tutorial meeting	33	58	8	1	0	3.2
7.		Explanation of the tutor is not comprehensive	3	18	45	34	0	1.9
8.		Explanation of the tutor is not systematic	3	20	32	42	3	1.8
9.	Question and answer and	Tutors conduct questions and answers with students	50	47	2	1	0	3.5
10.	discussion	Tutors facilitate discussions between students	34	59	3	1	3	3.2
11.		Tutors allow the students to ask questions	39	59	2	0	0	3.4

Table 1 shows that, in general, the tutors have carried out their duties properly according to the rules. No statements are getting a mean of less than 3, except for two: the explanation of the tutor who was considered incomplete and not systematic. The tutorial models used are exciting: as Englund, Olofsson, and Price (2017) mention, although novice teachers initially held more teacher-focused conceptions, they demonstrated more significant and more rapid change than experienced colleagues. Experienced teachers tended to exhibit little to no change in conceptions. This also happened in the selection of tutorial models used by HCSN tutors. New tutors are more eager to innovate using the latest teaching models, while senior tutors prefer to use the models they use from year to year.

Meanwhile, the discussions used by tutors are usually preferred by students rather than regular lectures, although it takes time. This is consistent with the opinion of Jang, Reeve, and Halusic (2016) that "students experienced an interesting advantage, an entertainment advantage, or a social engagement advantage through discussion."

In connection with the tutor's explanation, 23% of respondents said the explanation from the tutor was less systematic. While tutor explanations are considered incomplete according to 21% of respondents, there is a possibility that students cannot distinguish between ordinary lectures and face-to-face tutorials. In the tutorial, the tutor is not obliged to explain all the materials in the module. Tutors only guide students to learn and become facilitators if students experience difficulties actively. The students still expect tutors to act as ordinary lecturers in general. The main problem in the tutor's explanation is the transmission of information. Generally, we assume that students will remember the material precisely, as explained by the tutor. However, in reality, students will remember and keep instead their interpretations of what tutors have taught (Schmidt et al., 2015). This also happened to students during the HCSN tutorial when they were tested on the understanding they gained after listening to the tutor's explanation.

Table 2. Learning resources and media used.

No.	Aspects	SA	A	NAg	SNA	NA	Average
				Answers (%)			
1.	Package of teaching materials appropriate courses through a tutorial	41	54	4	0	1	3.3
2.	The material presented with a variety of media (handouts, BMP, etc.)	15	56	20	7	2	2.8

Based on Table 2, it can be seen that the learning resources and media used at the tutorial are good enough, but the diversity of media still needs be improved. There are still around 27% of respondents who think that the media used in the delivery of material is less diverse. In the case of the HCSN course, students got both printed teaching materials in the form of the module and non-printed teaching materials in the form of VCD, which they must learn independently. At the time of the tutorial, the most common media used by tutors other than books were the PowerPoint presentations for almost all courses taught. Ideally, the media used by tutors need to be adapted to five criteria: the conformity to learning goals; learning methods; the state of participants; availability; and efficiency (Ramdhani and Muhammadiyah, 2015). However, the use of printed books as the main media for students still needs to be supported with the recommendation of reading books or the task of making resume books. This is because "students tend to concentrate their learning activities based on the lecture notes rather than the more extensive book" (Schmidt, et al., 2015).

4 SUGGESTION

4.1 Conclusion

Based on the results and discussion above, it can be concluded that the implementation of the tutorial in terms of tutor performance, learning resources, and media used is good, with a value of more than 3. It only needs to be improved in several several, namely more systematic explanation of the tutor and the various media used.

4.2 Suggestion

Based on the conclusions above, it is suggested that tutors should get ongoing refreshment in the latest learning methods, including the use of media using technology. Therefore, face-to-face tutorials can be done well and meaningfully for students.

REFERENCES

Englund, C., Olofsson, A. D., and Price, L. (2017) Teaching with technology in higher education: Understanding conceptual change and development in practice. *Higher Education Research & Development*, 36(1), 73–87.

Hildayani, R. (2006) *Handling Children with Special Needs*. Jakarta: Universitas Terbuka.

Jang, H., Reeve, J. and Halusic, M. (2016) A new autonomy-supportive way of teaching that increases conceptual learning: Teaching in students' preferred ways. *The Journal Of Experimental Education*, 84(4), 686–701.

Ramdhani, M. A. & Muhammadiyah, H. (2015) The criteria of learning media selection for character education in higher education. Proceeding from the International Conference of Islamic Education: Reforms, Prospects and Challenges, Faculty of Tarbiyah and Teaching Training, Maulana Malik Ibrahim State Islamic University, Malang, December 2–3, 2015: 174–182.

Schmidt, H.G., et. al. (2015) On the use and misuse of lectures in higher education. *Health Professions Education*, 1, 12–18.

Universitas Terbuka. (2005) *Tutorial Guidelines For ECE Program Tutors*. Jakarta: Universitas Terbuka.

Wardani, I. G. A. K. 2005. Tutorial program in open and distance higher education systems. *Higher Education and Distance Learning Journal*, 1(2). http://simpen.lppm.ut.ac.id/ptjj/PTJJ%20Vol%201.2%20september%202000/12wardani.htm

Distance learning skills training (PKBJJ) for students

Agus Prastya
Universitas Terbuka, Tangerang Selatan, Indonesia

Devie Restia
Islamic University Pamekasan, Indonesia

Stefani
Universitas Terbuka, Tangerang Selatan, Indonesia

ABSTRACT: The process of increasing new students' understanding of IT (information technology) services in Open University (UT) needs to be improved because, if the distance program in UT is implemented and understood by students, the tutorial will run well in accordance with its objectives to improve students' learning outcomes. Failure in providing IT service materials and guidelines on how to study independently at UT was allegedly the main cause of the low students' learning outcomes. The purpose of the study is to describe the results of students' ability to operate IT better and no longer be technologically illiterate. By comprehending IT, students could better follow the tutorial because administrative and academic services in UT are IT-based. Data were collected using in-depth interviews, observation, and documentation on the research subjects, UT students, and managers. The results revealed that conducting PKBJJ training helped on the tutorial run effectively and smoothly, thus ensuring improvement in learning outcomes, students' enthusiasm, and motivation to study.

1 INTRODUCTION

According to Weber (Jones, 2009), methods used in appreciating the social environment mean that people should be able to understand their actions in accordance with objectives. Weber stated that the type of social action is classified into four categories: traditional social actions, meaning that actions are controlled by traditions, such as a habit or ordinary action on does daily; affective social actions, those determined by one's specific affections and emotional state, in which people tend not to think about consequences and base their actions on attractiveness and motivation; value rational social actions, those determined by a conscious belief in the inherent value of a type of behavior; and instrumental-rational social action, those carried out to achieve a certain goal, such as something a person does because it leads to a result.

The implementation of distance learning in universities such as in the Open University (UT) is in accordance with policies based on the needs of students. They need to understand and master the internet to follow the tutorial provided by the institution because all student activities, both academically and administratively, employ digital technology – distance learning relates to the students' need to receive IT training. Thus, before the tutorial starts, all students are required to take part in introduction activities for two days. Students should carry out social actions for this **PKBM** activity because participating in students' activities can result in increased IT and internet ability, meaning tutorial activities will run better in the future.

However, in reality, there are problems faced by new students. For instance, they may not obtain knowledge about digital technology/internet right away, so it is difficult to master the tutorial materials. Before the tutorial begins, students should receive training on academic and administrative services using IT; thus, at the beginning of their studies at UT, students can already understand how to learn independently. Using the internet and mastering the IT tutorials will impact positively on their exam results.

To overcome these issues, it is immensely important to conduct distance learning skills training (PJJ); new students can understand IT services, comprehend the UT academic and administrative service system, and access academic and administrative service systems at UT through PKB training.

2   METHODS

The subjects of the study were 50 new students from batch 2018, 10 regional managers, and 10 tutors (BBLBA UT Surabaya, 2018). Data was collected using observation ofPKB activities carried out in five districts and one city. Also, interviews with the 50 students were conducted to inquire about the benefits of PKB, as well as documentation to acquire data of PKB participants both in UT and at the managerial level (Bogdan and Taylor, 2012: 341).

Furthermore, the data was processed using the Interactive model of Miles and Huberman, as follows: data collection, which is the presentation of data after being collected; data reduction, i.e. the current data which was reduced as thoroughly as possible; display data, meaning that the data presented are grouped according to their characteristics; and conclusion/or verification, i.e. the present data is concluded with one conclusion sentence (Miles and Huberman, 2010: 201).

3   RESULTS AND DISCUSSION

Table 1.   Interview.

No.	Student	Total	Major	Interview result
1.	Grumpy	50	PGSD	Agree PKB-UT
2.	Non-Squeezing	50	Management	Agree PKB-UT
3.	Manager	10	Area	Agree PK BJJ
4.	Tutor	10	PTN/PTS	Agree PKB-UT

Social Action carried out by students is an expression that comes from this study to take responsibility for the next generation's path. UT students agreed that PKB implementation improves the quality of teachers as human resources by studying at UT. The results of interviews of 50 students (Part UT Registration, 2018) and of teachers and employees depicted that teachers are willing to become students, not because of material and financial problems but because of a call to improve the quality of the nation's generation. For example, students who are teachers will be called students even though they are teachers, whether in playgroup, kindergarten, and elementary schools. Hence, the similarity of professions creates a sense of unity. Now, the teachers become students at UT. Even though their salary is high, their current activity interrupts them in their required tasks and responsibilities.

The social actions of managers and tutors agreed that PKBJJ is significantly useful for students to take part in the UT tutorial training. It is worth praise, even though the social action is a normal thing to do in educating students. Hence, the social actions of managers and tutors are something that is commonly done by the community and its members, such as UT students in their duties as educators in schools. According to Weber, this action is categorized as a traditional social action because it is usually performed by the community members. UT students should master IT well, because if they do not they will be open to criticism from the public over how a teacher could not understand IT.

4 CONCLUSION

1. To improve the performance of UT students, IT training is needed by students through the UPBJJ-UT program by implementing the provisions and procedures for the proper application of UPBJJ-UT. Students followed and listened to the instructors' presentations; afterward, they understood particularly, regarding IT services at UT.
2. Motivation and good intentions of the manager are necessary. Furthermore, the tutors are satisfied and agree that PKBJJ is highly useful for students before participating in the tutorial.

REFERENCES

Bogdan, R. and Taylor, S. J. (1993) *Qualitative Research Basics*. Translation A. Khozin Affandis.
Creswell, J. W. (1988) *Quality Inquiry and Research Design*. California: Sage.
Jones, Pip (2010). *Social Theories*. Yogyakarta: Torch.
Moleong, L. J. (2007) Bandung Youth Qualitative Research Methodology Rosdakarya.
Ritzer, G. and Godmann, G. J. (2004) *Modern Social Theory*. Translation Alimandan. Jakarta: Prenada Media.
UT Catalog (2018). *Open University Basic Education Curriculum*. UT Publishers.

Emerging Perspectives and Trends in Innovative Technology
for Quality Education 4.0 – Kusmawan et al (eds)
© 2020 Taylor & Francis Group, London, ISBN 978-0-367-25803-0

Development of educational traffic signs game based on android for elementary school students

A. Rachman, D.A. Fatimah & H. Nugroho
Informatics Engineering, Institut Teknologi Adhi Tama Surabaya, Indonesia

Sulistyowati
Information Systems, Institut Teknologi Adhi Tama Surabaya, Indonesia

W. Widodo
Informatics Engineering, Institut Teknologi Adhi Tama Surabaya, Indonesia

W.A. Surasmi
Universitas Terbuka, Tangerang Selatan, Indonesia

ABSTRACT: Traffic signs are a tool used to provide road users with information about the rules that must be followed. Traffic signs must be introduced early on to children; this is because many drivers do not obey traffic signs when driving. We developed an Android-based educational game for elementary schools that provides an introduction to traffic signs in a fun way that doesn't feel like learning in the true sense. Our application test was carried out on 30 elementary school students in grade 5. The method we used to measure the cognitive abilities of students was to carry out a pre-test, play a game, and then carry out a post-test. The pre-test and post-test process provide multiple-choice questions. The process of playing games lasts for 10–20 minutes, every day for two months. At the end, we do a post-test. From the results of the post-test conducted by researchers, it was found that the educational traffic sign game developed by the researchers is capable of helping elementary school students better understand traffic signs and their function. Researchers used ISO 9126-3 usability factors, and, based on the test results, the educational game application is very useful as a medium for primary school student learning aids.

Keywords: Application, Educational Games, Iso-9126-3, Traffic Signs, Android

1 INTRODUCTION

Data from the World Health Organization (WHO) found that 90% of worldwide deaths caused by traffic accidents occur in developing and low-income countries. Indonesia has one of the highest accident rates in Asia. In accidents in Indonesia involving motorized vehicles, the average age of the driver is between 15 and 29 years (Nastiti, 2017). Among the leading victims of traffic accidents are children, who do not understand traffic signs. The introduction of traffic signs early on is very important for the safety of children on the road (Sugiyanto and Santi, 2015).

Elementary school in Indonesia is carried out for six years, from grade 1 to grade 6. Elementary schools play a role in student abilities as they have the potential to assist student development. To achieve its goals, teaching in elementary schools must pay attention to the cognitive, affective, and psychomotor aspects of students (Sarinengsih, Nur'aeni, and Pranata, 2018). In previous studies, the use of technology in teaching and learning processes made the learning process more innovative. Previous studies also found that the use of digital technology in the learning process can reduce the cost of school spending (Rachman, 2011).

Traffic signs can take the form of symbols, letters, numbers, sentences, or combinations that serve as warnings, restrictions, orders, or instructions for road users (Nugroho, 2009). Traffic signs are intended to regulate driving conditions on the road to avoid traffic accidents, but many drivers still violate traffic signs due to a lack of driver knowledge or a lack of willingness to obey existing traffic signs (Syahiruddin, 2016). Four types of traffic signs must be recognized and understood by motorists on the highway, namely Wednesday traffic orders, prohibited traffic signs, warning traffic signs, and directional traffic signs. Command traffic signs are marked in blue on traffic signs, prohibited traffic signs are marked in red, warning traffic signs are marked in yellow signs, and manual traffic signs are marked in green (Sari, 2019).

Software is a tool created to solve existing problems – starting from problems in offices, hospitals, schools, and even community groups (Mujumdar, Masiwal, and Chawan, 2012). The processes involved in software engineering work sequentially, from simple activities to complex activities (Mirza and Datta, 2019). To create software, the developers must rely on the model of the software development process: waterfall, incremental, rapid application development, spiral, v-shaped, even agile frameworks (Kaur and Boparai, 2015; Reddy and Kumar, 2019).

Digital games are counted as a type of application software (Rachman, Purwanto, and Nugroho, 2019). Digital games of various types of learning media are in demand in many countries, being considered effective learning media. The use of digital games in classes varies by country. Previous studies on digital games suggested that the use of digital games can stimulate cognitive functions and improve the skills of players. In clinical trials, games can improve the structure and function of neuropsychology in modifying the structure and function of brain architecture (Pereira and Valentin, 2018). Game-based learning is a very competitive approach in motivating students (Mokeddem, Plaisent, and Prosper, 2019), whereby playing games students try to win the game but learn indirectly as well.

As there is still a lack of awareness among road users about the importance of obeying traffic signs, and due to the power of digital games, the researchers developed an educational game for the introduction of traffic signs, with the target user being a grade 5 elementary school student. Because the planting of knowledge early on among elementary school students helps develop their character, such a game will have an impact on student behavior when they become adults (Hadisi, 2015).

2   RESEARCH METHODS

In this study, researchers used an incremental software engineering development model. In the incremental development model, there are five main activities carried out by researchers, namely engineering needs, design stages, application development stages, testing stages, and implementation stages. To obtain research objectives, researchers used a research method consisting of literature study activities, application development, pre-test, playing games, post-test, and application feasibility test.

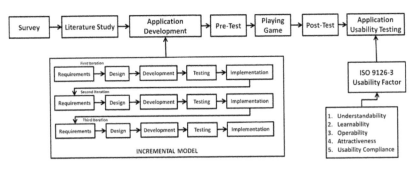

Figure 1.   Research methods of traffic sign game application.

The first activity carried out by researchers was to conduct a survey. Researchers surveyed one of the elementary schools in Surabaya, the Klampis Ngasem I Elementary School. The researchers met with the principal to discuss the research topic. For traffic signs material, researchers reviewed several web pages with the keyword "traffic signs" to get more information about them. In application development, researchers use the incremental model as a software development model. This model is very suitable for developing the subject of traffic signs into an educational game application. Researchers conducted two iterations in application development: learning material and game material. The game design, based on instructions, will instruct the player to take action by existing orders. The second activity carried out by researchers was to conduct a literature study on the material of traffic signs and software development models that match the research undertaken. The third activity was application development, carried out by researchers using the incremental process model software. In this study, researchers conducted two iterations where each iteration consisted of five main activities: requirements, design, development, testing, and implementation. The fourth stage of making an application was testing: the researchers carried out the testing process in a functional way, according to the objectives of the game-making process, especially the learning function. If there were functional errors, the researchers immediately fixed them. The fifth activity of this study was to have grade 5 elementary school students play an educational game application we made twice a week.

Our educational game application was made by using the game engine construct 2.0 and with the help of some graphics tools (Corel Draw and Adobe Illustrator). There are 140 kinds of picture signs in the game. In this section, the researchers conducted a test with a team of five people as quality control. Five teachers tried our application and played this game for 25 minutes, each using smartphones. For elementary students, 10 students first tried out the game for 10 minutes. The researchers then instructed the teacher that a total of 30 students could play the educational game twice during the week for a duration of 10–15 minutes. The researchers conducted a pre-test of the 30 students before they played the educational game. This pre-test was conducted by researchers to determine students' knowledge of current traffic signs. The questions given were in the form of multiple choices of 20 questions. The pre-test results found 25% of elementary school students made a lot of mistakes in answering questions on traffic signs.

3 RESULT

In this study, researchers developed an Android-based educational game application that was tested by 30 grade 5 students and five teachers. After playing the game, elementary school students were tested with the same questions as they were during the pre-test. The post-test results showed that students who answered questions wrongly decreased from 25% to 3%. From Figure 2, it can seen that overall the students, after playing the educational traffic signs game, showed better capabilities than before.

The last activity of this research was to conduct a feasibility test on the application made by researchers. Researchers used the Usability Factor from ISO 9126-3. From this Usability

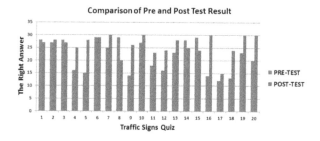

Figure 2. Comparison of test results in grade 5 students.

Factor, there are five things to be assessed: understandability, learnability, operability, attractiveness, and usability compliance sub-factors. The feasibility test of the educational game application was also assessed by the teacher. Researchers use the Likert scale to obtain test results. The results of the application assessment based on the usability factor were understandability (96%), learnability (96%), operability (84%), attractiveness (84%), and usability compliance (96%), with an average value of 91.2%. This means that the traffic signs game is very useful as digital learning media.

4 CONCLUSION

Researchers have developed an educational traffic signs game application using an incremental model, and this application can help elementary school students improve their understanding of traffic signs. This application can feasibly be used as a learning medium for elementary school students.

REFERENCES

Hadisi, L., (2015) Pendidikan Karakter Pada Anak Usia Dini. *Jurnal Al-Ta'dib*, 8, 50–69.

Kaur, S. and Boparai, A K, (2015). Process of moving from traditional to agile software development: A review. *International Journal of Advanced Research in Computer Science and Software Engineering*, 5, 586–591.

Mirza, M. S. and Datta, S. (2019). Strengths and weakness of traditional and agile processes: A systematic review. *Journal of Software* 14, 209–219.

Mokeddem, A., Plaisent, M., and Prosper, B. (2019) Learning with the games: A competitive environment based on knowledge. *Journal of e-Learning and Higher Education* 2019, 1–6. https://doi.org/10.5171/2019.133016.

Mujumdar, A., Masiwal, G., and Chawan, P. M. (2012) Analysis of various software process models. *International Journal of Engineering Research and Applications* 2, 2015–2021.

Nugroho, S. S. (2009). Undang-Undang Republik Indonesia Nomor 22 Tahun 2009 Tentang Lalu Lintas Dan Angkutan Jalan.

Pereira, V. F. A. and Valentin, L. S. S., (2018) The MentalPlusD digital game might be an accessible open source tool to evaluate cognitive dysfunction in heart failure with preserved ejection fraction in hypertensive patients: A pilot exploratory study. *International Journal of Hypertension*. https://doi.org/10.1155/2018/6028534.

Rachman, A. (2011).Optimalisasi VRML Sebagai Media Pembelajaran Playgroup Berbasiskan Web 3D. The 12th Seminar on Intelligent Technology and Its Applications.

Rachman, A., Purwanto, M. Y., and Nugroho, H. (2019) Development of educational games for the introduction of fruits and vitamins. *Journal of Educational Science and Technology* 5(1),76–81. https://doi.org/10.26858/est.v5i1.

Reddy, K. S. M., and Kumar, V. V., (2019). A review of conventional SDLC process models. *International Journal of Scientific Research and Reviews* 8, 4186–4191.

Sari, N. M. (2019) Tanda Rambu Lalu Lintas yang Perlu Dipahami dan Ditaati untuk Kurangi Kecelakaan. Available at: https://hot.liputan6.com/read/4012293/tanda-rambu-lalu-lintas-yang-perlu-dipahami-dan-ditaati-untuk-kurangi-kecelakaan.

Sarinengsih, S. S., Nur'aeni, L. E., and Pranata, O. H. (2018) Peningkatan Pemahaman Konsep Materi Simetri Lipat melalui Penerapan Model Pembelajaran learning cycle 5E. *Jurnal Ilmiah Pendidikan Guru Sekolah Dasar* 5, 9–20.

Sugiyanto, G., and Santi, M. Y. (2015) Pendidikan Kesalamatan Sejak Usia Dini Untuk Mengurangi Tingkat Fatalitas Pejalan Kaki. *Jurnal Teknik Sipil* 13, 104–123.

Syahiruddin, S. (2016) Aplikasi Pembelajaran Rambu-Rambu Lalu Lintas Serta Peraturan Berkendara.

Emerging Perspectives and Trends in Innovative Technology
for Quality Education 4.0 – Kusmawan et al (eds)
© 2020 Taylor & Francis Group, London, ISBN 978-0-367-25803-0

The benefits of Self-Regulated Learning (SRL) training for improvement of students' SRL

U. Rahayu, A. Sapriati & Y. Sudarso
Universitas Terbuka, Tangerang Selatan, Indonesia

ABSTRACT: Self-regulated learning (SRL) is a significant factor determining the success of student learning. Every student can be trained to be a self-regulated learner. This paper presents research relating the benefits of implementing SRL training in students at the Universitas Terbuka (UT), Indonesia. This research is mixed-method, employing one group design consisting of a pre-test and a post-test. The SRL training employed learning strategies integrated within an online tutorial. The tested population was made up of 65 students pursuing biology education at UT in academic year 2016. Although the increases seen were minimal, the results nevertheless indicated an improvement in students' SRL. This finding has applications relevant to university policy, particularly in integrating SRL training within other online courses.

Keywords: Distance Education, Training, Self-Regulated Learning

1 INTRODUCTION

Self-regulated learning (SRL) is one variable known to affect overall student learning. SRL is defined as a self-initiated, active, and constructive action in which students plan, organize, control, and evaluate their learning process. SRL includes cognitive, metacognitive, behavioral, motivational, and emotional aspects of learning (Panadero, 2017). Students mastering SRL learn faster and have better academic performance compared to ones that do not (Kizilcek, Sargustin, and Maldonado, 2016). To this end, an integrated and coherent SRL model and its relevant training supports improving student learning (Panadero, 2017).

Many studies researching the improvement of SRL in students have been conducted (Cheng and Chau, 2013; Lai and Hwang, 2015). However, investigation of certain methods to enhance SRL, especially for science students, is insufficient. Students lacking SRL skills show relatively low learning outcomes and tend to be inconsistent with their approach. Students at UT, primarily a distance-learning institution in Indonesia, are familiar with the online tutorial as a learning method. Therefore, training in this specific modality is predicted to improve the SRL for science students at UT.

2 RESEARCH METHOD

2.1 *Method*

The mixed-method used in this study consisted of a one-group design of pre-test and post-test (Cresswell and Clarck, 2007). To determine the effectiveness of online tutorials to improve the students' SRL, differences in ability were calculated by assessments done before and after training. Any increase of SRL was further delineated as low, medium, or high. Data collection occurred from January to April of 2016. The study population consisted of 65 students in a Biology Education study program at UT, Indonesia, in academic year 2016, in the courses of Assessment in Biology Learning, Strategy in Biology Learning, Human Anatomy and Physiology, and Animal Development.

2.2 Data collection

Data for students' SRL was collected using a modified Motivation for Learning Strategy (MLSQ), with four Likert scales, where 1 = never, 2 = rare, 3 = often, and 4 = very often (Rahayu and Widodo, 2017), and assessed both motivation and learning strategies. Measurement of motivation was further defined as both extrinsic and intrinsic and as self-efficacy. Learning strategies were measured in terms of goal setting, goal accomplishment, objectives in science, monitoring, time management, studying locations, effort regulation, self-evaluation, and self-reflection. The assessment contained 40 valid items, with a Cronbach Alpha reliability of (r) = 8.6 at α = 1%. Students were asked to fill out the MLSQ questionnaire both before and after taking an online tutorial.

The SRL tutorial techniques taught included study planning, self-monitoring, self-evaluation, mind mapping, and compiling questions and answers. Integrated training material in the form of a CERDAS book was provided, both in printed and online forms. One week prior to the online tutorial, students were required to monitor the implementation of their study plan and reflect on the learning process that occurred. The students were also asked to demonstrate an understanding of the concepts through the use of mind mapping and questions-and-answer formation. Students were interviewed for 10 weeks using guidelines that both explored SRL skills and reinforced and complemented other data.

2.3 Data analysis

Data collected was both quantitative and qualitative; SRL scores provided the quantitative data, while qualitative data were obtained from interviews. To measure the differences between SRL ability before and after training, parametric tests using a paired sample of t-test were carried out. The t-test is valuable when the data are distributed normally and have nonhomogeneous variance. Normality distribution testing was carried out using the Kolmogorov-Smirnov test. Homogeneity of variance testing was carried out by the Levene's test. To determine the improvement of the students' SRL, a normalized gain score (N-gain) analysis was carried out. The categories of gain values were defined as low (N-gain less than 3), moderate ($0.3 \leq$ N-gain ≤ 0.7), and high (N-gain ≥ 0.7) (Hake, 1998). Data analysis was carried out using SPSS 23 for Windows with a significance level of 5%.

3 FINDINGS AND DISCUSSION

Analysis of the mean difference showed that there was a significant difference between pre-SRL training scores and post-SRL training scores (p-value = 0.00 at α = 0.05). This is comparable to a previous study. Significant differences were shown between the students' SRL after training was evident, particularly in the self-monitoring parameter of the Biology Learning Strategies course (Rahayu, Widodo, and Redjeki, 2017), as well as the self-evaluation and self-reaction parameters of the Human Anatomy and Physiology (Rahayu, Widodo, and Damayanti, 2018). Table 1 shows the differences in the average score of students' SRL, before and after training.

Table 1. The result of the statistic using a t-test on students' SRL.

SRL	N	Mean	T	SD	Std. Error Mean	p-value
Pre-SRL	65	2.98	-4.19	.25	.031	<0.00
Post-SRL	65	3.21		.34	.042	

An N-gain test was carried out to identify the criteria for SRL improvement and compared the values of pre- and post-training SRL. The value of N-gain was 0.20, indeed showing an improvement of students' SRL, albeit a low level. Techniques taught in the training modules consisted of scheduling study time, time management, self-monitoring, self-evaluation, mind mapping, and devising questions and answers. These were shown to enhance students' SRL. However, the increases were not optimal. Presumably, several possible factors contributed to this. First, students were not yet familiar with the SRL training that was integrated within the online tutorials; indeed, the frequency and amount of student involvement in SRL training tended to decrease. Second, longitudinal research has shown that it takes four years to improve the study skills of students through SRL training (Wibrowsky, Matthews, and Kitsantas, 2016). SRL is a learned habitual behavior that requires time to acquire. As such, eight weeks was not long enough to transform the habits, attitudes, and behavior of the students. Therefore, consistency of the tutors and online tutorial programs, in addition to the time required to train, is required in order to improve aspects of SRL, which include motivation, self-efficacy, goal setting, goal accomplishment, self-monitoring, time management, studying location, effort regulation, self-evaluation, and self-reflection. Consistency of tutors as well as online tutorial programs are needed to build students' SRL.

Of the initial 65 study respondents, 56% remained involved in the overall online tutorial activities. Although the effect seen was low, the involvement of students in implementing the techniques learned from the tutorial training was key. In addition, participants interact with other students and/or instructors to enhance their SRL (Sun and Rueda, 2012). As supported by their interviews, it was shown that 92% of students found SRL training integrated into the online tutorials to be beneficial for them, and 99% of students stated that CERDAS as a material book helped and encouraged them to learn independently and systematically. Time management strategies learned with this systematic method were shown to increase motivation, particularly for distance education students. However, some students reported the tutorial experience to be difficult and found constraints in applying the learning strategies, namely in the subjects of time management, setting realistic short-term learning goals, self-monitoring, self-evaluation, and mind mapping.

Benefits seen from applied cognitive and metacognitive strategies did exist, however. During the training, students self-recorded their behavior and learning activities in a schedule. Specifically, they identified which materials had been mastered, the problems that were faced, and the possible solutions. The application of reflection and evaluation tasks can help students focus on the goal of building knowledge (Yang, Aalst, Chan, and Tian, 2016). It can also improve students' understanding, reflection, and critical thinking skills (Ghanizadeh, 2016). In this study, self-monitoring was shown to help students increase self-awareness of learning strategies.

4 CONCLUSIONS

The results of this research have shown that SRL training, when integrated within an online learning platform, improves students' SRL, although in the low category. University regulation is needed to support this training so that students can achieve SRL skills as well as modern-day self-assessment techniques. The implication of this study is that SRL training should be implemented in a broader range of courses, as it will lead to a gradual improvement of learning readiness and self-regulated learning skills in student populations.

REFERENCES

Cheng, G. and Chau, J. (2013) Exploring the relationship between students' self-regulated learning ability and their ePortfolio achievement. *Internet and Higher Education*, 17(1), 9–15.
Cresswell, J. W. and Clarck, P. V. L. (2007) *Designing and Conducting*. New Delhi: Mixed Methods Research.

Ghanizadeh, A. (2016). The interplay between reflective thinking, critical thinking, self-monitoring, and academic achievement in higher education. *Journal of Higher Education*, 74(1), 101–114.

Hake, R. (1998). Interactive engagement vs traditional methods: A six-thousand student survey of mechanics tests data for an introductory physics course. *American Journal of Physics* 66(1), 64–74.

Kizilcek, R. R., Sargustin, M. P., and Maldonado, J. J. (2016) Self-regulated learning strategies predict learner behavior and goal attainment in massive open online courses. *Computers & Education*, 104, 18–33.

Lai, C. L., and Hwang, G. J. (2015) A self-regulated flipped-classroom approach to improving students' learning performance in a mathematics course. *Computers & Education*, 100, 126–140.

Panadero, E. (2017). A review of self-regulated learning: Six models and four directions for research. *Frontiers in Psychology*, 8(422), 1–28.

Rahayu, U. & Widodo, A. (2017) The development of online tutorial integrated with learning strategy guide to practice self-regulated learning. *Jurnal Pendidikan & Kebudayaan*, 2(2), 201–210.

Rahayu, U., Widodo, A., and Damayanti, T. (2018) Enhancing students' self-regulated learning and achievement through metacognitive and cognitive strategy training. *Advance Science Letter*, 24(11), 8414–8417.

Rahayu, U., Widodo, A., and Redjeki, S. (2017) *The Enhancement of Self-Regulated Learning and Achievement of Open Distance Learning Students through Online Tutorials: Ideas for 21st Century Education*. London: Francis Group.

Sun, J. C. Y. & Rueda, R. (2012) Situational interest, computer self-efficacy and self-regulation: Their impact on student engagement in distance education. *British Journal of Educational Technology*, 43(2), 191–204.

Wibrowsky, C. R., Matthews, W. K., and Kitsantas, A. (2016) The role of skills learning support program on first-generation college students' self-regulation, motivation, and academic achievement: A longitudinal study. *Journal of College Student Retention: Research, Theory & Practice*, 19(3), 317–332.

Yang, Y., Aalst, J., Chan, C., & Tian, W. (2016) Reflective assessment in knowledge building by students with low academic achievement. *International Journal of Computer-Supported Collaborative Learning*, 11(3), 282–311.

Emerging Perspectives and Trends in Innovative Technology
for Quality Education 4.0 – Kusmawan et al (eds)
© 2020 Taylor & Francis Group, London, ISBN 978-0-367-25803-0

Initial research on online-enriched microteaching in higher teacher's open and distance education

Udan Kusmawan, Sri Sumiyati & Della Raymena Jovanka
Universitas Terbuka, Tangerang Selatan, Indonesia

ABSTRACT: Online-enriched Microteaching (OMT) is a series of learning processes that are carefully designed and intensely implemented in order to achieve a positive impact on the basic teaching skills of student teachers. In Indonesia, the OMT emphasizes its training strongly on integrating skills of the 21st Century into the traditional field of microteaching. The appropriate number of practitioners and the proper time involved constitutes another variable to be considered while harvesting the dividend of the practices. This initial research recommends a total embrace of strong practices of the OMT in order for the country to achieve quality ICT enrichment while applying learning theory and didactics, pedagogical, methodical, and andragogical learning into its practices.

1 INTRODUCTION

Traditional microteaching has been practiced worldwide and has been seen to have equipped and improved the nature of educating and learning of programs (Remesh, 2013; Anthonia, O, 2014). As referenced above, normally, microteaching is acquainted with accomplished higher nature of instruction and learning. Different courses of action and media have been utilized to reinforce the intensity of the microteaching. Advancement came with the combination of video-recordings into the conventional microteaching system. Video-recordings were ordinarily circulated to more extensive clients (understudies) to empower them to work outside research hall exercises with the point of giving the clients more chances to adapt altogether. The additional estimation of this technique was to give understudies (clients) the chance to think about their own recorded presentations, notwithstanding having extra time for learning and giving comments on their partners' presentations. As Kourieos (2016) demonstrated, this strategy elevates a higher method to draw in understudies in reflective activities during their learning.

Advancements in information and communication technologies (ICT) have contributed to the involvement of the media in the quality of the microteaching program. Several studies suggested that microteaching may be practiced through online media(Yorke, 1975; Bell, 2007; Kusmawan et al., 2009; Kusmawan, 2013, 2017; Remesh, 2013). They suggested that online microteaching has strengthened teachers' abilities to develop more extensive critical thinking and reflective actions while practicing quality teaching. These critical and reflective thinking and actions constitute the essential teaching skills for which the microteaching may have a greater impact on quality practices of teaching preparations.

Based on the proposition, research entitled "Reinventing concepts and models of online-enriched practical courses for teacher education programs in open and distance higher education system in Indonesia" was conducted under UT Research Grand no.32691/UN31.LPPM/PM/2019, dated 23 Juli 2019. This research aimed at generally developing an online platform where students are facilitated to build their academic portfolio regarding their continuous learning achievements and authentic reports of practical-course activities during their study at UT. We hope to provide clinical assistance for the students considering the quality of their practices. This is critical so that the teacher students can receive appropriate training regarding the basic skills that they need to acquire for efficient online teaching careers. As for the initial goal of the research, this draws on a renewed course outcome in learning microteaching, as one of the practical courses provided by UT. This research essentially has intense information and communication technologies that grant access to current learning and teaching processes and strategies, as well as aligning with the current student teachers' learning styles.

While the traditional microteaching activities are conducted in face-to-face (F2F) settings, online-enriched microteaching promotes thorough interactions for a student-teacher who aims at improving his/her teaching quality through distance education modes(Kusmawan, 2018). Further, Kusmawan suggested that while practicing the microteaching online, experts are required to provide advice for the practitioners and colleagues in a group discussion, who are involved in discussion forums, share ideas and reflective comments on the teaching practices. These interactions are organized through asynchronous online activities and controlled by a moderator assigned to the forum. In the context of community online learning, Lambert and Fisher (2013)indicated that students do not manage themselves well for conducting synthesis phases of inquiry without some degree of supports. Therefore, regulated online learning is critical in leading students' learning. Collaborative and communicative skills are essentials in a 21st Century global world; therefore, it is important to offer opportunities to students to put these skills into practice.

In the context of gaining meaning through discussion as part of online learning activities that are emulated from behavior, Wang et al. (2015) identified some significant associations between those of discourse behavior and learning activities. This e-learning structure is to facilitate students with high and proper possibilities of gaining directly with the lecturers or supervisors on their necessities. This direct interaction mostly occurs through online, both synchronous and asynchronous. In his research, Özpolat and Akar (2009) assessed the exhibition of the proposed computerized student demonstrating the approach. They observed that the proportion between the acquired student's learning style of utilizing the proposed student's model and those gotten by the polls that were generally utilized for learning style evaluation is reliable for a large portion of the elements of Felder-Silverman learning's style. Felder-Silverman's learning style model is developed to depict different learning styles and preferences. Further research in this area was reported, including the research by Hawk and Shah (2007) and El-Bishouty et al. (2019), which suggested that lecturers should create courses to reflect their individual teaching strategies which do not always fit with various learning styles of students. Further, they concluded that a course structured with certain learning styles could improve the learning of the students with those learning styles.

As a result, online learning, which is programmed on a certain pattern of recognition and communication, will describe the classes in the language of the signs; hence, the recognition system can be built with automatic training. In case of, otherwise, a resource for initial information is limited when describing the classes, and for some reason, it is less appropriate to compile such a description since pattern recognition system may be shaped by means of training. With this online-enriched microteaching conception, automatic learning may become part of the teaching and learning system to a certain adequate proportion.

2 PRELIMINARY RESULTS AND DISCUSSION

As mention previously, this article describes five issues regarding the basic ideas of microteaching (MT) with functions that are enriched through online services. First of all, as shown in Table 1, this research reported that above 98% of the respondents gave positive responses to the proposed conception of microteaching. Strong agreement was indicated by all leaders of the institutions, especially the deans and vice deans of the faculties of education. Out of all the responses, there was no additional comment from the respondents regarding the concept of microteaching. As a result, we assumed that the concept of this online-enriched microteaching (OMT) being used in this research had been widely accepted. The OMT is defined as a sequence of practical rehearses that are planned and implemented through face-to-face activities and at the same time are monitored and supervised by teaching professionals through the online platform so that its entire quality practices can be assured, in particular in the aspects of readiness to provide professional teachers.

Table 1. conception and learning outcome.

Agreement	Conception		Learning Outcome	
Strongly Agree	16	36%	14	31%
Agree	28	62%	30	67%
Disagree	1	2%	1	2%
Strongly Disagree	0	0%	0	0%
TOTAL	45	100%	45	100%

Secondly, the point of the survey is concerned with the course Learning Outcome (LO) of the microteaching scheme. A statement regarding the LO was shared with all respondents who asked them the extent of their agreement to the LO statement. As shown in the above Table 1, the research observes that about 98% of the respondents expressed their positive agreement towards the proposed statement. Among them, a respondent suggested that there should not be an explicit involvement of the ICT in the description of the LO. By reason of the vast number of respondents, this research was not able to feature the responses of all the respondents, but it looked through these responses and thus concluded that the proposed statement is widely accepted by most of the respondents. This research accepts that the LO of the microteaching program is to provide opportunities for students to practice basic teaching skills in a constructive, supportive, and conducive atmosphere, with strong enrichment on various ICT-based teaching and learning media and strategies. As a result, the course of the MT needs to elevate targets of the program into preparing students not only for those who are skillful in teaching and learning but also those that are skillful in implementing the 21st-century skills into their teaching and learning processes.

The third point of this research is to investigate more specific learning outcomes (SLO) of the MT program required to be enhanced through the online-enriched microteaching.15 statements regarding the SLO were given to all the respondents. All the respondents were advised to select more than one statement from the list that was provided. As illustrated in, this research observed that only one aspect was appreciated by below 25% of the respondents, namely the aspect of no.4. This means that only a few of the respondents did agree that the students need to be capable of explaining the meaning of the OMT. This fact might indicate that the MT program should not require the student practitioners to interpret MT as editorially. Instead, the respondents tend to suggest that the MT program should directly focus more on OMT basic skills training. This result is associated with the previous conclusion on the LO statement, which suggested that MT programs should provide wider opportunities for students to practice basic teaching skills in a constructive, supportive and conducive atmosphere.

The following result discusses the basic teaching skills that must be rehearsed by student teachers while working on the OMT program. Figure 2. illustrates the result of the survey. It

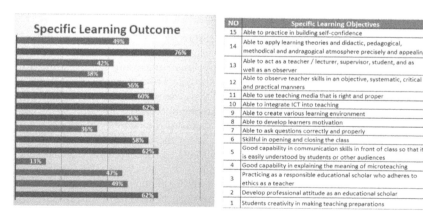

Figure 1. Specific learning outcomes of the OMT.

176

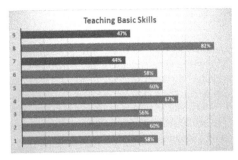

No.	Teaching Basic Skills
9	Skills of organizing small group discussion
8	Skills of classroom management
7	Skills of teaching small group and individual
6	Skills of opening and closing the class
5	Skills of providing explanation
4	Skills of integrating ICT into teaching
3	Skills of creating various teaching environment
2	Skills of giving reinforcement
1	Skills of asking questions

Figure 2.   Teaching basic skills of the OMT.

is indicated that two of the basic skills, namely the skills of No.7 and No.9, were considered less imperative to be deliberately considered as basic skills of the OMT. On the contrary, this research observed that being skillful in managing the classroom is appreciated as the most important skill for the OMT practitioners (suggested by 82% of the respondents). The second priority of the basic skills for the OMT is the skill of integrating ICT into teaching (No.4). This priority was followed by skills of giving reinforcement (No.2) and providing explanations (No.5). This finding may associate with the role of the teacher expected in the digital era, as implied in the assented learning outcome of the OMT. The teachers nowadays are required to emphasize their teaching strategies on facilitating students with more explanative reinforcement so that the 21st Century skills, such as collaborative and participative learning, and flipped teaching, can be orchestrated by the current teachers.

The following result concerns the environment which needs to be prepared for student teachers while practicing their online-enriched microteaching. Figure 3 illustrates the results of the survey. Inconsistent performance with the previous results, the environment of the OMT should be given priority in order to facilitate practical training, especially those that involve the limited number of practitioners (5-10 peoples), to facilitate the provision of the appropriate length of time (10-15 minutes), and to facilitate proper feedbacks. Though with less emphasis, but still above 25% of the respondents expected that the OMT would pay attention to quality interactions among the practitioners that are involved, including low-threat and low-risk situations so that rehearsals will function to motivate the student teachers to improve their teaching skills and secure the participants that are involved in the practices.

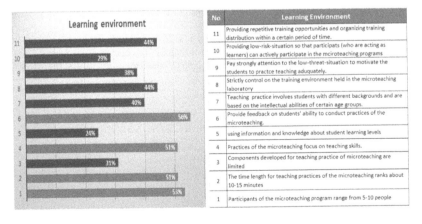

Figure 3.   Learning environment of the OMT.

3 CONCLUSION

We discovered from this initial research that online-enriched microteaching is designed to facilitate greater opportunities for continual upskilling while the student teachers are practicing the basic skills of teaching. Quality practices are projected through intensive rehearsals in classroom management. The second priority of the basic skills for the OMT is the skill of integrating ICT into teaching (No.4). This priority was followed by skills of giving reinforcement (No.2) and providing explanations (No.5). These findings may associate with the role that is expected of the student teachers in the digital era, as implied in the assented learning outcome of the OMT. Nowadays, student teachers are required to emphasize their teaching strategies in order to facilitate students with more explanative reinforcement so that the 21st Century skills, such as collaborative and participative learning, and flipped teaching, can be orchestrated by the current teachers.

REFERENCES

Anthonia, O, I. (2014) 'No Title', An International Multidisciplinary Journal, 8 (4)(35), pp. 183–190. doi: http://dx.doi.org/10.4314/afrrev.v8i4.15.

Bell, N. D. (2007) 'Microteaching: What is it that is going on here?', Linguistics and Education. doi: 10.1016/j.linged.2007.04.002.

El-Bishouty, M. M. et al. (2019) 'Use of Felder and Silverman learning style model for online course design', Educational Technology Research and Development. doi: 10.1007/s11423-018-9634-6.

Hawk, T. F. and Shah, A. J. (2007) 'to Enhance Student Learning', Decision Sciences Journal of Innovative Education.

Kourieos, S. (2016) 'No Title', Australian Journal of Teacher Education, 41(1), pp. 65–80. Available at: https://ro.ecu.edu.au/cgi/viewcontent.cgi?article=2907&context=ajte.

Kusmawan, U. et al. (2009) 'Beliefs, attitudes, intentions and locality: the impact of different teaching approaches on the ecological affinity of Indonesian secondary school students', International Research in Geographical and Environmental Education. doi: 10.1080/10382040903053927.

Kusmawan, U. (2013) 'Teachers' online forum: an online interactive forum for sustaining teacher professional development, by Universitas Terbuka', in Perspectives on open and distance learning: Open Educational Resources: an Asian perspective.

Kusmawan, U. (2017) 'Online microteaching: A multifaceted approach to teacher professional development', Journal of Interactive Online Learning.

Kusmawan, U. (2018) 'Online Microteaching: a Multifaceted Approach to Teacher Professional Development', Journal of Interactive Online Learning.

Lambert, J. L. and Fisher, J. L. (2013) 'Community of inquiry framework: Establishing community in an online course', Journal of Interactive Online Learning, 12(1), pp. 1–16.

Özpolat, E. and Akar, G. B. (2009) 'Automatic detection of learning styles for an e-learning system', Computers and Education. doi: 10.1016/j.compedu.2009.02.018.

Remesh, A. (2013) 'Microteaching, an efficient technique for learning effective teaching', Journal of Research in Medical Sciences.

Wang, X. et al. (2015) 'Investigating how student ' s cognitive behavior in MOOC discussion forums affect learning gains', Proceedings of the 8th International Conference on Educational Data Mining.

Yorke, D. M. (1975) 'Microteaching', Journal of Educational Television. doi: 10.1080/1358165750010304.

Emerging Perspectives and Trends in Innovative Technology
for Quality Education 4.0 – Kusmawan et al (eds)
© 2020 Taylor & Francis Group, London, ISBN 978-0-367-25803-0

Students' perceptions of learning media video blogs in online tutorials

Binti Muflikah, Yusak Suharno & Vica Ananta Kusuma
Universitas Terbuka, Tangerang Selatan, Indonesia

ABSTRACT: Online tutorials are part of the online learning assistance services used at Open University to facilitate education. Effective tutorials happen where there is good interaction between a tutor and students, as well as among learners themselves. Improving the success of an online tutorial ensures that learning is carried out more successfully. However, tutorials require creativity and innovation in their presentation of material to keep students active and engaged. Educational institutions should adopt and innovate on the appropriate learning media due to the growth of internet users, among them the large number of Generation Y in Indonesia. Vlogs, or video blogs, are a learning media innovation that needs to be developed further. The purpose of this study was to determine students' perceptions of the use of vlogs as a learning media tutorial for three courses in the Open University English Education program. The results of the study show that vlogs play a significant role in the online tutorial.

1 INTRODUCTION

According to Mennheim (1923), Generation Y consists of those people born between 1981 and 1995. These individuals are also referred to as the millennial generation, since they were the group who were children and adolescents at the start of the new millennium. In general, members of Generation Y shows some similar characteristics. Howe and Strauss (2011) explained that Generation Y members have high levels of self-confidence and motivation, are optimistic, focus on goals, and tend to be more group-oriented rather than working individually. Though these features are standard in Generation Y adults' work environment today, the prominent characteristic of these individuals is their ability with and dependence on digital devices. According to Ruth N. Bolton (2013), Generation Y includes those who spend time in a digital environment where information technology plays an essential role in their lives and work. The generation actively contributes, shares, searches, and consumes content, including knowledge, work, and games, through social media.

In 2018, members of Generation Y were estimated to be between those 23 and 38 years of age. Based on Indonesian statistical data, these individuals constitute around 85,531,500 people out of a population of 265,000,000 (Databoks, 2018). Since this is a large proportion, experts think Generation Y is starting to dominate the market and have a significant effect on the country's sectors. Generation Y are now the main focus in secondary and tertiary learning. All educational institutions are faced with the challenge of testing their ability to adapt and innovate in order to deliver quality education for their students following changing times. The same challenge is also faced by distance learning institutions (PJJ), including Open Universities.

One of the principles of distance learning is using up-to-date technology or modern learning methods, utilizing advances in information and communication technology (ICT). Therefore, both teachers and students need to learn in order to keep pace with progress in this sophisticated era.

Video blogs, or vlogs, are widely used by Generation Y and continue to gain more popularity with time. Burgess and Green (2009) define vlogs as a collection of videos that function both as audiovisual documentaries and a means of communication and interaction on the internet. The act of making continuous vlogs is referred to as vlogging, while people who make them are called vloggers. YouTube is considered the most critical platform for vlogs, with nearly 40% of them uploaded and featuring in the sections of most viewed, most discussed, most favorites, and most responded to by users.

Vlogs, which were initially used as a medium for expressing an opinion, are now used as a means of learning. Various channels about topics such as, for example, appropriate dress, cooking a variety of foods, and how to educate children also appear on YouTube. This trend is attributed to the advantages of the vlog platofrm itself. Educause Learning Initiative notes several advantages of vlogs over other content forms, including being natural to make, more dynamic than text-based content, developing communication options, potentially becoming the latest commercial tools, and facilitating self-expression.

Open University uses an online-based distance learning media called Tutorial Online, held every semester for eight weeks or approximately two months before the final exams. Typically, tutors first upload the material in the form of text in PowerPoint, Word, and Excel files to the leading site before giving time for students' responses. The process of collecting material and holding discussions between tutors and students is called initiation. Online tutorials have a maximum contribution of 30% toward the final grade of courses taken.

The purpose of this study, therefore, was to examine students' perceptions of the use of vlogs as an online learning media tutorial on Reading 3 subjects. It is beneficial for the development (innovation) of distance learning media, which advocates for the principle of recency or a condition where both teachers and students tend to use modern learning methods.

2 METHODS

The research was conducted for one year using the Open University online tutorial media. The subjects include Generation Y students enrolling for Reading 3 courses, and the object is a video blog (vlog), a learning medium in the online tutorial. Since a large number of students in Generation Y are the potential target of distance learning in Open University, a study on the students' perceptions of using vlogs as a learning medium in online tutorial courses for Reading 3 was conducted. The online tutorial of Reading 3, held October–December 2018, was attended by 30 students. However, only 18 returned the research questionnaires distributed during the seventh week of Tutorial Online: expressed as a percentage, 60% of respondents returned the questionnaire, while 40% did not. This 60% of respondents became the focus of the research object and as research respondents.

Therefore, the framework of thinking in this study is more or less formulated as follows:

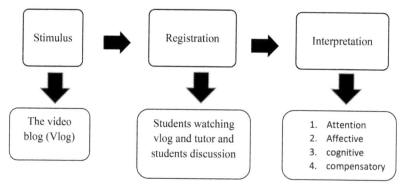

Figure 1. The framework of the study.

Primary data on the perception of using vlogs in online tutorials were obtained through a questionnaire collected at the seventh meeting. The data obtained were analyzed using perception measurement criteria using the Formula Index %.

$$Formula\ Index\% = T/Y \times 100 \qquad (1)$$

T = total score of the answers given by respondents.
Y = highest score Likert x number of respondents

The Formula Index % from each statement was matched with the value perception table and concluded.

3 RESULT AND DISCUSSION

Students' perceptions were divided into four categories – attention, affective, cognitive, and compensatory – as shown in Table 1.

Table 1. Students' perceptions.

Number	Perception	Average Perception Results	Perception interval	Category
1	Attention	77.78%	75% - 100%	Agree
2	Affective	78.89%	75% - 100%	Agree
3	Cognitive	77.78%	75% - 100%	Agree
4	Compensatory	71.11%	50% - 74.99%	Neutral

Table 1 implied that tutors need to encourage and direct students to watch the vlog given each week. Second, Generation Y's show a strong preference on the use of vlogs as learning media in online tutorials, as seen in the number of respondents who recommend the use of vlogs from Generation Y. Generation Y individuals embrace change and are familiar with the internet, apart from being active in social networking media. This is likely the reason why they recommended vlogs. The use of video versus text is considered a new and refreshing challenge. This is reflected in the comments of one respondent who claimed that the vlog learning media help to understand the material instantly and gave the impression of "getting closer" to the tutor-and-student relationships in cyberspace.

The video quality and techniques used should be carefully considered. Several respondents in this study admitted that they had difficulty hearing the tutor due to background noise. Therefore, tutors need to think carefully about the settings used in each vlog and come up with ways of preventing distraction. For instance, they can consider using a microphone with noise-canceling or sound suppression features to manage the background noise.

4 CONCLUSION

Vlogs have the potential to improve learning through online tutorials. This study tested the students' perception of the use of vlogs as a learning medium in the online tutorial of Reading 3. The results show the respondents' perception is positive, and therefore vlogs should be developed as learning media for future Open University students.

REFERENCES

Biel, J.-I. (2013) *Voices of Vlogging*. Proceedings of the AAAI International Conference in Social Media and Weblogs.

Central Bureau of Statistics (BPS). (2013) *Indonesia Population Projection 2010-2035*. Jakarta: BPS Catalog.

Bolton, R. N. (2013) Understanding Generation Y and their use of social media: A review and research agenda. *Journal of Service Management*, 24(3), 245–267.

David, E. R. (2017) Effect of vlog content on YouTube on formation of communication student attitudes at the Faculty of Social and Political Sciences at Sam Ratulangi University,.Medan. *Acta Diurna E-journal*, 6(1).

Databoks (2018). Available at: https://translate.google.com/translate?hl=en&prev=_t&sl=auto&tl=en&u=https://databoks.katadata.co.id/datapublish/2018/05/18/2018-jumlah-penduduk-indonesia-mencapai-265-juta-jiwa.

Kecksemeti, P (ed.) (1952) *Karl Mannheim: Essays*. London: Routledge.

Parker, C. (2005) *Video Blogging: Content to the Max*. Washington, DC: IEEE Computer Society.

Soekartawi. (2006) *Blended E-Learning: Alternative Distance Learning in Indonesia*. Yogyakarta: Journal of the National Seminar on Information Technology Applications.

Margono, S. (2007) *Educational Research Methods*. Jakarta: Pt Rineka Cipta.

VanMeter, R. A., Grisaffe, D., Chonko, L., and Roberts, J. (2013) Generation Y's ethical ideology and its potential workplace implications. *Journal of Business Ethics*, 117(1), 93–109.

Yaya, S. and Tedi, P. (2008) *Educational Research Methods*. Bandung: PT. Azkia Main Library.

Emerging Perspectives and Trends in Innovative Technology
for Quality Education 4.0 – Kusmawan et al (eds)
© 2020 Taylor & Francis Group, London, ISBN 978-0-367-25803-0

Reality and public perception of the implementation of Islamic sharia laws in Banda Acèh

Abdul Manan, Gunawan & Muhibbuthabry
The State Islamic University of Ar-Raniry Banda Acèh, Acèh, Indonesia

ABSTRACT: The present research was conducted to critically analyze contemporary perspectives on Islamic sharia laws in Banda Acèh, the capital city of Acèh province, Indonesia. Special attention is paid to the reality as well as perception and valuation of implementation. This field research has been conducted by means of participant observation as the principle method. Besides meticulous observation, in-depth discussions with informants were also carried out. The results of the research show that implementation of sharia has tended to focus on physical things; it is based on matters relating to women's clothing, seclusion, and writing Arabic names for government institutions and so forth, while the substance of the application of sharia has not yet fully appeared. The implementation of sharia is still more on symbolic discourses, while the goals and objectives of sharia have not yet been fully understood and implemented. Not all the applications of sharia law have been fully carried out in accordance with the rules, procedures, and objectives. Meanwhile, what exists today can be said in general to be somewhat "surface-level" only. However, many informants said that the implementation of Islamic sharia laws has been relatively successful compared to the years before the laws were legalized and implemented.

Keywords: Islam, sharia, perception, implementation of Islamic sharia laws

1 INTRODUCTION

Acèh had applied Islamic sharia laws since the Kingdom Acèh Darussalam led by Sultan Iskandar Muda (Manan, 2017); it is the only province in Indonesia authorized by the central government to implement the laws on the whole. To realize it, several legislations, especially on the implementation of *sharia*, were brought into force, such as law no. 44 of 1999 on the implementation of the Acèh privilege and law no. 18 of 2001 on the exclusive autonomy (in the era of President Megawati) for Acèh province, later reinforced by law no. 11 of 2006, the Acèh Local Government Law, known as UUPA. This latter granted full authority to the Acèh government in applying Islamic sharia laws comprehensively in Acèhnese people's lives.

At some levels, the implementation and practice of sharia in Acèh province has gone on for more than seventeen years. At this time, several institutions and some *qanun*, as vehicles in implementing sharia laws, were established. These institutions are sharia courts, sharia police called *Wilayatul Hisbah* (a task force for implementing *qanun*), and the assembly of scholars as well as a number of *qanuns*, which became the foundation in the practice of sharia laws in Acèh. Still, the absence of legal laws has led to problems of enforcement of Islamic sharia laws in Acèh; thus, the *qanunjinayat*, meant to strengthen the implementation of sharia in Acèh, have to be legalized immediately.

The implementation of Islamic sharia laws on the first five years (2002 to 2007) since it was declared went well, but from 2008 until today the reverberation (*gaung/gema*) of its implementation decreased (Isa, 2013). Nowadays, most of the current reality of society, as admitted by many informants, is no longer in accordance with the customs and culture of the Islamic sharia laws. This fact can be seen from the behavior of the social interaction of the people in everyday life, unconsciously changing and shifting the cultural values and the customs of the Acèhnese.

Several local scholars have conducted studies about the implementation of sharia laws (Isa, 2013; Ikhwan, 2013; Abubakar, 2005, 2013; Salim, 2008, 2015; Ismail, 2007). No specific attention, however, has been paid fully to the modalities of sharia laws in the capital city of Acèh province, and no up-to-date studies exist about the current problems involved in its implementation. Foreign researchers have conducted studies on the Acèhnese sharia laws, too (Morries, 1983; Aspinall, 2007; Feener, 2013). Their attention was mostly focused on the historical and political dynamics leading to the adoption of sharia laws in Acèh as part of the central government's endeavor to bring Acèh province under its control and suppress the local support for the Free Acèh Movement. Therefore, this research deeply focuses on the perception and valuation of the people of the Autonomous Province of Acèh, particularly in the capital city of Acèh, Banda Acèh. This city was chosen due to the complexity and homogeny of the population. This research is focused only on the formalization of Islamic sharia and ways of upholding Islamic laws in order to know the strengths and the weaknesses of implementation in the public perception.

2 RESEARCH METHOD

The research was conducted by means of participant observation as the principal research method. It entailed meticulous observation of the implementation of Islamic sharia laws, including modes of conduct deviating from the sharia rules. The researcher conducted an interview and in-depth discussions with religious leaders, community leaders, *adat* leaders, community members, and so on relating to the formalization and implementation of sharia laws. In addition, a systematic survey of relevant published sources has been made. The data collected during field and library research were subjected to qualitative analysis involving the stages of data deduction, data display, conclusion, and verification as standardized by Miles and Huberman (1994).

3 DISCUSSION

3.1 *The reality of the implementation of Islamic sharia laws and public perception of their formalization*

The application of Islamic sharia has caused numerous polemics among some communities in Acèh, but pro and contra. Attention was given not only to the formulation of the rules of law in a number of *qanun* but also to the law enforcement process of sharia law. The expectation of a number of people toward the implementation of sharia in Acèh was that it could serve as a model for the application of sharia law in Indonesia more widely. Many countries appeared to have failed in implementing the sharia laws controlled by provisions of the State. Nurcholish Majid stated that countries such as Pakistan, Afghanistan, and Sudan failed in implementing *sharia* law, so the only hope was with Acèh. The arguments in favour of and against the implementation of Islamic sharia in Acèh can be seen in the context of efforts to sensitize the authorities to realize *sharia* as a whole and to apply it without favoritism through humanism and sociological approaches (Salim, 2008). This means that the full implementation of Islamic sharia, which covers all facets of life, has yet to happen.

Arguments against the implementation of Islamic law can be identified views expressed in mass media or in various meetings such as seminars, workshops, and other gatherings. For instance, the application of flogging was criticized as a violation of human rights. The implementation of sharia was said to be very discriminatory and does not pay attention to or guarantee the interests of vulnerable groups such as women and children. Wilayatul Hisbah (municipal police) were said to be improper in performing their duties, sometimes acting beyond their powers of arrest, investigation, searches, and so forth. These were among the views disagreeing with formal implementation of Islamic sharia law in Acèh.

On the other side, views in favour of Islamic sharia implementation state that it is a grace, for which people should be grateful that the juridical base is strong enough to implement Law no. 44 of 1999 and Law no. 11 of 2006. These laws increased expectation that sharia will be realized as a whole: without these laws, implementation could be limited to the areas of faith and worship only, and public aspects could not be implemented at all. Criticisms of flogging were answered by the fact that, according to this view, it is in conformity with the provisions of the Al-Quran and Al-Hadith, in which flogging is one form of punishment that can be applied to perpetrators of a crime, called *hudud* or *ta'zir*.

Furthermore, in Banda Acèh today some people even dare to violate sharia against officers during a raid. The levels of vice in the city of Banda Acèh is now very worrying. Meanwhile, Banda Acèh should provide an example for other regions. Nevertheless, public awareness of Islamic sharia law seems still very low. In fact, people often fight the implementation of Islamic sharia openly in Banda Acèh. This is caused, among other reasons, by limited facilities and infrastructure, limited human resources, insufficient rules and regulations (especially in the enforcement of *sharia*), and limited authority in the resolution of criminal cases, as well as the lack of seriousness and commitment of all parties including the government in efforts to accelerate the implementation of Islamic law in Acèh. This implies that the government, communities, and provincial parliament as a legislative body should work hand in hand in guarding and taking responsibility for the implementation of Islamic sharia law in Acèh, especially in Banda Acèh. People expect that the commitment to the implementation of Islamic sharia should not only be mere wishful thinking and that its implementation can be accomplished in reality in accordance with the objectives of sharia itself. In fact, the pessimistic responses toward the implementation of Islamic sharia in Acèh have not run in accordance with the rules, procedures, and objectives of sharia itself. This could have been caused by the absence of systematic procedural clarity in the implementation of sharia in Acèh.

There are five answers to the question of why sharia is applied in Acèh. First, on religious (theological) grounds, the implementation of *sharia* is a religious commandment for a Muslim to be more perfect – that is closer to Allah. Second, for psychological reasons, people will feel safe and secure because of what prevails around them; they lead activities in education and in everyday life appropriately and in line with their conscience. The third goal is for people to live in a system that is more in line with the rule-of-law awareness, sense of justice, and values that grow and develop in society. The fourth reason relates to economic and social welfare reasons, value-added economic activities, and social solidarity in the form of helping economic activity or social activities to be more easily formed and solid. The fifth goal of implementing sharia is to build mentors who can meet the needs of the people of Acèh today – transition toward modernization without the need to feel there is a gap between religious guidance and worldly prosperity. As mentioned, the implementation of Islamic sharia not only refers to the model and ideas of scholars of the past but also to formulating a new understanding so that people feel more secure with the guidance of the Qur'an and the sunnah of the Prophet. Thus, sharia will refer to the future in order to meet the demands and needs of the times, which are more complicated and complex. A further goal is in the field of law and justice, which is good implementation of sharia covering all aspects of life, implemented honestly and, sincerely, to bring about justice and order in accordance with the legal awareness of the people of Acèh itself.

4 CONCLUSION

Strategies for the continuation and deeper implementation of sharia law in Acèh will require continued refresher training of the coaching personnel in the means of successful implementation of awareness and understanding of *sharia* law amongst the public. Although there have been some continuing public discussions about the institutionalization of sharia law in Acèh, especially in the capital, Banda Acèh, in general *sharia* law has been accepted by the Acèhnese, especially since it has tended to reinforce the concept of a unique Acèhnese identity distinctive from the more general Indonesian identity; although the Acèhnese did not win independence, they gained a special identity, something they had wanted since independence. In reality, however, after more than seventeen years, the application of sharia law in Acèh – what we can call Acèhnese sharia law – is still a work in progress, with its most obvious implementation limited to physical implementation. Implementation in the fields such as health, education, finance, and the economy is still continuing, step by step. Certainly, the realization of the Islamic ideal of peaceful, principled, prosperous people living on earth in Islam will require continuing education at all levels of society.

REFERENCES

Abubakar, I. (2005) *Syariat Islam di NAD, Paradigma Kebijakan dan Kegiatan*. Banda Acèh: Dinas Syari'at Islam.

Aspinall, E. (2007) Association for Asian Studies Annual Meeting. In *The Politics of Islamic Law in Acèh*. Boston, MA: Boston University.

Feener, R. M. (2013) *Sharia and Social Engineering: The Implementation of Islamic Law in Contemporary Acèh, Indonesia*. Oxford: Oxford University Press.

Isa, A. G. (2013) *Formalisasi Shariat Islam di Acèh: Pendekatan Adat, Budaya dan Hukum*. Banda Acèh: Yayasan PeNa.

Ismail, A. (2007) *Shariat Islam di Nanggroe Acèh Darussalam*. Banda Acèh: DinasSyari'at Islam.

Manan, A. (2017) The social fact of the implementation of Islamic sharia laws in West Aceh, Indonesia. In *Proceedings from International Collaboration of ASEAN Researchers*. Jakarta: Humboldt University, p. 25.

Miles, M. B. and Huberman, A. M. (1994) *Qualitative Data Analysis: A Source Book of New Methods*. Beverly, NJ: Sage.

Morries, E. E. (1983) *Islam and Politics in Acèh: A Study of Center–Periphery Relations in Indonesia (Unpublished dissertation)*. New York: Cornell University Press.

Salim, A. (2008) *Challenging the Secular State: The Islamization of Law in Modern Indonesia*. Honolulu: University of Hawaii Press.

Salim, A. (2015) *Contemporary Islamic Law in Indonesia: Sharia and Legal Pluralism*. Edinburgh: Edinburgh University Press.

*Emerging Perspectives and Trends in Innovative Technology
for Quality Education 4.0 – Kusmawan et al (eds)*
© 2020 Taylor Francis Group, London, ISBN 978-0-367-25803-0

Improving children's emotional intelligence through cooperative learning methods

Sri Sukatmi
Universitas Terbuka, Tangerang Selatan, Indonesia

ABSTRACT: It has been observed that children are prone to emotional problems when they are not well taught with cooperative learning methods in kindergarten. Therefore, the objective of this research, which was conducted at MutiaraInsani Kindergarten, Mustika Jaya, Bekasi, is to obtain a clearer picture of children's emotional intelligence through cooperative learning. Cooperative learning helps children to understand the fundamentals of manners such as the development of attitudes, behavior, religion, social-emotional language, knowledge of physical motor skills, and art skills needed to adjust to their environment. Problems found in the field are related to children's emotional development, and the results showed that there was a significant increase in the development of children's emotional intelligence through the application of cooperative learning methods in each learning cycle.

Keywords: Cooperative, emotional, intelligence

1 INTRODUCTION

Kindergarten institutions are a place where a child's personality can be molded toward the development of attitudes, behaviors, social-emotional skills, language, knowledge of physical motor skills, and arts. In order to develop a well-rounded personality, children must possess individuality, be independent, be confident, have a good manner, have a high social spirit, and be able to work together in a team. Educators can arrange educative learning that is packaged into fun, effective, and creative learning for children, using a variety of relevant media, so that the children's involvement can be guaranteed.

Based on the results of the observations conducted, there were some emotional deviations that mostly came from financially capable families who had problems in fostering active communication with their children. The children lacked the attention of parents, because there was often insufficient time for parents to respond to the problems their children are faced with, and so children were not given the opportunity to express their opinions or ideas. These are factors that affect early childhood development and can cause problems or disorders that are worrisome to educators and parents.

One learning technique that can be applied to improve children's abilities to develop emotional intelligence is cooperative learning. Cooperative learning methods are conducted by providing opportunities for students to carry out various activities in small groups. During such activities, students can interact, communicate, and discuss with fellow group members with guidance from their teachers. The teacher is obliged to be active in responding to children's opinions in order for the children to reach an agreement with their team. Early childhood education institutions are expected to be able to implement and develop cooperative learning methods with various activities according to the kindergarten curriculum.

1.1 Problem formulation

Based on the background of the problems mentioned, this research describes problem formulation as follows:

1. What is the condition of children's emotional intelligence before applying cooperative learning?
2. What is the process of applying cooperative learning methods in an effort to improve children's emotional intelligence?
3. How much does children's emotional intelligence increase after cooperative learning is applied?

2 LITERATURE REVIEW

2.1 Early childhood education

The National Association for the Education of Children (NAEYC) has argued that early childhood education is very important because mental development, which includes the development of intelligence, personality, and social behavior, takes place quickly at an early age. Early childhood education therefore includes the development of various aspects of life that are tailored to the needs of children.

2.2 The principle of intelligence

David Wechsler argued that intelligence is the ability to act according to the rules, to think in a rational way, and to deal with the environment effectively. In addition, Piaget stated that children's intelligence develops through active learning activities. Based on these opinions about intelligence, for a human being to carry out all activities successfully he or she requires the ability to think so that all actions that will be carried out in a directed, systematic, rational manner in order to produce the expected result.

2.3 Emotional intelligence

Daniel Goleman stated that the factors that influence success are determined not only by the level of intellectual intelligence but also emotional stability, called the emotional quotient or emotional intelligence. Emotion is basically the impulse to act, usually a reaction to external and internal stimuli.

2.4 Cooperative learning

According to Balcony, cooperative learning is a learning strategy carried out in small groups of students who have different abilities, in which the group seeks collectively to develop the ability to learn something. Each group member cooperates in improving the learning progress and increasing the success of all group members. For Slavin, cooperative learning refers to a variety of teaching methods in which students work in small groups to help each other in learning the subject matter.

Furthermore, the Ministry of National Education has argued that cooperative learning is carried out with the aim of improving learning outcomes and providing opportunities for students to receive friends who have different backgrounds to develop their social skills. Anita Lie stated that the special characteristics of cooperative learning embrace the elements that must be applied, including positive interdependence, individual responsibility, face-to-face communication between members, and group process evaluation.

Based on some of the opinions already outlined about the role of cooperative learning in developing and implementing cooperative learning strategies in early childhood, the preparation should be designed to the maximum extent possible so that it can bring children into the learning process and help them to carry out various activities in groups, negotiating, among other challenges, group members acting as liabilities when resolving problems that must be solved together.

3 RESEARCH METHODOLOGY

The population in this study was kindergarten teachers at MutiaraInsani Kindergarten, Mustika Jaya, Bekasi. The method used was action research to improve the effectiveness and efficiency of educational practices. Furthermore, Ebbut, as quoted by Wiriatmaja, stated that action research is a systemic study of efforts used to improve the implementation of educational practices by a group of teachers by taking actions in learning based on their reflection on the results of these actions.

Based on this method, it became necessary to have appropriate data collection methods and techniques. The techniques used in action monitoring were non-testing-based, using field records, interview notes, observation guides, and documentation carried out by collaborators. Field notes were carried out directly, using a camera for documentation. Observations were carried out using an observation guide, both for the teachers while carrying out the learning process and for students in the development of their emotional intelligence.

3.1 Research design

The intervention design of the action in this study used the Kemmis and Taggart models. This action plan consisted of four stages: planning, acting, observing, and reflecting. After the stages in cycle 1 were completed, it was followed by re-planning, acting, observing, and reflecting for the next cycle, and so on, thereby forming a spiral. A step taken by investigators in this research was to perform activities in a cycle of learning as undertaken by students of S1-ECD Open University, which include:

1. Planning: Arrange action planning in the form of a Daily Activity Plan.
2. Acting: In this activity, the researcher and collaborator carried out the Daily Activity Plan, namely cooperative learning.
3. Observation of actions conducted toward student learning in the class and observations of students' behavior during the learning process.
4. Reflection: After planning, acting, and observing, researchers, together with collaborators, conducted a reflection of actions taken, to analyze achievement of given actions and also to process and analyze the causes of actions not achieved.

3.2 Results and discussion

3.2.1 Results of cycle 1
The results of observation from the first cycle showed various abilities in different children. Some children were enthusiastic and showed a lot of passion, but some were shy and were not able to carry out these activities. This does not mean that the children were unable to complete the activity but that they needed some motivation and guidance in order to be more confident and brave. At the first meeting, the children needed support through the use of some appealing pictures to be able to listen and retell the contents of the storybook "Kinds of Vehicles." They also needed encouragement to tell the contents of the storybook "Sea Ship," even though the story was not the same as what was expressed. Furthermore, some children began to enjoy speaking activities involving sorting and telling the contents of a series of images, but this did not happen to all, as some still did not want to share their experiences.

From the data obtained, the results of evaluation in cycle 1 showed that, out of 17 children who scored well, there were eight (47%) whose grades were good, nine children (53%) whose grade were adequate, and two (12%) who still had lesser grades. This shows that the mastery level of the children in the first cycle was 47%.

Table 1. Success rates of Cycle 1 and Cycle 2.

Cycle	Good	Adequate	Less
Cycle I	47%	53%	12%
Cycle II	76%	24%	0%

3.2.2 *Results of the implementation of cycle 2*

The results from the second cycle of activities showed that some children were successful and were enthusiastic in carrying out the activities given, while others who had not succeeded still needed to be given motivation and guidance by either their teachers or parents in order to get better results. Some children were able to capture, receive, and also complete the tasks that were given to them. However, at the second meeting (in cycle 2), the children were getting used to collaborative activities and sharing experiences in a simple way. Children were able to follow the activities of communicating well through group activities with given themes.

The data obtained showed from the evaluation results in cycle 2 that, out of 17 children who scored well, there were 13 (76%) whose grades were good, 4 (24%) who were adequate, and none who scored less (0%). This shows that the mastery level of children in the second cycle was 76%.

4 CONCLUSION AND SUGGESTIONS

Efforts to improve emotional intelligence were carried out in MutiaraInsani Kindergarten, Mustika Jaya, Bekasi, through cooperative learning with various indicators – i.e., attitudinal/behavioral, social, and emotional. At the end of cycle I and cycle II, field notes included: greetings when children meet their teachers, friends, or known people; speaking well and politely when talking with others; apologizing if feeling guilty; being willing to forgive if friends asked for forgiveness or apologized for wrongdoing; being friendly; being polite to friends and teachers; being more in control; trying to help friends who are struggling; and being honest in their actions. There is a need for the development of habitual attitudes and behaviors and fulfillment of social and emotional needs from an early age, so that future generations will have discipline be virtuous, honest, independent, and skilled, and have a true spirit of fair play.

REFERENCES

Bowler, P. and Linke, P. (1996) *Your Child from One to Ten*. Victoria: Impac.
Bromley, K. D. (1992) *Language Arts: Exploring Connections*. Boston, MA: Allyn and Bacon.
Daniel Goleman (1996). *Emotional Intelligence*. Jakarta: Gramedia Main Library.
David, I. K. (1987), Learning Management. Jakarta: David McCoy.
Díaz, V. M. (2017) The augmented reality in the educational sphere of student of degree in childhood education: Case study. *Píxel-Bit: Revista de Medios y Educación*, (51), 7–19.
Isenberg, J. P. and RenckJalongo M. (1993) *Creative Expression and Play in the Early Childhood Curriculum*. New York: Macmillan.
Isjoni (2007) *Cooperative Learning Effectiveness of Group Learning*. Bandung: Alfabeta.
Mathis, J. A., Fairchild, L., and Cannon, T. M. (1980). Psychodrama and Sociodrama in Primary and Secondary Education. *Psychology in the Schools*, 17(1), 96–101.
Moleong, J. L. (2014) *Qualitative Research Methodology* (revised edition). Bandung: Youth Rosdakarya.
Papalia, D. (1990). *A Child's World: Infancy through Adolescence*. NewYork: McGraw-Hill.
Santrock, J. W. (2001) *Educational Psychology*. New York: McGraw-Hill.
Tadkirotun Musfiroh (2010) *Multiple Intelligence Development*. Jakarta: Open University.
Vygotsky, L. (1962) *Thought and Language*. London: The MIT Press.
Winda G., Lilis S., and Azisah M. (2008). *Methods of Developing Behavior and a Basic Ability for Early Childhood*. Jakarta: Open University.

*Emerging Perspectives and Trends in Innovative Technology
for Quality Education 4.0 – Kusmawan et al (eds)*
© 2020 Taylor & Francis Group, London, ISBN 978-0-367-25803-0

Construction of Islamic education in the education system in Indonesia

Hayati, Z.A. Tabrani & Syahril
Serambi Mekkah University, Banda Aceh, Indonesia

Saifullah Idris & Ramzi Murziqin
Ar-Raniry State Islamic University, Banda Aceh, Indonesia

ABSTRACT: This article aims to map the implementation of Islamic education in Indonesia in realizing the generation of a nation of superior quality without ignoring the national education system. The approach used in this discussion is qualitative by using descriptive methods. The data were analyzed using the inductive method. Islamic education in Indonesia has a legitimate need to exist to meet the educational needs of Muslims. The Islamic system of educational institutions has a strong foundation on which to develop. However, the problem is more complicated when compared with general education. Most Islamic educational institutions are considered less able to meet learners' needs, so many people think that Islamic education is second-class and not a suitable alternative to general education. Conceptually speaking, the integration of Islamic education into learning refers to the understanding that any knowledge is a means to God, provided humans realize early on that life in the world requires it to attain the afterlife.

1   INTRODUCTION

Islam views humans as God's creatures who have a certain uniqueness and special features. As one of God's creatures, the characteristics of human existence must be sought concerning the Creator and other creatures of God (Djumana, 1995: 54). The education process is an inseparable series of human creation processes. In order to understand the nature of education, an understanding of human nature is needed (Muhaimin, 2004: 27). Langeveld (in Pratiwi, 2010: 1) states that humans are "animal educandum" – creatures that need to be educated.

One of the teachings of Islam is to oblige its people to carry out education to attain better provisions and direction in life. Education is conscious guidance by educators to guide the physical and spiritual development of the educated towards a better personality, which essentially leads to the formation of an ideal human being. The ideal human is a perfectly moral man. What appears and is in line with the apostolic mission of the Prophet Muhammad is to perfect the noble character. Islam is a universal religion that teaches mankind about various aspects of life, both worldly and *ukhrawi* (in the afterlife).

The essence of education is the formation of humans in the direction they aspire to. By seeing the current reality, with the increasingly slack Islamic values and Eastern customs that become the identity of our country – for example, the increasing number of people who take actions that deviate from religious rules – problems of education are among the problems that need to be overcome or at least minimized. Therefore, the Islamic education system is very much needed and must be reconstructed in the process of implementing learning. The Indonesian nation is a religious nation; this attitude of religious life has been possessed by the Indonesian people since time immemorial. The beliefs of the ancestors of our nations came in the form of animism and dynamism that developed in Indonesian society, later converting to

Hinduism and Buddhism to Indonesia, accompanied by the entry of Islam, and later the entry of Christianity prove that Indonesian society is a religious society.

The two education systems in Indonesia – the National Education System and Islamic Education – at the beginning of the independence period were often considered to be in conflict, and they developed separately from each other. National Education, at first, could only be reached by the upper classes of society. In contrast, Islamic Education grew and developed independently among the people and had deep roots in the community (Tabrani, 2014).

In accordance with the aspirations of the nation, the problem that we need to discuss is how to implement religious education so that it is more useful in realizing a generation of people of superior quality, outwardly and inwardly, capable of high life in *aqliyah* and *aqidah* and weighted in *amaliyah* and *muamalah* behavior, so that these qualities survive in the current dynamic of sociocultural change.

2 METHOD

The approach used in this discussion is qualitative and uses descriptive methods. This discussion also explores ideas related to the topic of study and is supported by data or information obtained from literature sources. The authors chose and philosophically study materials related to the study material. The method used to analyze the data in this discussion is the inductive method of thinking that moves from specific facts and events to draw general conclusions.

3 RESEARCH AND DISCUSSION

3.1 *Integration of Islamic education curriculum in learning*

According to Sanusi (1987: 11), integration refers to a unified whole, not divided or divorced. Integration includes the needs or completeness of members who form a unity with a close, harmonious, and intimate relationship between the members of that unit.

In the conceptual level, the integration of Islamic education in learning refers to the understanding that any knowledge is a means to God if humans realize early on that life in the world requires it to attain the afterlife. In the end, all kinds of knowledge that provide good in the world and the hereafter are important to learn. Al-Ghazali stressed the need for humans to scale education priorities by placing religious knowledge in the most important position (Asrorum Ni'am Sholeh, 2006).

The Islamic education curriculum is still faced with difficulties in integrating the two poles of the dualistic scientific paradigm. On the one hand, it must deal with "secular subjects" and on the other hand with "religious subjects." Subjects that are considered secular usually consist of general scientific matters such as mathematics, physics, biology, medicine, sociology, economics, politics, botany, zoology, and so on. Religious subjects consist of types of knowledge related to revelation such as the Qur'an, al-Hadith, Fiqh, Theology, Sufism, and others. From the dichotomy above, the general education curriculum and Islamic education curriculum are still in their respective regions, so the learning process is partial and fragmented between the science of divine revelation and the natural sciences.

Ideally, there is no need for ambivalence and dichotomy problems in educational orientation. Understanding the integration of Islamic values in learning is implicit in the Qur'an and does not contradict science and religion. The Qur'an states that science, including that relating to human life, is an integral part of religion. Science teaches humans about how to manage nature, carry out various processes, and produce something for the needs of life. Meanwhile, religion teaches humans about the value system.

In implementing Islamic education in schools, we can refer to the references offered by Bagir et al. in Sauri (2009: 11), dividing them into four levels of implementation: conceptual, institutional, operational, and architectural levels.

At the conceptual level, Yusoff and Hamzah (2015: 119–132) explained that the integration of value education could be realized through the formulation of vision, mission, goals, and school programs (school strategic plans). Institutionally, integration can be realized through the formation of an institutional culture that reflects a blend of values and learning. At the operational level, according to Steinbach and Afroozeh (2016), curriculum design and extracurricular activities must be formulated in such a way that the fundamental values of religion and sciences are coherently integrated. Architecturally, integration can be realized through the formation of a physical and science-based physical environment and *imtaq*, such as complete religious facilities, adequate laboratory facilities, and libraries that provide religious books and general science in full.

3.2 *The actualization of the integration of Islamic education in the Indonesian education sytem*

The most important basis of education in Indonesia is the national philosophical theory of Pancasila as well as the 1945 Constitution. Basic education indirectly requires us to carry out a national education process that is consistent in moving toward the achievement of the final goal: the formation of fully qualified Indonesian people who develop and grow on a balanced life pattern between *lahiriyah* and *bathiniyyah*. The road to that goal is none other than through an educational process that is oriented toward a three-way relationship: the relationship of students with their Lord, with their communities, and with their natural surroundings (Daulay, 2007: 210).

In Indonesia, the problems of Islamic education are more complicated than those of general education. For example, from a small manual device that can be used for madrasah diniyah, we have not been able to meet these needs. As a result, most Islamic educational institutions are considered less able to meet needs, and many people think that Islamic education is a second-class education and cannot be a viable alternative to general education. This perception is caused by several factors, as explained by Ismail (2007: 173–174) and Tabrani (2014).

First, there are internal barriers: 1) the absence of a standardized curriculum as a borderline to other education systems; 2) the lack of a standard teaching methodology; and 3) the absence of reliable measuring tools in assessing educational outcomes.

Second, there are external obstacles: 1) the Islamic education system still depends on the education patterns outlined by the government, namely education to support development; 2) lack of funds and facilities, so that Islamic education is oriented to the tastes of consumers and helps the marginal; 3) the national education system is still unstable; 4) cultural development mean that Islamic education is increasingly powerless to compete with the pace of change in society; 5) public appreciation of Islamic educational institutions have not been encouraging; 6) social factors based on materialistic measures cause people to race to favorite educational institutions, without regard to their ideological aspects; and 7) mismanagement tendencies lead to such as unhealthy competition between leaders and closed leadership (Tabrani, 2014: 250–270).

Although Islamic education cannot compete with general education in Indonesia, its presence is still welcomed. Educational institutions under the Ministry of National Education cannot accommodate all students who need education. Islamic education institutions – mostly in rural areas – offer relatively inexpensive tuition fees. Some people still feel attached to Islamic education or feel obliged to provide religious education for their children. And finally, certain regions do not have public education institutions that can be reached by the community. In this way, Islamic education has indeed become a necessary alternative for some.

To develop the thoughts and feelings of students undergoing religious education, it is necessary to design a dynamic curriculum model in substance/material that mobilizes educators and students politically, as explained by Tabrani (2013). In such a system:

a. Substance/subject matter is more focused on the problems of today's sociocultural life and perspectives toward the future, which encourages students' interest and attention to conceptualize the goals and values inherent to God's guidance.
b. Education must be able to create a dialogical community situation that contains interdependence between God and students.
c. Students in the teaching-learning process must actively engage in dialogical communication with educators, peers, and natural surroundings. The curriculum model outlined above does require a critical formulation of the existing religious education curriculum that is implemented based on an approach oriented to the efficiency and effectiveness of teaching and learning.

Muhaimin (2009: 106) and Tabrani (2014b: 211–234) mapped the four problem areas in Islamic education. First, the dichotomy of science led to the problem of Islamization of science (education). Second, the quality of Islamic religious education in schools and public tertiary institutions is insufficiently strong. In principle, this problem concerns the internal and external problems of Islamic religious education. Another aspect is the narrow understanding of religion teachers/lecturers of the essence of Islamic teachings; the design and preparation of Islamic Religious Education materials which are not quite right; the conventional-traditional methodology; and so on. The third problem concerns efforts to build Islamic education in an integrated manner to develop Indonesian people as a whole. The fourth problem concerns the exploration of philosophical concepts of Islamic education as well as the thoughts of Islamic education figures from the classical period to the modern period, from within or outside the country (Tabrani, 2014: 250–270; 2013b).

Of the four categories of problems of Islamic education as described above, the problem of integration of Islamic education in the national education system falls into the first category, namely the problem of the dichotomy of science. It has been mentioned earlier that Islamic education that has taken place since the entry of Islam into Indonesia, and it is an inseparable part of the national education system, both explicitly and implicitly.

4 CONCLUSION

Education is a cultural process to improve human dignity and status, lasting throughout life and carried out in the family, school, and community. Therefore, education is a shared responsibility between family, community, and government.

Education is the process of achieving objectives and managing needs in an integrated and harmonious system. Thus the religious education strategy in all educational environments is not only tasked with motivating life and eliminating the negative impacts of development but also must be able to internalize the absolute values of God into a whole person who is able to filter and select, as well as serve as an antidote to the negative impacts from within and from outside the national development process. For this purpose, religious education is directed at the formation of Indonesian people with Pancasila identity and personality, conducive religious morality, and personal persistence and determination in facing the ups and downs of national development.

REFERENCES

Abdullah, A., and Tabrani, Z. A. (2018) Orientation of education in shaping the intellectual intelligence of children. *Advanced Science Letters*, 24(11), 8200–8204. Available at: https://doi.org/10.1166/asl.2018.12523.

Daulay, H. P. (2007) *Sejarah Pertumbuhan dan Pembaharuan Pendidikan Islan di Indonesia*, Jakarta: Kencana.

Djumana, H. (1995) *IntegrasiSpikologidengan Islam*. Yogyakarta: Pustaka Pelajar.

Idris, S., Tabrani, Z. A., and Sulaiman, F. (2018) Critical education paradigm in the perspective of Islamic education. *Advanced Science Letters*, 24(11), 8226–8230. Available at: https://doi.org/10.1166/asl.2018.12529.

Ismail, A Kholiq dan Nurul Huda (2001) *Paradigma Pendidikan Islam*. Semarang: Putaka Pelajar.

Muhaimin (2009) *Rekonstruksi Pendidikan Islam: dari Paradigma Pelembagaan, Manajemen Kelmbagaan, Kurikulum, hingga Strategi Pembelajaran*. Jakarta: PT. Raja Grafindo Persada.

Sauri, S. (2009) *Implementasi Pendidikan Nilai dalam Pedagogik dan Penyusunan Unsur-unsurnya*. Bandung: SPs PU UPI.

Steinbach, M. and Afroozeh, S. (2016) Comparative education in the educational systems and problems in likenesses and differences between regions of the world. *Jurnal Ilmiah Peuradeun*, 4(3), 333–346.

Tabrani, Z. A. (2013) Modernisasi Pengembangan Pendidikan Islam (Suatu Telaah Epistemologi Pendidikan). *Serambi Tarbawi*, 1(1), 65–84.

Tabrani, Z. A. (2014) Islamic Studies dalam Pendekatan Multidisipliner (Suatu Kajian Gradual Menuju Paradigma Global). *Jurnal Ilmiah Peuradeun*, 2(2), 211–234.

Tabrani, Z. A. (2015). *Persuit Epistemology of Islamic Studies (Buku 2 Arah Baru Metodologi Studi Islam)*. Yogyakarta: Penerbit Ombak.

Walidin, W. (2016) Informal education as a projected improvement of the professional skills of employees of organizations. *Jurnal Ilmiah Peuradeun*, 4(3), 281–294.

Yusoff, M. Z. M. and Hamzah, A. (2015) Direction of moral education teacher to enrich character education. *Jurnal Ilmiah Peuradeun*, 3(1), 119–132.

Emerging Perspectives and Trends in Innovative Technology
for Quality Education 4.0 – Kusmawan et al (eds)
© 2020 Taylor & Francis Group, London, ISBN 978-0-367-25803-0

The local governance system based on the special autonomy law in Indonesia

Mukhsin Nyak Umar, Ramzi Murziqin & Baihaqi As
Ar-Raniry State Islamic University, Banda Aceh, Indonesia

Sanusi
Syiah Kuala University, Banda Aceh, Indonesia

Fauza Andriyadi
STAI Al-Washliyah, Banda Aceh, Indonesia

Sulaiman
STAI-PTIQ Aceh, Indonesia

Syahril & Z.A. Tabrani
Serambi Mekkah University, Banda Aceh, Indonesia

ABSTRACT: Regional autonomy provides flexibility for a region to discover the potential of and manage its own natural resources. This eliminates jealousy or injustice between the central and local governments. Aceh, DKI-Jakarta, and DIY-Yogyakarta are the three provinces in Indonesia specially recognized as autonomous regions, with privileges different from those of other provinces. This study is normative-empirical legal research (applied normative law) with positive law researches written with regards to the behavior of citizens due to normative legal enforceability. The results showed that autonomy was implemented in Indonesia through Law No. 22 of 1999 on Regional Government. In 2004, this law was considered incompatible, and, due to the development of the state, public administration, and the demands of regional autonomy, it was replaced with Law No. 32. Furthermore, Law No. 32 of 2004 on Regional Government has been amended several times, with the most recent being Law No. 12 of 2008. This is an excellent opportunity for local government to prove its ability to implement the rights of local authorities and create freedom of expression in order to develop the area and avoid violating the law.

1 INTRODUCTION

The implementation of decentralization and regional autonomy in Indonesia is believed to have the ability to attract better public services, improve people's welfare, and foster local democracy. Indonesia as a nation (Bhineka Tunggal Ika) is composed of thousands of islands, with hundreds of cultures and subcultures spread throughout the archipelago. Historically, some regional administrative areas have special autonomy, with different designations in accordance with their historical background and the formation of regional autonomy policy governing its time. This relates to the adoption of Law No. 5 of 1974 concerning Governance Principles in the Special Regions of Aceh and Yogyakarta, while the term "Jakarta" refers to the special status of Jakarta as the country's capital.

Grants in the form of specificity to the Aceh, DKI Jakarta, and DIY Yogyakarta provinces are due to the fact that they possess special historical, cultural, and special regions compared to other provinces in Indonesia.

These special areas are privileged under Law No. 23 of 2014 on Regional Government, the legal basis of local government in Indonesia.

2 METHOD

This study used the normative-empirical research method, which is a written positive law regarding the behavior of every citizen as a result of normative legal enforceability.

According to Abdul Kadir Muhammad (2004: 66), the normative approach: identifies the subjects (topical subject) and sub-principal (subtopical subject) based on the formulation of research; identifies the legal provisions of the normative to become the benchmark and applied sourced in line with the sub-principal discussion; and the implementation of the legal provisions applied in the normative benchmark pertinent legal events, which resulted in the application of appropriate or inappropriate behavior.

Legal materials were derived from interviews, library, regional regulations, *qanun*, and law to obtain more systematic research to address the formulated issues. Qualitative analysis was used to analyze the data obtained, which are descriptively presented in the form of sentences that are true, complete, and systematic, to avoid creating a variety of interpretations.

2.1 *Research approach*

The normative-empirical approach was the research strategy utilized. Researchers, in formulating problems and objectives, need to be detailed, clear, and accurate.

2.2 *Location of the study*

In accordance with the type of research and the approach used, the locations of the study were the Provincial Government of Aceh; the Provincial Government of DKI Jakarta; and the Provincial Government of DIY Yogyakarta.

3 RESEARCH AND DISCUSSION

3.1 *The existence of special autonomy in the system of local governance*

3.3.1 *Aceh special autonomy*

Compared to other regions, Aceh acquired three attributes of special autonomy. First, the Law of the Republic of Indonesia Number 44 for the Year 1999 confirmed that Aceh is free to implement the law of God for all its people. This means that no one will feel hurt, disappointed, and resentful toward the implementation of sharia law. Any tribe or religion should be aware that Aceh applies sharia law and is therefore not the same as other provinces in Indonesia.

Secondly, Law No. 18 of 2001 on Special Autonomy for Aceh Province states that the considerations of granting special autonomy are: (1) the administration system under the Act of 1945 recognizes and respects the units of local government regulated by the Act; (2) the historic distinctive character of the people's struggle is in line with their durability and high fighting spirit, which is based on their view of life, social character, and community with Islamic culture. This made the Aceh Region a strong capital region, which seized and maintained independence for the Unitary State of the Republic of Indonesia; (3) in order to give broad authority to run the government for the Province of Aceh, it is necessary to provide special autonomy; (4) Law No. 22 of 1999 on Regional Government and Law No. 25 of 1999 on Financial Balance between Central and Local Government does not fully accommodate the origin of the rights and privileges of the Province of Aceh; and (5) the implementation of Law No. 44 for the Year 1999 concerning the province's features needs to be harmonized in governance in the Province of Aceh as Nanggroe Aceh Darussalam (Djojosoekarto et al., 2008).

Thirdly, through Law No. 11 for the Year 2006, the primary considerations are: (1) the system of government of the Republic of Indonesia, according to the Constitution of the Republic of Indonesia Year 1945, recognizes and respects the local government units that are special; (2) based on constitutional law, Aceh is a unit of regional government that is specifically related to one of the distinctive characteristics of the people's history of struggle with resilience and perseverance; (3) the resilience and perseverance are sourced from a view of life that is based on Islamic law that gave birth to a strong culture, which made it possible for them to struggle for, seize, and defend the independence of the Republic of Indonesia; (4) the administration and development in Aceh cannot fully realize the people's welfare, justice, and promotion, fulfillment, and protection of human rights, and therefore the government must develop and implement principles of good governance; (5) the earthquake and tsunami has grown the solidarity of the nation in rebuilding the communities and regions to resolve conflicts in a peaceful, comprehensive, sustainable, and dignified manner (Djojosoekarto et al., 2008).

In an effort to determine the political direction of law as outlined by the 1999–2004 guidelines, the release of Act No. 18 of 2001, and Act No. 11 of 2006 for the province of Aceh as part of the Unitary Republic of Indonesia declared the Sharia Courts as part of the national judicial system (article 25 paragraph (1) of Law No. 18 of 2001; article 128 paragraph (1) of Law No. 11 2006). The Indonesian Justice environments that run the judiciary is free from any party and consists of the following:

1) General justice;
2) Religious courts/Appellate Syar'iyah;
3) The military judiciary; and
4) The judiciary of the country (Cik Basir, 2011).

3.3.2 DKI Jakarta special autonomy

Single autonomy in Jakarta means the ability to nurture and cultivate services in a unified manner. It is expected to be able to deliver services more quickly and precisely and to be more integrated into the community. In addition, a single autonomous nation is closer to the granting of autonomy to one level of the province, for the autonomous region to have its own provincial government. This has consequences for the institutional governance of the DKI Jakarta area, because the municipality and District Administration is part of a device called the region by the provincial government, not a separate government.

In granting special status to Jakarta (DKI Jakarta), emphasis is placed on its historical aspects. Jakarta, due to the history of the Indonesian independence struggle, is always chosen as a place to host major events. It was once known as Batavia, Jakarta, and is also a central place in the birth of the movement of Boedi Utomo, the Youth Pledge, and Proclamation of independence till 1945. The concentration of the government serves to make Jakarta its state capital.

The privileged status of the Jakarta administration came into existence through Presidential Decree No. 2 of 1961 on the Regional Government of DKI Jakarta Raya by President Sukarno. Huda further revealed that the bases for this are:

1) Jakarta as the capital of the country should be used as indoctrination, as an exemplary city and ideal town for Indonesia;
2) As the country's capital, Jakarta Raya region needs to meet the minimum terms of an international city in the shortest possible time;
3) To create the above objective, the Jakarta Raya should be given a special position as the area directly controlled by the President/Leader of the Revolution (Ni'matul Huda, 2014: 168).
4) The implementations of asymmetric decentralization in the context of the evaluation are totally different from other areas, especially Aceh, which also received the status of "special." However, granting more privileges to Jakarta is historical.

3.3.3 DIY Yogyakarta special autonomy

Yogyakarta, on the other side of the autonomy arrangement, possesses no difference in its provincial autonomy that is not enjoyed by others with no privileged status. This is different from the status currently enjoyed by Aceh and Papua. Assessing Yogyakarta involved just looked at the position of the head and deputy of the executive in Yogyakarta, which is occupied by the Sultan/Pakualam/the royal family and authority in the land sector (known as the sultan grond) and also culture.

There is some resemblance, however, to the granting of privileged status to the city of Jakarta. Historical aspects are considered vital for granting privileges, especially for Yogyakarta. Before the independence of Indonesia, Yogyakarta already had full sovereignty as part of an empire led by Sri Sultan Hamengkubuwono IX (HB IX) and Sri Paku Alaman XIII (PA XIII), so the establishment of the Homeland could not be released without the establishment of DIY Yogyakarta.

The de facto status was noted by HB IX mandate and PA XIII, which was a mandate on September 5, 1945, with each region of the Sultanate and Pakualaman declared as a special region. Furthermore, on October 30, 1945, there was a published work with only one special region in the Republic of Indonesia, Yogyakarta. HB IX and PA XIII, and all the people of Yogyakarta, responded through a series of heroic struggles to regain independence. Yogyakarta is believed to be the Capital of Indonesia due to the emergency situation in Jakarta at the time. Legally, recognition as a special region was born March 3, 1950, with the issuance of Law No. 3 1950. It is a form of recognition and legalization of the de jure privilege of Yogyakarta. DIY privilege was then further recognized by the 1945 Constitution and regulated by Law No. 13 of 2012 on Privileges in DIY Yogyakarta.

4 CONCLUSION

The implementation of regional autonomy is an important focal point in improving people's welfare. The development of an area tends to be adjusted by the local government given the potential and specificity of each region. Regional autonomy was implemented in Indonesia through Law No. 22 of 1999 on Regional Government. In 2004, Law No. 22 for 1999 on Regional Government was deemed no longer appropriate in the circumstances, and was replaced by law No. 32 of 2004. Furthermore, Law No. 32 of 2004 concerning the government has been amended several times, with Law No. 12 of 2008 regarding the Second Amendment to Law No. 32 of the Year 2004 on Regional Government as the most current.

This is an excellent opportunity for local government to prove its ability to implement the rights of local authorities. Progress or setbacks in a region are determined by the ability and willingness to carry out local government activities enabling them to create freely and express themselves in order to develop their regions, provided they do not violate the law.

Conceptually, Indonesia is based on three main objectives: political, administrative, and economic. Political objectives in the implementation of regional autonomy attempt to achieve democratization through parties and the Regional Representatives Council. Administrative purposes to be achieved through the implementation of regional autonomy are the division of affairs between central and local government, which includes financial resources, as well as management reform in local government bureaucracy. The economic objectives to be achieved by the implementation of regional autonomy in Indonesia include improved scores on the human development index as an indicator of the welfare of the Indonesian people.

REFERENCES

Ahmad R. (1995). *Hukum Islam di Indonesia*. Graha Ilmu, Jakarta.
Bambang S. (2006). Metodologi Penelitian Hukum, Raja Grafindo Persada, Jakarta.
Cik Basir, 03 April 2011, Kedudukan, Kewenangan, dan prospek Mahkamah Syar'iyah dalam Judicial Power, diakses dari. Available at: http://www.badilag.net/artikel/publikasi/artikel/kedudukan-kewenan gan-danprospek-mahkamah-syariyah-sebagai-judicial-power-oleh-drs-cik-basir-shmhi–227.
Djojosoekarto, A., Rudiarto S., dan Cucu S. (2008) Kebijakan Otonomi Khusus di Indonesia. Kemitraan bagi Pembaruan Tata Pemerintahan di Indonesia. Jakarta, 2008. Available at: www.kemitraan.or.id, Agung Djojosoekarto, Rudiarto Sumarwono, Kebijakan otonomi khusus di Indonesia, diakses 28 Januari 2017.

Fabretti, V. and Paola, N. (2017). The relations between religion and politics in European education systems. *Jurnal Ilmiah Peuradeun* 5(2), 225–236. doi:10.26811/peuradeun.v5i2.148.

Handoyo, S. (2018). The role of public governance in environmental sustainability. *Jurnal Ilmiah Peuradeun* 6(2), 161–178. doi:10.26811/peuradeun.v6i2.255.

Kadir, M. A. (2004). Hukum dan Penelitian Hukum, Aditya Citra Bakti, Bandung.

Moh, M. (1987). Pokok-Pokok Hukum Administrasi Negara, Penerbit Liberty, Yogyakarta.

Ni'matul H. (2014). *Desentralisasi Asimetris Dalam NKRI, Kajian Terhadap Daerah Istimewa, Daerah Khusus, dan Otsus*. Bandung: Nusa Media.

Penjelasan Umum Undang-Undang Nomor 29 Tahun 2007 tentang Pemerintahan Provinsi Daerah Khusus Ibukota Jakarta Sebagai Ibukota Negara Kesatuan Republik Indonesia (Lembaran Negara Republik Indonesia Tahun 2007 Nomor 93, Tambahan Lembaran Negara Republik Indonesia 4744).

Pasal 7 ayat (2) Undang-Undang Nomor 13 Tahun 2012 tentang Keistimewaan Yogyakarta (Lembaran Negara Republik Indonesia Tahun 2012 Nomor 170, Tambahan Lembaran Negara Republik Indonesia Nomor 5339).

Sulaiman (2018). *Studi Syariat Islam di Aceh*. Banda Aceh: Madani.

Tabrani Z. A. (2015). *Persuit Epistemology of Islamic Studies (Buku 2 Arah Baru Metodologi Studi Islam)*. Yogyakarta: Penerbit Ombak.

Undang-Undang Republik Indonesia Nomor 44 Tahun 1999 tentang Penyelenggaraan Daerah Istimewa Aceh.

Emerging Perspectives and Trends in Innovative Technology
for Quality Education 4.0 – Kusmawan et al (eds)
© *2020 Taylor & Francis Group, London, ISBN 978-0-367-25803-0*

Behavior of millennials in using the internet for learning

Jaka Warsihna
Universitas Terbuka, Tangerang Selatan, Indonesia

ABSTRACT: This study aims to determine the behavior of the millennials in using the Internet for learning. The study was conducted by performing a survey of the millennials in Jakarta, Bandung, Yogyakarta, Denpasar, and Banda Aceh. As a result, the general behavior of the millennials in the use of the Internet for learning was relatively good, even though those who have ever used the Internet for learning was less than half. Based on the next-generation data regarding the millennials' internet use, it was found that the intensity of the internet use is in the moderate category; they mostly access the Internet to do the assignments given by teachers; most of the access is at school; they access the Internet to learn as ordered by their teachers; most of the materials are interactive multimedia; and nearly all of the students who never use the internet for learning spend no more than two hours a day; the rest use of internet is for other purposes. Based on the correlation test, it was revealed that the students' intensity in using the Internet to learn was significantly and positively associated with students' attitudes towards the internet, the intensity of the tasks or lessons given by the teachers, their search for interactive materials, and the encouragement from their teachers. To improve these variables, the teachers' encouragement to use the Internet for the learning needs to be improved, particularly in classroom assignments, materials enrichment, and finding learning resources.

Keywords: Behavior, Generation, Millennials, Internet, Learn

1 BACKGROUND

The development of information and communication technology (ICT) has changed human civilization. Currently, changes occur rapidly, and the competition is intense. In these circumstances, the pace of innovation and creativity are the advantages. Speed and innovation are some of the characteristics of the millennial generation. Millennials refer to generation aged 15-39 years (Ali, Aji and Ghazali, 2019; Waliyuddin, 2019). Meanwhile, William Strauss and Neil (2000) stated that millennials are those who were born in 1982 and above (Howe and Strauss, 2000). The millennial generation is also known as generation Y. However, a different idea was conveyed by Putra (2016), his study on the theory of generational differences provides a breakdown of generation based on their behavior in consuming media (one of which is the millennial generation). The research results of Alvara Research Center (2019) found that a greater emphasis is on the aspect of behavior. He stated that the millennial generation is a creative generation who is thinking out of the box and rich in ideas and concepts. Furthermore, they are able to communicate and connect good ideas. Besides, the millennials are individuals who are good at socializing, especially in the communities that they are involved in, active in social media and the internet. They are people who are confident and dare to express opinions on the online media and in front of the public.

When viewed by age and behavior, the millennial generation is the generation that actively uses the Internet and social media. Research concerning the use of the Internet for learning by Sherlyanita & Rakhmawati (2016) found a positive impact on the use of the internet.

Millennials in junior high school were able to determine the use of the Internet, both for the means of learning and entertainment, such as accessing social media. They knew how to divide their time well to have a direct interaction in daily life and to have interaction on social media. Furthermore, they were aware of basic security, such as privacy restriction and data sharing only to people they know.

Various findings above demonstrate that millennials are young generations aged between 13-39 years who actively utilize social media and the internet; those who are confident and creative, think out of the box; and learn through social media and the Internet. In such a condition, it is good to understand the millennials' behavior in learning via the Internet thoroughly. Several factors may encourage and influence their learning via the Internet. The research on millennials and internet use in adolescents has grown in number and impacts. However, the study on the millennials' behavior aged 13-18 years on the use of the internet for learning is limited. This is the unique nature of this research, which is to investigate the behavior of millennials in learning through the internet and factors which encourage and influence their learning via the internet.

2 RESEARCH METHODS

This study employed a survey of millennials in five cities: Jakarta, Bandung, Yogyakarta, Denpasar, and Banda Aceh. The population of this research is the millennials aged 13 to 18 years. Stratified purposive sampling was used by utilizing a random sampling of 535 respondents. Data collection was conducted in September 2018.

Data were analyzed using descriptive statistics and Spearman correlation statistics. To facilitate the processing of the data, SPSS version 24 was employed. The data analysis was expected to answer the research questions.

3 RESULTS AND DISCUSSION

The millennials' behavior of the use of the internet for learning purposes is measured by how frequent they use the internet. Data of respondents in this study were collected from 535 respondents (Table 1). It was found that only a few millennials have never used the Internet for learning (6.5%); nearly half stated that they rarely use the Internet for learning (47.1%); approximately a quarter of them posited that they sometimes use the internet for learning (28.6%); and the rest always use the internet for the learning purpose (21.3%).

Furthermore, the results were developed via interviews with the millennials. Based on the data on the frequent use of the internet for learning, it was revealed that 6.5% of the millennials did not use the internet for learning because of financial constraint like they do not have an internet data on their mobile phone or wifi at home. Furthermore, the number of those who sometimes use the internet to learn was 47.1%, depending on whether or not their assignments need internet access. Next, the millennials who frequently access the internet for learning accounted for 28.6% because their teachers often give them homework or extra materials over the internet. In such circumstances, the millennials inevitably need to acquire internet access. Finally, the millennials who invariably have internet access reached 21.3%.

Based on the reasons for the internet use for learning the purpose, it was revealed that most of the millennials (46.2%) use the internet to do assignments, while 5% of them to seek a referral. Also, 19.8% was found to use the internet for exploring their school's material, while 29% to find information. The data above indicate that, in general, the millennials utilize the internet because of the tasks given by their teachers, and only a few of them use it to find references or to deepen their subjects. This study found that the use of the internet by the millennials needs to be guided and directed towards productivity.

In addition, it was revealed that Internet use is largely done at school (43.3%). The millennials usually access the internet at school to complete their assignments from their teachers or to find information during their spare time. However, some schools also ban the use of the smartphone by students. Additionally, internet usage at home is also significant (37.9%) but lower than at school because the internet access at home is limited. The millennials who access the internet in other places, such as libraries, Internet cafes, and public facilities are noticeably small.

The majority of the millennials utilizing the internet because of their teachers' requests for 65.9%. This depicts that this generation's behavior is greatly influenced by who is the referral. Additionally, parents' request obtained 16.4%. In this regard, in-depth interviews were performed to investigate whether the use of the internet for learning was greatly influenced by teachers because they give homework and assignments which allow students to conduct their own learning via the internet. On the other hand, the parents' request was due to the learning environment at home, especially where one of the parents is a college graduate or where wifi is available. The rests were influenced by friends, own willpower, and idle time.

According to the result of the use of learning media on the Internet, it was found that over 45.1% of the millennials prefer interactive multimedia; 5.0% of them like using audio; 35.5% preferably choose video; 12.1% prefer textbooks; and the remainder (2.3%) like to learn from augmented reality. From the data above, most of the millennials highly preferred to use interactive multimedia, such as games, simulations, animations, and video for learning.

Based on the result of learning styles, it was depicted that the majority of the millennials (46.2%) prefer games or interactive games, while only 7.1% of them choose audio. Those who prefer watching the video are accounted for 34.6%, and the rest (12.1%) prefer reading. The data indicated that most of the millennials preferred playing interactive games, and only a few of them liked to listen to the audio.

Millennials' interest is a crucial factor in improving learning achievement. This notion is explained in the studies conducted by Maghfiroh (2016) and Ariastuti et al. (2014). Hence, the low internet proficiency among students is caused by the low interest of the students towards the lessons conducted via the internet.

The majority (40.6%) of the millennials access the internet for learning for the duration of the half to 1 hour. Furthermore, 34.6% of them spend 1 to 2 hours, 20.5% access the internet to study for less than ½ hour, and 4.3% spend more than 2 hours. This result indicates that a maximum duration for accessing the internet for learning is only two hours, while a few spend over two hours for the same purpose. After interviews, the result showed that, in fact, the Millenials generally access the internet for more than two hours for the purpose of study briefly, while the rests are for entertainment, searching information, and accessing social media like Instagram, facebook, twitter, and youtube.

According to the national survey report by CSIS (Center for Strategis and International Studies, 2017), each online media found a striking difference between the millennial and non-millennial generations. Approximately 54.3% of the millennials admitted to reading online media daily, while merely 11.9% of the non-millennials perform the same activity.

Based on the results from the correlation test, factors related to the intensity of the internet use for learning showed that the correlation was 0.261 with a significance value of 0.000. This indicates that the ability of students to access the internet for learning relates positively and significantly with their interest in learning with the internet at the 99% confidence level. It means that the more frequent the students access the internet, the higher their interest. Conversely, the less frequent they access the internet for learning, the less their interest in accessing the internet.

In general, millennials sometimes use the internet for learning (47.1%). According to table 2, generally, the millennials use the internet only for work assignments (46.2%), tasks given by the teachers (65.9%), and assignments at school (43.3). The most preferred content, according to the millennials, is interactive multimedia.

4 CONCLUSION

The frequency of the millennials' use of the internet for learning is still low. Their internet utilization to learn is mostly to complete their assignments. The millennials prefer to find materials on the internet because of interactive multimedia and a variety of learning styles in the form of interactive games. These results suggest that the millennials prefer to learn from inter-active, challenging, and instant materials which can be easily found on the internet.

The frequency of the millennials in accessing the internet for learning is still small. There-fore, to increase their interest in learning through the internet, it is advisable for them to access the internet for learning frequently. Learning via the internet can be conducted in vari-ous ways, such as e-learning, blended learning, Massive Open Online Course (MOOC), online distance learning, searching for learning resources, references, and latest journals, and deepen-ing their subjects. Hence, the internet can be used in an interactive and communicative way to improve the Millenials' competence, updating and increasing their current knowledge.

To raise the students' interest in learning through the internet, there are numerous assign-ments for the teachers. Material developed and uploaded to the internet should be short, con-cise, and interactive. In addition, the given assignments should be in the form of group projects which can challenge students to explore the internet.

REFERENCES

Ali, H., Aji, I. and Ghazali, M. H. (2019) "Desain Pendidikan Islam di Pondok Pesantren Sindangsari Al-Jawami Cileunyi Bandung Dalam Menghadapi Generasi Milenial," Tarbawi: Jurnal Pendidikan Islam, 16(1).

Center for Strategis and International Studies (2017) Laporan Survai Nasional "Orientasi Sosial, Eko-nomi, dan Politik Generasi Milenial, CSIS. Available at: https://www.csis.or.id/uploaded_file/event/ ada_apa_dengan_milenial____paparan_survei_nasional_csis_mengenai_orientasi_ekonomi__sosial_ dan_politik_generasi_milenial_indonesia__notulen.pdf (Accessed: September 2, 2019).

Howe, N. and Strauss, W. (2000) Millennials rising: The next great generation. New York: Vintage.

Pusat Penelitian Alvara (2019) Memahami Milenial Indonesia, Pusat Penelitian Alvara. Available at: https://www.academia.edu/35915408/.

Putra, Y. S. (2016) "Theotrical Review: Teori Perbedaan Generasi," Among Makarti, 9(18), pp. 123–134.

Sherlyanita, A. K. and Rakhmawati, N. A. (2016) "Pengaruh dan pola aktivitas penggunaan internet serta media sosial pada siswa SMPN 52 Surabaya," Journal of Information Systems Engineering and Business Intelligence, 2(1), pp. 17–22.

Waliyuddin, M. N. (2019) "Religious Expression of Millenial Muslims within Collective Narcissism Dis-course in Digital Era," Wawasan: Jurnal Ilmiah Agama dan Sosial Budaya, 4(2).

Emerging Perspectives and Trends in Innovative Technology
for Quality Education 4.0 – Kusmawan et al (eds)
© 2020 Taylor & Francis Group, London, ISBN 978-0-367-25803-0

Employers' response to chemistry education graduates of Universitas Terbuka

Sandra Sukmaning Adji, Sri Hamda & Jamaludin
Universitas Terbuka, Tangerang Selatan, Indonesia

ABSTRACT: This study aims to obtain an overview of employers' responses to chemistry education graduates to assess their competitiveness compared to others. The study was conducted on a chemistry distance education program with a sample of 36 employers. Data were obtained from a questionnaire about the past four years inquiring about employers' satisfaction levels regarding certain skills required for the graduates as well as the perceived importance of those skills among employers. The questionnaire contained 29 rating-scale items and was content-validated by distance education experts. The results revealed that the majority of the items obtained high scores on satisfaction. However, ratings for their importance levels were still higher. The lowest score obtained is in English writing skills of X=3,36 and Y=3,77, while the highest is in calculation skills with X=4.27 and Y=4.55. Also, the quadrant analysis demonstrates that the time management item obtained a low score for satisfaction and high score for importance. This particular item is therefore the highest priority for improvement.

1 INTRODUCTION

Tracing studies aim to investigate the relationship between higher education and the professional world, assessing the relevance of higher education, information for stakeholders, and complete requirements for accreditation of higher education. This information is necessary to improve the program and to ensure graduates have the skills required in their workplace. Thus, it is necessary to analyze how employers respond to graduates from the chemistry education program. The purpose of this study is to examine: 1) employers' response to basic skills of graduates; 2) employers' response to their professional competencies; 3) employers' response to graduates' personal qualities; 4) employers' response to graduates' organization skills; 5) employers' response to graduates' specific technical knowledge; and 6) performance analysis from employers' responses.

2 METHODOLOGY

The study was conducted at the chemistry department of a distance education program. The data was obtained from employers' surveys of graduates. The instrument used was questionnaires with 36 high school principals (i.e., the graduates' employers) as respondents. The questionnaire contains five indicators with 29 question items adopted from the instrument developed by Universitas Terbuka. Even though it had been validated, the instrument was tested again via two distance education experts and two principals to assess the level of readability. The instrument employed a five-part scale, ranging from (5) highly satisfied or essential to (1) highly dissatisfied or not at all important (1). The data obtained were analyzed descriptively and presented in a bar chart.

3 RESULTS AND DISCUSSION

One of the crucial aspects of higher education is graduate employability. Harvey (2001) defines employability through the perspectives of various individuals and institutions. Graduates' ability to demonstrate their attributes to obtain jobs is defined as individual employability. Employers' responses to graduates are described as X, indicating satisfaction, with Y depicting its importance.

3.1 *Employers' response to graduates' necessary skills*

The success of chemistry graduate employees could be identified by their basic abilities, such as the ability to communicate actively, to write in English, to count, to possess ICT skills, and to have information literacy. Employers' responses to graduates' basic skills are illustrated in Figure 1.

Figure 1 shows that, for entire components, the values for importance were higher than those for satisfaction. Satisfaction scores ranged from X=3.36 to X=4.27 on a scale of 1 through 5. The ability to count (numeracy) had the highest score for employer satisfaction, while the lowest employer satisfaction score was for English writing skills. The graduates' numeracy ability received high scores for both satisfaction and importance, with X=4.27 and Y=4.55 respectively. These values were close to those relating to ICT skills, which were scored at X=4.05 and Y=4.50.

In education, ICT knowledge has become a highly essential skill and will only become increasingly so in the 21st century. The role of ICT is to transform teaching and learning; thus, it is important to explore how ICT will impact the way programs are offered and delivered in future universities and colleges (Sharma, 2012).

3.2 *Employers' response to professional competencies of graduates*

The ability to think critically is indicated by several other abilities, such as ability to read, take notes, or compose a piece of writing.

Figure 1. Employers' responses to basic skills of graduates.

Figure 2. Employers' responses to professional graduates' competencies.

206

Figure 2 reveals that graduates' personal qualities were scored as moderately good, including for the items of critical thinking and problem-solving (X=3.88; Y=4.38). Nearly all components in professional competencies obtained X scores that were above average (where averages were X=3.75 and Y=4.00). The highest satisfaction score was for graduates' ability to work in a team – i.e., cooperative skills with X score=4.08. In contrast, the highest importance score was for innovative and creative thinking, with Y =4.47. Furthermore, innovative and creative thinking had the most significant gap (0.58) between employers' satisfaction (X=3.88) and importance scores.

3.3 *Employers' response to personal quality of graduates*

Personal competencies include ethical normative competencies, balancing personal ethical scores and business objectives, realizing self-regulated and active involvement, and reflecting on personal experiences (Osagi and Wesselink, 2016). In this study, graduates' ability to accomplish general and professional ethics standards were given a score of X=4.05, while the importance was scored Y=4.44. As seen in Figure 3, employers give satisfaction and highly important scores with averages above 4.0, except for the score for employers' satisfaction with intercultural skills, at X=3.58. The highest scores were for self-reliance in job performance, at X=4.16 and Y=4.38. Graduates' ability to communicate and relate well with others were scored at X=4.16 and Y=4.41.

3.4 *Employers' responses to organization skills of graduates*

Intense communication is necessary to complete complex tasks. However, this type of communication is hard to achieve during teamwork. In an organization, leadership is needed, and someone has to be aware of forces that could bring disruption to their organization. According to Lurey and Raisinghani (2001), the virtual team efficacy is strongly influenced by good leadership.

Figure 3. Respondents' scores for personal quality.

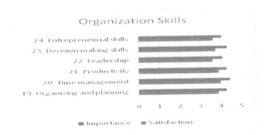

Figure 4. Employers' responses to organizational skills of graduates.

Figure 5 shows that all components of graduates' organizational ability were considered highly important, with Y scores averaging above 4.0. The employers expected the graduates to be able to make decisions and manage time well. The highest scores for both satisfaction and importance were for time management, with values of X=4.00 and Y=4.46. Furthermore, the lowest importance score was for entrepreneurial skills, at Y=4.05.

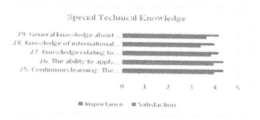

Figure 5.   Employers' response to graduates' technical knowledge.

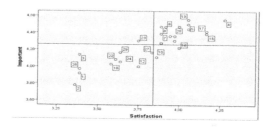

Figure 6.   The plot of performance in quadrant analysis.

3.5   *Employers' response to specific technical knowledge*

One determining factor in perceptions of graduates' skills is technical skills (Muilenburg and Berge, 2005). However, in this study technical ability components obtained average satisfaction scores below 4.0.

As shown in Figure 5, the technical knowledge importance scores averaged above 4, except for the capability component of international standards, at Y=3.94. The highest scores for satisfaction and importance were for the skill of continuous learning, at X=3.97 and Y=4.35. Ability to use knowledge and skills in the professional world scored lower, at X=3.94 and Y=4. Based on these scores, employers expect the graduates to be able to develop greater capacity in fieldwork.

3.6   *Analysis of the benefits of performance (importance of performance analysis)*

Quadrant analysis in Figure 6 reveals that most of employers' responses fit in quadrant 4, which means graduates were ranked as doing satisfying jobs in areas important to employers. However, number 23 (time management) is in quadrant 1, indicating low satisfaction with graduates but high importance for employers. Several studies on the importance of time management have been carried out, including its relation to job satisfaction, health, and academic performance (Claessens et al., 2007). In this study, results showed that graduates' time management skills did not satisfy employers, even though employers ranked time management as greatly necessary. Hence, it is important for Universitas Terbuka to improve the time management skills of students in their chemistry education program.

4   CONCLUSION

The results revealed that challenges still emerge despite graduates' many skills and knowledge. Graduates should have been more active and able to develop themselves in accordance with the demands of time management. Thus, employers posit that educational institutions should pay closer attention regarding this matter. Advanced skills, excellent soft skills, and technical skills are always in demand. Higher education, therefore, should focus highly on programs which could encourage and foster hard and soft skills, in addition to developing an understanding of the subject.

REFERENCES

Akkermans J, Schaufeli W., and Brennikmeijer, V. (2013) The role of career competencies in the Job Demands–Resources model. *Journal of Vocational Behavior*, 83(3),356–366.

Claessens, B., Erde, W; Rutte, C., Roe, G., and Robert, A. (2007) A review of the time management literature. *Personnel Review* 36(2),255–276.

Havey, L. (2001), Defining and measuring employability. *Quality in Higher Education* 7(2),97–109.

Lurey, J. S. and Raisinghani, M. S. (2001) An empirical study of best practices in virtual teams, *Information & Management*, 38(8),523–544.

Muilenburg L. Y. and Berge, Z. L. (2005) Student barriers to online learning: Factor analytic study, *Distance Education* 26(1),29–48.

Osagi, E. and Wesselink, R. (2016) Individual competencies for corporate social responsibility: A literature and practice perspective. *Journal of Business Ethics*. 135(2),233–252.

Sharma, K. (2012) Critical reflections on the benefits of ICT in education. *Oxford Review of Education*, 38(1),9–24. DOI: 10.1080/03054985.2011.577938.

Emerging Perspectives and Trends in Innovative Technology
for Quality Education 4.0 – Kusmawan et al (eds)
© 2020 Taylor & Francis Group, London, ISBN 978-0-367-25803-0

The use of hypermedia for Indonesian language learning in distance education

Brillianing Pratiwi
Universitas Terbuka, Tangerang Selatan, Indonesia

ABSTRACT: Hypermedia as digital technology innovation is relevant to education 4.0. It is a combination of various technologies. Hypermedia can be used for the learning of the Indonesian language. This paper aims to report on the design of hypermedia to contribute positively to the achievement of learning goals for students. The case study was conducted at the Open University, Indonesia. Open University (UT) is a long-distance tertiary education institution utilizing learning media. UT has students from all over the country and abroad. The students are expected to be able to study independently based on their initiatives. This paper discusses the definition of hypermedia, the use of hypermedia in Indonesia, and the advantages and disadvantages of hypermedia. The results of this paper indicate that hypermedia provides an opportunity for students to explore information in their ways. Students can interactively use hypermedia.

1 INTRODUCTION

Open University is a distance state university, which means the learning at the university is not done face-to-face; instead, the learning uses print media (modules) and non-print (audio/video, computer/internet, radio broadcasts, and television). One form of non-printed media is hypermedia. Hypermedia is different from multimedia. Multimedia refers to the use of various media. Meanwhile, hypermedia is a combination of several technologies that assist in organizing and accessing information.

At Open University, hypermedia is used in the process of teaching and learning. The purpose of this paper is to examine the use of hypermedia for learning the Indonesian Language at Open University, whether or not the user can contribute positively to the achievement of learning goals. This is in line with how to study at the Open University. UT students are expected to be able to study independently. If they face difficulties, they can request information from the study group, tutorial group, and the Open University Distance Learning Program Unit (UPBJJ-UT) spreading from Aceh to Jayapura.

Hypermedia, as a digital technology innovation connected to the internet, is very relevant to education 4.0. Discussions on education 4.0 are inseparable from the industrial revolution. According to (Prasetyo), the world experienced four stages of the revolution. The 1.0 industrial revolution took place in the 18th century through the invention of steam engines, facilitating mass production. The 2.0 industrial revolution occurred in the 19-20th century marked by the use of electricity, which made production costs low. The 3.0 industrial revolution occurred in the 1970s through computerization, and the industrial revolution 4.0 occurred in the 2010s through intelligence engineering and the internet of things as the backbone of the movement and connectivity of humans and machines.

The key to facing the industrial revolution 4.0 is preparing technological advances and developing human resources from the humanities side.

2 HYPERMEDIA DEFINITION

Hypermedia is derived from two words; the term hyper, which means nonlinear or random, and media that refers to information in many formats. According to Blanchard and Rotenberg (in Munir (2009), hypermedia is a combination of various media regulated by hypertext. Hypermedia includes various media such as video/visual, audio/music, text, animation, film, graphics, and images.

Kustandi (2011) defines hypermedia as an extension of hypertext which combines other media into text. Hypermedia systems can be used to make a corpus material that includes text, graphics, animated images, sound, video, music, etc.

Prasojo, Lantip Diat, and Riyanto (2011) propose that hypermedia is a data file containing much information that is sent via the internet to a computer and displayed graphically in a user-friendly way.

Landow (1992) stated that "electronic linking shifts the boundaries between one text and another as well as between the author and the reader and between the teacher and the student." In a similar vein, Dryden (1994) argues that hypermedia environments can indeed promote the appreciation of literature (and of texts in other disciplines) as they nurture the growth of the learner in intellect and spirit. Furthermore, hypermedia has the potential to transform the structure of both classrooms and entire institutions-schools and universities-and to make the teaching and practice of literate thinking and behavior a truly free enterprise that respects and serves the needs of both the individual learner and the broader community of learners.

From this view, it can be concluded that hypermedia is a media which includes not only texts, but also images, sounds, videos, animations, and graphics aimed at presenting interactive information. Besides, users are connected to the internet to store, search, retrieve, format, and display information.

3 LEARNING

The National Education System Law No. 20 of 2003 states that learning is the interaction between students and educators as well as learning resources in a learning environment. We are learning as a process built by teachers to develop creative thinking that can improve good mastery of the subject matter.

Learning, according to Sanjaya (2010), can be interpreted as a process of cooperation between teachers and students by using all the potentials from within the students themselves such as interests and talents of necessary abilities, including learning styles and potentials that exist outside the students such as the environment, a means of learning resources as an effort to achieve certain learning goals.

Arifin (2010) argues that learning is a process or activity that is systematic and systemic, that is, interactive and communicative between educators "teachers" with students, learning resources, and the environment to create a condition that allows the occurrence of student learning.

Meanwhile, according to Komalasari (2013), learning is a system or process of learning which is planned, implemented, and systematically evaluated so that learners can achieve the learning goals effectively and efficiently.

From the discussion, it can be concluded that learning is an activity carried out by teachers by utilizing all the potential and existing resources to achieve individual learning goals.

4 RESEARCH METHODOLOGY

This paper reports on descriptive qualitative research. Descriptive qualitative research is research that uses a case study method or approach. This paper focuses on one particular object, which is the case studied. Case study data were obtained from all parties involved. In other words, in this case, study, the data were collected from various sources Nawawi, (2003:1).

In summary, what distinguishes the case study method from other qualitative research methods is the depth of analysis in more specific cases (both specific events and phenomena).

The population this research is all types of hypermedia on websuplemen (online enrichment program, laboratory learning in junior high school, and online tutorial at Open University).

The sample of this research is hypermedia in Indonesia language learning.

5 THE USE OF HYPERMEDIA IN TEACHING OF INDONESIAN LANGUAGE

Open University is a distance and non-face-to-face higher education institution. Students do not have to study in a classroom. The learning process of students is conducted independently by using various printed and non-printed media. Printed products include subject matter books (BMP/modules). Meanwhile, non-printed products include BMP audio, interactive video, drylab, radio programs, television programs, and online-based enrichment materials. The non-printed programs are uploaded to the internet so that students can access them any-time and anywhere.

The Use of Hypermedia in Teaching of Indonesian Language, for example in:

5.1 Online based enrichment program (web supplement)

The online-based enrichment program is a unity with the Basic Material Book (module), but not all BMPs are equipped with additional teaching materials. The following are examples of courses that are complemented by online-based enrichment programs. Students can access the online-based enrichment program on page http://web-suplemen.ut.ac.id/.

Online based enrichment material consisting of text, images, audio, video, and animation that aim to present interactive information. Interactivity can be seen from the way users store, search, retrieve, format, and display information.

5.2 Indonesian language learning series

The Indonesian learning series is part of the SMP Learning Laboratory. The SMP learning laboratory is part of the educational laboratory. Educational laboratories provide enrichment materials (web supplement), which are materials designed to broaden teachers' insights and knowledge regarding learning and education in general. The enrichment material on the guru-pintar.ut.ac.id page is presented in two material formats, namely video and article forms. Students can access the materials from home or their workplace without the constraints of time and space.

5.3 Online tutorial

Online Tutorial is an internet-based service offered by the Open University and attended by students through the internet network. Other forms of tutorials are face to face tutorials, radio tutoring, and television tutorials.

The purposes of the online tutorials are:

1. Optimizing the use of the internet network to provide learning assistance services to students.
2. Allowing the distance learning process to be more communicative and interactive
3. Providing choices for students who have access to the internet network to obtain optimal learning assistance services.

Before participating in the online tutorial, students activate their accounts on the site http://elearning.ut.ac.id. After this process is carried out, the students will obtain an account password to use the online tutorial features. The students must read the online tutorial guide available on the website before accessing the system so that the login process can be quickly done.

The online tutorial is held every semester for 8 weeks or approximately 2 months before implementing the UAS each semester. The contribution of the tuton value to the final grade is a maximum of 30%. The maximum value can be obtained if the student becomes an active participant in the tuton. Active participants are participants who read each session, respond by asking questions or responses, discuss and work on assignments in sessions 3, 5, and 7. Passive participants are participants who only read sessions without giving questions, responses, and answers to assignments.

6 EXCESSES AND LACKS OF HYPERMEDIA

All media, including hypermedia, has advantages and disadvantages.
The advantages of hypermedia are:

1) Hypermedia allows for adaptation between students
2) Hypermedia provides support mechanisms aimed at reinforcing ideas and concepts related to teaching materials.
3) Hypermedia is connected to the internet, so students are freer.
4) Hypermedia combines sound and images with text to be kept in the brain.
5) Regarding the link, students can connect ideas from different media sources so that control is on the students. This makes students more motivated.
6) Students can find their information individually.
7) Building mental structures of students based on their exploration.
8) Hypermedia enables interactive communication and creates a collaborative learning environment. Students can develop complex cognitive abilities such as breaking a topic into subtopics.

Meanwhile, the drawbacks of hypermedia are:

1) Differences in student backgrounds. Students do not have the expertise to operate computers.
2) The environmental freedom offered by hypermedia can cause students not to focus on learning goals.
3) Information obtained by many students sometimes makes students easily tired.
4) The mismatch of interactions makes students bored.

7 CONCLUSION

Hypermedia allows students to explore various information in their way. Hypermedia learning media can also make learning fun, and students will be more interested.

Hypermedia provides many benefits for distance learning because students can access, read, download, ask questions, provide responses, and work on questions.

Hypermedia can be improved by increasing the use of animation, increasing practice questions, displaying comprehensive explanations, adding sounds to the material, and clarifying navigation.

REFERENCES

Arifin (2010) Evaluasi Pembelajaran. Bandung: remaja Rosdakarya.
Dryden, L. M. (1994) Literature, Student-Centered Classrooms, and Hypermedia Environments. New York: Modem Language Association of America.

Komalasari, K. (2013) Pembelajaran Kontekstual: Konsep dan Aplikasi. Bandung: Refika Adiatama.

Kustandi, C. d. S., Bambang (2011) Media Pembelajaran Manual dan Digital. Bogor: Ghalia Indonesia.

Landow, G. (1992) HYpertext: The Convergence of Contemporary Critical Theory and Technology. Baltimore: Johns Hopkins University Press.

Munir (2009) Pembelajaran Jarak Jauh Berbasis Teknologi Informasidan Komunikasi. Bandung: ALFABETA.

Nawawi (2003) Metode Penelitian Bidang Sosial. Yogyakarta: Gadjah Mada University Press.

Prasetyo, B. a. T., Umi 2018. Revolusi Industri 4.0 dan Tantangan Perubahan Sosial dimuat dalam Prosiding Semateksos 3 Strategi Pengembangan Nasional Menghadapi Revolusi Industri 4.0.

Prasojo, L. D. d. R. (2011) Teknologi Informasi Pendidikan. Yogyakarta: Gava Media.

Sanjaya and Wina (2010) Perencanaan dan Desain Sistem Pembelajaran. Jakarta: Kencana.

Social behavior of students in the "seize the ball" game

U.Z. Mikdar & Roso Sugiyanto
FKIP Palangka Raya University, Indonesia

ABSTRACT: The purpose of this study was to assess the social behavior of students in the game "seize the ball." A descriptive qualitative method was used, with the subjects and the informants being students undertaking physical education courses in the PGSD Study Program FKIP Palangkaraya University. The results showed that those participating in the game displayed attitudes and social behavior, such as excitement (cheerfulness), competiveness, cooperation, conflict, sincerity, honesty, respect, and enthusiasm. Additionally, 11 out of 30 students (36.66%) had relatively high morale compared to other players. However, there were still a few techniques or movements employed by players that were dangerous or negative, which necessitates further changes to the game.

1   INTRODUCTION

In educational and sports philosophy, four questions can be asked of a game of "seize the ball": (1) Is the game in line with education and sports? (2) What do you want to achieve (objectives)? (3) How are objectives implemented? (4) What are the benefits?

Ontologically speaking, the game is played by two teams, each consisting of five players, attempting to move a rubber ball past the opponent's finish line. It can be played on a closed or open field, indoor or on grass, of a size of 10 x 20 meters. This rectangular shape is bounded by 2 left/right sidelines, two end lines, and a midline, which includes a midpoint.

The game requires physical values such as intensity and agility but can also spur uplifting feelings and involve qualities such as honesty. The game fosters relationships and social interactions between individuals, groups, and other community members. Epistemologically speaking, it aims to foster the pursuit of physical, intellectual, social, moral, emotional, and spiritual potential. For this reason, this research aims to uncover the social behavior involved in a game of "seize the ball."

2   METHOD

This study involved descriptive qualitative research, carried out via a survey polling the behavior, thoughts, or feelings of respondents. The researcher conducted observations, polls, and interviews about the social behavior of students in the Faculty of Teacher Training and Education of the University of Palangkaraya, who played the game.

A total of 30 students were divided into three groups of 10 members each, with each group playing a five-on-five game for two periods of 5 minutes and then 2 minutes of a hospitality break. During the game, the social behavior of students, who had previously been told how to play, was observed.

After the game, all players were interviewed and asked to fill in an open essay observation sheet on social behavior during the game.

2.1 Data collection technique

In this study, data collection was carried out directly, in real life, though observations and interviews were also used. Importantly, data analysis techniques involved in-depth interviews and observation techniques. The data were analyzed in an interconnected way to arrive at a provisional guess, which was used as a basis for gathering the following data, then confirmed with the informants continuously in triangulation.

3 RESULTS AND DISCUSSION

Some social behaviors observed during the game of "seize the ball" include (1) excitement, which arises during a process of carrying and winning the ball, scoring goals, being crushed, and cooperation after winning the game; (2) competitive behavior, which is encouraged by urge to win, knock each other down, and win and maintain the ball; (3) cooperation, which occurs while compiling tactics, dividing tasks, guarding opponents, helping each other, passing, and scoring goals; (4) disagreement, which happens when there is a violation and protests against it; (5) seriousness, which occurs when the ball is out of reach while carrying and winning it; (6) honesty, displayed when winning a ball, scoring goals, accepting defeat, and receiving punishment, although some do not admit to committing an offense; (7) respect, evident when apologizing for mistakes and respecting losers and winners; and (8) enthusiasm, evident in the process of scoring goals and winning the ball or the game, although there are still people who lack this quality due to lack of understanding of the rules.

In this game, what appeals to the player is (1) running with the ball while being chased by an opponent; (2) taking the ball; (3) teamwork (cohesiveness); (4) getting the ball at the beginning of the game; (5) running swiftly past the opponent; and (6) defending the ball and bypassing the opponent.

These behavioral findings reinforced the theory of stimulus-organism-response (S-O-R), where organisms intervene between stimulus and response. Watson stated that the objectivity of individual conduct only applies to overt behavior. Every behavior is primarily a response or response to a stimulus, since it activates the activities of stimuli. The intervention of organisms toward stimuli involves social cognition and perception, values, and concepts (Soelaeman, 2008: 47). Therefore, the game "seize the ball" raises social behaviors such as excitement, cooperation, sincerity, respect, respect, and spirit. Other social behaviors that appear in the game include competition, cooperation, and opposition. In Simmel's interaction theory, interaction shows (1) the community is formed from a network of relations between people constituting a unity – in the network of relations, there are many actions and reactions, and therefore the community is a dynamic process determined by the behavior of its members; (2) the network of relations is not the same – this means that, from the network of relations, community associations might be formed; (3) integration does not always form a harmonious network, but criticism, opposition, and conflict, among other reactions, often occur (Soelaeman, 2008: 56). Therefore, in the game of "seize the ball" relatively dangerous acts such as kicking, clothes-pulling, scratches, pressing, squeezing, elbowing, pushing, hair-pulling, pushing, sliding, neck-holding, crashing, nail-scratching, and slamming are human behaviors, which occur even though they are contrary to existing rules.

4 CONCLUSION

PGSD FKIP students at Palangkaraya University playing a game of "seize the ball" display social behaviors such as excitement, cheerfulness, competition, cooperation, conflict, sincerity, honesty, respect, respect, and enthusiasm. Further, 11 (36.66%) out of 30 students had higher spirits compared to others.

Dangerous movements and attitudes displayed during the game include kicking, clothes-pulling, scratching, pressing, squeezing, elbowing, pushing, hair-pulling, sliding, holding the neck, crashing, nail-scratching, slamming, and knocking off-balance.

These relatively dangerous movements need to be reduced by tightening and perfecting the rules and regulations to ensure the safety of the players while maintaining the excitement of the game.

REFERENCES

Nawawie, A. H. (2009) Perilaku Sosial Mayarakat Terhadap Pekerja Anak. Disertasi. Universitas Merdeka Malang. Program Pascasarjana.

Beilharz, P. (2005) *Teori-teori Sosial, Observasi Kritis Terhadap Filsof Terkemuka.* Yogyakarta: Pustaka Pelajar.

Burhan Bungin (2003) *Analisis Data Penelitian Kualitatif.* Jakarta: PT. Raja Grafindo Persada.

Djunaidi Dhony (1997) *Dasar-Dasar Penelitian Kualitatif: Prosedur, Teknik, dan Teori Grounded disadur dari [Basics of Qualitative Research: Grounded Theory Procedures and Techniques].* Suarabaya: Bina Ilmu.

Koentjaraningrat (1982) *Seri Teori-teori Antropologi-Sosial, Sejarah Teori Antropologi.* Jakarta: UI Press.

Lauer, R. H. (2003) *Perspektif Tentang Perubahan Sosial.* Jakarta: Rieneka Cipta.

Moeleong, L. J. (2002) *Metodologi Penelitian Kualitatif.* Bandung: Remaja Rosdakarya.

Soelaeman, M. (2008). *Ilmu Sosial Dasar: Teori dan Konsep Ilmu Sosial.* Bandung: PT. Rafika Aditama.

Soerjono Soekanto (1999) *Sosiologi Suatu Pengantar.* Jakarta: Rajawali Press.

Strauss, A. and Corbin, J. (2007). Dasar-dasar Penelitian Kualitatif: Tata Langkah dan Teknik-teknik Teoritisasi Data. Yogyakarta: Pusataka Pelajar.

Sugiyono (2009) *Metode Penelitian Kuantitatif, Kualitatif dan R & D.* Jakarta: CV. Alfabeta. Available at: https://jurnalmanajemen.com/teori-hierarki-kebutuhan-maslow/.

Emerging Perspectives and Trends in Innovative Technology
for Quality Education 4.0 – Kusmawan et al (eds)
© 2020 Taylor & Francis Group, London, ISBN 978-0-367-25803-0

Assessment of practical work in the biology education program, distance education

A. Sapriati, M. Sekarwinahyu, U. Rahayu & S. Suroyo
Universitas Terbuka, Tangerang Selatan, Indonesia

ABSTRACT: This paper analyzes the results of a descriptive study of practical work carried out as part of the Biology Education Program. It was based on distance learning and showed the student's perception of the practical work component. Data were collected using document analysis sheets, observations, and questionnaires. The results of the study indicated that assessment is conducted before, during, and after practical work by instructors, lab assistants, or lecturers. The assessment of the skills and abilities was carried out using an observation sheet and document analyses form. Also, an assessment of readiness to carry out the practicum examined the purpose of lab work, the description or explanation of equipment and materials used, and the work procedures or steps. This study showed that students had positive attitudes toward the practical work component of the program and felt that all the activities involved, including writing reports, were easy, enjoyable, and useful. In general, practical work enhances science knowledge and improve skills.

Keywords: Practical Work, Assessment, Biology Education, Distance Education

1 INTRODUCTION

Practical work is an integral part of science education, especially for theoretical verification and for inquiry activities (Abrahams and Millar, 2008; Abrahams and Reiss, 2012; Almroth, 2015; Hofstein, 2017; Mamlok-Naaman and Barnea, 2012). As part of the curriculum, practical work components, or practicums, confirm and illustrate the theories, developing students' knowledge and skills (Ferreira and Morais, 2013; Katchevich, Hofstein, and Mamlok-Naaman, 2011; Sapriati, Rahayu, and Kurniawati, 2013). Without exception, the significance of practicus also applies even to in-service training biology education in the distance learning context (Shaw and Carmichael, 2010).

In supporting education activities, including distance learning, practicum implementation faces several obstacles, including type and tasks involved, conformity of the learning objective with the assessment, time adequacy and supporting facilities, and so on (Hofstein, 2004; Hofstein and Lunetta, 2003; Jordan et al., 2011; Shaw and Carmichael, 2010). For this reason, it is necessary to understand and explore the implementation of practicums and how students perceive them. A teaching-learning program needs to ensure that students gain experience and confidence in a lab (Abrahams and Reiss, 2012; Roberts and Reading, 2015). Through practical work, lecturers teach and demonstrate various abilities and skills that will be implemented in further education and daily life (Hofstein, 2015; Pellegrino, 2013). Therefore, it is vital to identify the skills that are more important and worthy of assessment.

In science, practical skills might be directly modelled, through the manipulation of real objects to demonstrate skills, or indirectly shown, with data and reports assessed (Abrahams, Reiss, and Sharpe, 2013). Evaluation of a practical task showed the analytical framework offered a means of assessing the learning demand of practical tasks. It also helped in identifying tasks that require specific support for students' thinking and learning to be effective.

Assessment is critical in students' learning experiences and is an essential part of an aligned curriculum (Biggs and Tang, 2007; Boud and Falchikov, 2006; Care et al., 2018; Guerrero-Roldán & Noguera, 2018). Science assessment uses the system approach, designed to support and monitor learning and to evaluate the effectiveness of science education (Pellegrino, 2013). The assessment should be carried out on both practical work skills and lab reports (Fadzil and Saat, 2013; Hofstein, 2004). It should evaluate understanding of work procedures and safety, practicum readiness and ability (practice skills/process performance), observational outcomes, attitudes and behavior, and results, including practicum reports (Abrahams and Reiss, 2012; Abrahams et al., 2013; Gobaw and Atagana, 2016; Sapriati et al., 2013). Teaching how to assess practical work skills is an essential component in science education, including, in biology, understanding how to understand and explore practical work and how students' perceptions and attitudes toward practical work help to achieve learning objectives. The research contributes to the literature providing information on practical work assessment on biology courses in open and distance education, as well as in presenting students' abilities, skills, attitudes, and perceptions toward biology education programs in open and distance education.

2 METHODOLOGY

This was descriptive research on the implementation of a practical work assessment component of the Biology Practicum Course at the Biology Education Program in Universitas Terbuka. The study used a purposive data collection sampling method, with a sample of 52 students, from July to October 2017.

The data was collected using document analysis sheets and questionnaires; it included student perceptions of readiness and performance and profiles of practicum learning assessments and reports. The valid data comprised 45 of 52 students and were analyzed descriptively and quantitatively. Quantitative analysis was performed by calculating the percentage of student responses.

The purpose of the study is to explore the implementation of practical work assessments and analyze how students perceive and carry out practicums. On this basis, research questions consider (a) the description of practical work biology assessment, (b) students' perceptions of practical work, and (c) how students in their practical work show self-confidence.

3 FINDINGS AND DISCUSSION

3.1 *Skills and abilities in practical work*

Assessment is conducted before, during, and after practical work (Practicum Report) by instructors, lab assistants, or lecturers. These individuals provide technical assistance to students during practical sessions. The assessment of the skills and abilities is carried out using an observation sheet and document analyses form. The study showed that assessment of readiness to carry out the practicum involves pre-tests about the purpose of lab work, description of equipment and materials used, and explanation of work procedures or steps. Importantly, students' abilities assessed in science practicums consist of readiness, manipulating and using the tools/materials, conducting experiments, improvisation, observing accurately, recording data/information, reporting the result, cleanliness, neatness, and work safety (Sapriati et al., 2013).

In this case, the score for student performance used a scale between 0 (very bad) and 4 (very good). Moreover, assessment aspects of student practical report include (a) Recording of observations, (b) Discussion, (c) Formulating conclusions appropriate to the objectives and observation, and (d) Description/explanation of answers to the given questions. The skills and abilities are assessed during preparation and before, during, and after practical work through the session reports.

The assessment involves the analysis of performance in order to demonstrate skills (to plan, implement, and deliver practical results) and knowledge (Hofstein and Mamlok-Naaman, 2007). Variables to be assessed include physical skills such as measuring, observing, experimental design; (b) thinking skills and logic, for instance, formulating conclusions and choosing the method; and (c) knowledge of science concepts and materials. Other essential elements include skills to plan, implement, and deliver practical results (Hofstein and Mamlok-Naaman, 2007). The assessment might be carried out for learning outcomes during practice and practicum products (e.g., observations, reports, or both). In more detail, it might be performed on (1) understanding of work procedures and safety; (2) students' readiness for the practicum; (3) work process during lab sessions; (4) the result of observation; (5) behavior; and (6) practical reports. According to experts, almost all practicum activities need to involve assessment of the results of work processes, observations, and reports.

3.2 Students' perception of practical work

The results showed that the students conducted hands-on labs and practical works to confirm and illustrate the theory of science and to use and present their experiences for future teaching. Additionally, the results also indicated that facilities, materials, and instructors were available during the practical sessions.

Moreover, students had positive attitudes toward practical works and assumed that the activities involved, including report-writing, were easy, enjoyable, and useful for them. Consequently, they acknowledge the fact that practical works could enhance their science knowledge and improve their practical work skills. According to students, practical work develops their understanding of scientific concepts and skills, their curiosity, and their ability to work carefully and systematically (Sapriati et al., 2013). Therefore, practical work attempts to develop students' understanding of concepts, skills, attitudes, and interest in science (Sapriati et al., 2013).

Furthermore, students felt they were confident in preparing practical work, using online and physical learning materials, conducting observations and experiments, measuring and recording data, drawing conclusions, writing reports, and carrying out lab work. However, they reported that they were less confident in working independently using the Guidelines, conducting multiple experiments, making accurate measurement, performing accurate calculation, analyzing data or information, making a hypothesis, planning the experiment, using a microscope, monitoring biological processes, calculating the concentration of a particular solution, and recognizing bacteria accurately.

Overall, however, students felt that they gained valuable experience by participating in the practicum. These experiences help them obtain additional knowledge and skills, reinforcing and strengthening biological practices. The experiences were also reported to be useful in supporting school learning and career development, providing inspiration in daily life, and improving understanding practical and personal understanding. Additionally, students gained experience in supporting the implementation of work; writing practicum reports; obtaining a satisfactory grade; and building networks, cooperation, and excellent partnerships among fellow professionals in future.

4 CONCLUSIONS

Assessments were conducted on the skills and behavior of students in terms of readiness, using tools, making observations, data/information recording and communicating, work safety, accuracy, cleanliness, and neatness. The students had positive attitudes toward practicums and felt that the activities involved, including writing reports, were easy, enjoyable, and useful. The practical work could enhance their science knowledge and improve practical work skills. Therefore, practical work helps to develop students' understanding of concepts, skills, attitudes, and interest in science.

REFERENCES

Abrahams, I. and Millar, R. (2008) Does practical work really work? A study of the effectiveness of practical work as a teaching and learning method in school science. *International Journal of Science Education*, 30(14),1945–1969. Available at: https://doi.org/10.1080/09500690701749305.

Abrahams, I., and Reiss, M. J. (2012) Practical work: Its effectiveness in primary and secondary schools in England. *Journal of Research in Science Teaching*, 49(8), 1035–1055. Available at: https://doi.org/10.1002/tea.21036.

Abrahams, I., Reiss, M. J., and Sharpe, R. M. (2013) The assessment of practical work in school science. *Studies in Science Education*, 49(2), 209–251. Available at: https://doi.org/10.1080/03057267.2013.858496.

Almroth, B. C. (2015). The importance of laboratory exercises in biology teaching: Case study in an ecotoxicology course. *Pedagogical Development And Interactive Learning*, (September), 1–11.

Biggs, J., and Tang, C. (2007). *Teaching for Quality Learning at University*. Berkshire: Open University Press.

Boud, D., & Falchikov, N. (2006) Aligning assessment with long-term learning. *Assessment & Evaluation in Higher Education*, 31(4), 399–413. Available at: https://doi.org/10.1080/02602930600679050

Care, E., Kim, H., Vista, A., and Anderson, K. (2018) *Education System Alignment for 21st Century Skills: Focus on Assessment*. Brookings: Center for Universal Education at The Brookings Institution.

Fadzil, H. M., and Saat, R. M. (2013) Phenomenographic Study of Students' Manipulative Skills during Transition from Primary to Secondary School. *Jurnal Teknologi*, 63(2), 71–75. Available at: https://doi.org/10.11113/jt.v63.2013

Ferreira, S., and Morais, A. M. (2013) Conceptual demand of practical work in science curricula. *Research in Science Education*, 44(1), 53–80. Available at: https://doi.org/10.1007/s11165-013-9377-7.

Gobaw, G. F., and Atagana, H. I. (2016) Assessing laboratory skills performance in undergraduate biology students. *Academic Journal of Interdisciplinary Studies*, 5(3), 113–122. Available at: https://doi.org/10.5901/ajis.2016.v5n3p113

Guerrero-Roldán, A.-E., and Noguera, I. (2018) A model for aligning assessment with competences and learning activities in online courses. *The Internet and Higher Education*, 38, 36–46. Available at: https://doi.org/10.1016/j.iheduc.2018.04.005.

Hofstein, A. (2004) The laboratory in chemistry education: Thirty years of experience with developments, implementation, and research. *Chemistry Education Research and Practice*, 5(3), 247–264. Available at: https://doi.org/10.1039/B4RP90027H.

Hofstein, A. (2015) The development of high-order learning skills in high school chemistry laboratory: Skills for life. In J. García-Martínez and E. Serrano-Torregrosa (eds.), *Chemistry Education*. New York: Wiley, pp. 517–538. Available at: https://doi.org/10.1002/9783527679300.ch21.

Hofstein, A. (2017) The role of laboratory in science teaching and learning. In K. S. Taber and B. Akpan (eds.), *Science Education*. Rotterdam: Sense, pp. 357–368. Available at: https://doi.org/10.1007/978-94-6300-749-8_26.

Hofstein, A., and Lunetta, V. N. (2003) The laboratory in science education: Foundations for the twenty-first century. *Science Education*, 88(1), 28–54. Available at: https://doi.org/10.1002/sce.10106.

Hofstein, A., and Mamlok-Naaman, R. (2007) The laboratory in science education: The state of the art. *Chemistry Education Research and Practise*, 8(2), 105–107. Available at: https://doi.org/10.1039/b7rp90003a

Jordan, R. C., Ruibal-Villasenor, M., Hmelo-Silver, C. E., and Etkina, E. (2011) Laboratory materials: Affordances or constraints? *Journal of Research in Science Teaching*, 48(9), 1010–1025. Available at: https://doi.org/10.1002/tea.20418.

Katchevich, D., Hofstein, A., and Mamlok-Naaman, R. (2011) Argumentation in the chemistry laboratory: Inquiry and confirmatory experiments. *Research in Science Education*, 43(1), 317–345. Available at: https://doi.org/10.1007/s11165-011-9267-9.

Mamlok-Naaman, R., and Barnea, N. (2012) Laboratory activities in Israel. *EURASIA Journal of Mathematics, Science and Technology Education*, 8(1), 49–57. Available at: https://doi.org/10.12973/eurasia.2012.816a.

Pellegrino, J. W. (2013) Proficiency in science: Assessment challenges and opportunities. *Science*, 340(6130), 320–323. Available at: https://doi.org/10.1126/science.1232065.

Roberts, R., and Reading, C. (2015) The practical work challenge: Incorporating the explicit teaching of evidence in subject content. *School Science Review*, (357).

Sapriati, A., Rahayu, U., and Kurniawati, Y. (2013). Implementation of Science Practical Work at Faculty of Teacher Training and Educational Science, Universitas Terbuka, Indonesia. International Conference on Education and Language (ICEL), 2.

Shaw, L., and Carmichael, R. 2010. (2010). Needs, costs, and accessibility of the science lab programs. In D. Kennepohl and L. Shaw (eds.), *Accessible elements: teaching science online and at a distance*. Edmonton: AU Press.

Emerging Perspectives and Trends in Innovative Technology
for Quality Education 4.0 – Kusmawan et al (eds)
© 2020 Taylor & Francis Group, London, ISBN 978-0-367-25803-0

Are digital books in android applications and web-based virtual books effective for distance learning? A lesson from Universitas Terbuka Indonesia

Yasir Riady
Universitas Terbuka, Tangerang Selatan, Indonesia

ABSTRACT: In order to provide greater ease of access, better facilities, and a more user-oriented experience for tutorials, registrations, and examination, Universitas Terbuka (an Indonesian distance learning university) creates learning material based on Android device applications as well as a virtual web-based reading room. These provide a new platform in the distance education system and are in line with increasing usage of smartphones in Indonesia, not only for communication but also for learning purposes. As of the beginning of 2017, more than 10,000 users had downloaded the application. This article aims to explore the implementation of Android applications in digital teaching material by evaluating the platform's effectiveness and efficiency through six measurements: system quality, number of users, purposes, simplicity, user-orientedness of application, and time of use. A sample of 55 respondents was employed, using a questionnaire and interviews and applying a descriptive conceptual approach. It concludes that the application of Android-based digital teaching materials really helps students at the Universitas Terbuka in Indonesia to better understand materials in the areas of online learning, face-to-face tutorials, and also final examinations.

Keywords: Android, eBook, Access, Application

1 INTRODUCTION

Currently there are many applications developed by universities such as Open Universities in Indonesia, allowing easy access, greater facilities, and more needs-oriented approaches to tutorials, registration, and examinations. In general, applications are developed to assist students in accessing facilities and services in completing their academic tasks. The digital transformation supported by the internet and computer technology makes things easier to comprehend and achieve and also increases competition among users and institutions (Wilson-Higgins, 2018).

One of the applications developed by the Universitas Terbuka is Digital Books, launched on January 30, 2017 (Universitas Terbuka, 2017) and accessed via Android devices. Bahan Ajar Digital (Digital Printed Material or BA Digital) is adapted from the UT Basic Printed Materials (BMP) that had previously been the main source material for the students.

This study discusses the details of the digital books application developed by the Universitas Terbuka, focusing on the activities and guidelines and the use of Android applications based on aspects of usability, effectiveness, and efficiency (Universitas Terbuka, 2017).

Figure 1. Home page of Bahan Ajar Digital UT/UT Digital Library, in Android (play.google.com).

Along with the Android application, Universitas Terbuka offers the Ruang Baca Virtual or Virtual Reading Room, which can be seen by students who access the website. The Digital Library app and the Virtual Reading Room offer a variety of information services and objects that support access to information through digital devices (Sismanto, 2008). This service is expected to facilitate the search for information in collections of information objects such as books, documents, images, and databases in digital format quickly, precisely, and accurately. The Digital Library does not stand alone but is linked to other sources, and its information services are open to users around the world.

STATISTIK HIT WEB UPT PERPUSTAKAAN

Statistik	2012	2013	2014	2015	2016	2017	2018	2019
Ruang Baca Virtual								
Januari	1382	9968	26761	26000	32756	32182	86443	46010
Februari	11699	16352	41779	40057	51744	61041	163398	79490
Maret	23318	28794	4438	91716	94265	143177	403894	184171
April	10809	26305		66032	68458	103844	263247	608241
Mei	12012	30050		43919	73115	97172	260773	345290
Juni	2359	8965		15414	12138	17022	36018	217365
Juli	3837	19691		12537	20661	76213	49011	53098
Agustus	6390	27660		48814	74136	232906	85193	77386
September	25189	68778		91311	91079	487557	192227	
Oktober	18835	60551		87359	101587	379472	692996	
Nopember	15222	44566		53017	46443	146043	415908	
Desember	4637	15379		16947	15778	40079	285779	
	135,589	355,859	72,978	593,123	682,160	1,796,708	2,534,786	1,510,051

Figure 2. Ruang Baca virtual hits in the last 10 years.

2 DIGITAL BOOKS FOR LEARNING MATERIAL IN ANDROID

Digital books are one of the most important parts of the learning process. It would be a great thing if institutions offered electronic books to support every student in accessing materials (Park, Kim, and Lim, 2019).

Widodo and Jasmadi (in Lestari, 2013) state that teaching materials are a set of learning tools or tools systematically designed and made interesting in order to achieve expected goals of competence.

3 CONCLUSIONS

Various digital applications and online services developed by universities such as Open Universities in Indonesia offer students easy access, better facilities, and more needs-oriented approaches to tutorials, registration, and examinations. Applications and facilities developed are very helpful for students in completing their academic tasks. Evaluation of the Android-based material shows user numbers exceeding 10,000 just four months after the January 2017luanch. (Universitas Terbuka, 2017). Use of the Virtual Reading Rooms (RBV) likewise increased over the past 10 years (with more than 20,000 users on average each month).

Students at the Universitas Terbuka find the application of Android-based digital teaching materials aids in their learning of material, as shown by the number of downloads (Chau and Jung, 2018). It is expected that the program will continue to expand, so that Apple iOS users will also be able to access the Digital Teaching Facilities at the Universitas Terbuka in future.

REFERENCES

Android (2018) Available at: www.android.com.about/.

Chau, N.-T. and Jung, S. (2018) Dynamic analysis with Android container: Challenges and opportunities. *Digital Investigation*, 27, 38–46. Available at: https://doi.org/10.1016/j.diin.2018.09.007.

International Organization for Standardization (2018) Available at https://www.iso.org.

Mulyasa (2006). Menjadi Guru Profesional Menciptakan Pembelajaran.

Park, J. H., Kim, H.-Y., and Lim, S.-B. (2019) Development of an electronic book accessibility standard for physically challenged individuals and deduction of a production guideline. *Computer Standards & Interfaces*, 64, 78–84. Available at: https://doi.org/10.1016/j.csi.2018.12.004.

Pendit, P. L. (ed.) (2007) *Perpustakaan Digital: Sebuah Impian dan Kerja Bersama*. Jakarta: Sagung Seto.

Razak, M. F. A., Anuar, N. B., Salleh, R., Firdaus, A., Faiz, M., and Alamri, H. S. (2019) "Less Gives More": Evaluate and zoning Android applications. *Measurement*, 133, 396–411. https://doi.org/10.1016/j.measurement.2018.10.034.

Riady, Y. (2014) Assisted learning through Facebook: A case study of Universitas Terbuka's students group communities in Jakarta, Taiwan and Hong Kong. *Turkish Online Journal of Distance Education*, 15(2),227–238. Available at: https://doi.org/10.17718/tojde.71656.

Zhang, C., & Li, L. (2006) Digital teaching reference book service: A case study on knowledge-object-based microstructure of digital resources. *The International Information & Library Review*, 38 (3),110–116. https://doi.org/10.1016/j.iilr.2006.06.002.

Emerging Perspectives and Trends in Innovative Technology for Quality Education 4.0 – Kusmawan et al (eds)
© 2020 Taylor & Francis Group, London, ISBN 978-0-367-25803-0

The effect of learning style and motivation on students' learning achievement at UPBJJ-UT Makassar (study on undergraduate public administration students)

Jamil & Nuraziza Aliah
Universitas Terbuka, Tangerang Selatan, Indonesia

ABSTRACT: Distance learning at the Open University calls for the adoption of an appropriate learning method, pattern, and style from students. This adoption assists them in navigating toward tutorial sessions (e.g., face-to-face and online tutorials) and leverage online learning resources (e.g., Independent Online Practices), which holds the key to an optimal learning outcome. Capturing the notion of students' preference to learning style and their motivation to activate the learning style can be best used to demonstrate the attainment of that outcome. This study is survey research that seeks to measure the effect of learning style and motivation on learning achievement at the Open University of Makassar. Multiple regression modeling was adopted to analyze the data with a population of undergraduate students of the Public Administration program in East Luwu Regency. The results were favorable; learning style and student motivation had a positive and significant effect on learning achievement, both simultaneously and partially.

Keywords: Learning Style, Learning Motivation, Learning Achievement

1 INTRODUCTION

The notion of learning style speaks to the understanding that each individual student learns in a different manner, and in most cases, students do not know how to identify their own learning preferences. Learning style entails the combination of absorbing, organizing, and processing information DePorter and Hernacki (2005). In the distance learning environment, students are called upon to demonstrate independent learning by processing, absorbing, and regulating their own understanding of Subject-Matter Book (Buku Materi Pokok), both print and online form.

The core tenet of independent learning lies in students' ability to showcase meaningful individual understanding. Independent learning calls for the skill of reading fluency (speed reading) to grasp reading better and faster. To optimize it, students need the motivation to drive such learning behavior and show how self-regulated learning can be activated to enhance mastery in a particular subject matter. The learning achievement among the students of Public Administration in East Luwu can set the direction toward the understanding of learning style and motivation in a way that can be implemented in a self-regulated setting.

While it is common that students have a particular way in which they learn best, they may fit into a mix of learning style categories. Nevertheless, some students may find a dominant style of learning, with far less use of the other styles. Some may find auditory learning to fit them better than visual learning, while others learn best by a hands-on approach to learning. This study looks at these dominant styles among the students of Public Administration in East Luwu across their learning trajectories in a distance environment and measures their effect on how these students achieve optimal learning.

2 LITERATURE REVIEW

2.1 *Learning style*

Gunawan (2004) defines learning style as the preferential way in which learners process and understand information. Learners who recognize a better way of how self-regulated learning is approached are able to identify the underlying dynamics of academic success. Distinguishing the different learning styles will enable them to understand learning with a particular style that suits them best, which ultimately guides them to optimal learning.

DePorter (2009) breaks down learning styles into three common categories, i.e., visual, auditory, and kinesthetic styles. The indicators for visual learning are organized; learners best remember colors and images, read and spell words without distraction, and understand artistic work. They are typically weak listeners but tend to observe teacher's attitudes and movements, including lip movements during teaching. Auditory learners get through active engagement with sounds but are easily distracted by them. They are good at remembering what they hear as they learn, are typically talkative, and work best with verbal instructions. Kinesthetic learning deals with the use of visual cues such as body language and movement, and physical contact to retain information and remember it. Kinesthetic learners often speak slowly, accessing their feeling as they talk. They absorb information best using a sense of touch by doing, experiencing, moving, and being active in some way.

2.2 *Motivation for self-study*

Motivation in learning initiates and guides the process to maintain goal-oriented behavior in learning. Key characteristics in motivation include decision-making responsibility, risk-taking skills, meaningful work completion with desirable results, and passion for gaining mastery in one's field (Schultheiss, Campbell, & McClelland, 1999). Learning is an activity that brings about behavioral and mental changes in an individual. These changes occur because there is an effort from within each individual and are relatively permanent as a form of response to the situation or as a result of experience and interaction with the environment.

Self-study, or self-regulated learning, manifest from the tendency to learn without direct supervision or in-class attendance. Barnadib highlights the notion of self-study behavior in the following characteristics: 1) taking the initiative, 2) problem-solving skills, 3) a strong sense of self-confidence, 4) the ability to do something with minimum assistance from others, and 5) competitiveness.

When it comes to motivation for self-study practices, learners motivate themselves to take ownership of learning. These learners actively process information while capitalizing on wide-ranging resources available to them in order to obtain new knowledge and skills in a self-regulated way. Motivation, in this sense, is part of students' independence to take on self-management and self-monitoring of their own learning. Looking at one's motivation in self-study is hence observing one's willingness to meet the demands of academic workload as an effort to meet the needs and requirements to achieve optimal learning.

2.3 *Learning achievement*

Zainal Arifin (2009) defines the clear distinction between learning achievement and learning outcome; while the former concerns with the aspects of knowledge, the latter deals with the formation of students' characters. Learning achievement reflects on the result of individual student performance after experiencing learning processes within a certain period. On this basis, measuring learning achievement means assessing the actual skills obtained by an individual student after learning. Learning achievement involves a process in which proof of an individual learning milestone is recorded.

3 RESEARCH METHODOLOGY

Multiple regression analysis was adopted to measure the effect of learning style, and motivation on learning achievement within a population of 40 undergraduate students of Public Administration in East Luwu Regency sampled using a saturation point. The study was conducted in 2018 in Tomoni District, East Luwu. Data collection techniques included observation, interview, and questionnaire.

4 RESULT AND DISCUSSION

4.1 *Results*

4.1.1 *Coefficient of determination*

Table 1. Calculation output of correlation and determination coefficient.

Model Summary

Model	R	R Square	Adjusted R	Std Error of the Estimate
1	.727[a]	.528	.503	.830

a Predictors: (Constant), motivas belajar (X2), gaya belajar (X1)

The resulting R is 0.72, indicating that learning style (X1) and motivation (X2) are strongly correlated with learning achievement. R2 of 0.52 represents 52% of the total effect of both learning style and motivation on learning achievement. The remaining 48% accounts for unknown variables.

4.1.2 *Simultaneous test*

Table 2. Calculation output of f-test.

ANOVA[b]

Model	Sum of Square	Df	Mean Square	F	sig
1 Regression	28.507	2	14.254	20.708	.000[a]
Residual	25.468	37	.688		
Total	53.975	39			

a Predictors (Constant), motivasi belajar (X2), gaya belajar (X1)
b Dependent Variable prestasi belajar (Y1)

The resulting F value of 20.708 with sig 0.00 validates the positive and significant effect of learning style (X1) and motivation (X2) on learning achievement.

Table 3. Calculation output of t-test.

Coefficients[a]

Model	Unstandardized coefficients		Standardized coefficients		
	B	Std Error	Beta	1	Sig
1 (Constant)	7.203	2.178	.704	3.307	.002
gaya belajar (X1)	.426	.069	.329	6.153	.000
motivasi belajar (X2)	.342	.119		2.874	.007

a Dependent Variable: Prestasi belajar (Y1)

T value for X1 is 6.15 at a significance of 0.00, indicating a positive and significant effect of learning style on learning achievement. Similarly, the T value for X2 is 2.87 at a significance of 0.007, indicating a positive and significant effect of motivation on learning achievement. The output from Table 3 settles down to the following model:

$$Y = 7.203 + 0.426 \, X1 + 0.342 \, X2$$

The model further confirms that the partial effect of learning style and motivation is positive and significant on learning achievement, ultimately suggesting that when X increases, Y increases. The statistical significance simply indicates that the improvement of learning style brings about the improvement of learning achievement; and that the improvement of motivation brings about the improvement of learning achievement. Between the two Xs, the learning style turns out to demonstrate a more dominant effect on learning achievement with a coefficient magnitude of 0.426.

5 DISCUSSIONS

As results have indicated, the preference the students exhibit in the learning process, be it visual, auditory, or kinesthetic preference, was statistically significant in the learning achievement in Public Administration in East Luwu. The same significance applies between motivation and learning achievement. When these students are able to recognize the difference between these preferences, adjust their learning methods, and activate self-motivation, increased achievement across their academic trajectories is not too far behind. Being aware of these learning styles and initiating self-motivation gives a tremendous advantage in the process of self-relatedness in learning.

This finding is markedly consistent with that of Baker et al. (2003), who identified better achievement and academic satisfaction among students who incorporated specific styles into learning. In a similar sense, Geiger (1992) observed a group of students who performed better with a particular learning style during a lecture of a particular course and found a set of relevant evidence that attested to higher student satisfaction.

It is worth noting, however, that learning styles are not really concerned with what these students learn, but rather how they learn it. The idea of utilizing a particular learning style does not necessarily justify the claim about making students universally smarter; instead, this particular style enables them to work better and more effectively on academic tasks. Due to different strengths and preferences, some prefer to work with an auditory mode; some work better with visual skills, while others are more comfortable with kinesthetic experience. In some cases, these students are less consistent with one particular style, are therefore not exclusively visual students, auditory students, or kinesthetic students, and eventually accommodate the amalgamation of different learning modes.

6 CONCLUSIONS

When jointly assessed, the overall significance of learning style and motivation was positive and strong. The indication is obvious; when self-awareness about learning styles and motivation to take on a learning strategy that best fits students' strengths improves, the development of learning achievement also improves. Similar significance applies in the partial assessment of learning style and motivation in relation to learning achievement.

REFERENCES

Arylien Ludji Bire(1*), Josua Bire(2), Jurnal Kependidikan PPs Universitas Nusa Cendana.

Baker, J. A., Dilly, L. J., Aupperlee, J. L., & Patil, S. A. (2003). The developmental context of school satisfaction: Schools as psychologically healthy environments. School Psychology Quarterly, 18(2), 206.

De Porter, B., & Hemacki, M. (2005). Quantung Teaching, Mempraktekkan Quantum Learning Membiasakan Belajar Nyaman dan Menyenangkan (A. Abdurraman, trans.). Bandung: Kaifa PT. Misan Pustaka.

De Potter, B. (2009). Quantum Teaching, Mempraktekkan Quantum Learning di Ruang Kelas. Bandung: Kaifa.

Geiger, M. A. (1992). Learning styles of introductory accounting students: An extension to course performance and satisfaction. The Accounting Educators' Journal, 4(1), 22.

Gunawan. (2004). Genius Learning Strategy. Jakarta: PT. Gramedia Pustaka Utama.

Schultheiss, O. C., Campbell, K. L., & McClelland, D. C. (1999). Implicit power motivation moderates men's testosterone responses to imagined and real dominance success. Hormones and Behavior, 36(3), 234–241.

Zainal, A. (2009). Evaluasi Pembelajaran. Bandung: PT Remaja Rosdakarya.

Emerging Perspectives and Trends in Innovative Technology for Quality Education 4.0 – Kusmawan et al (eds)
© 2020 Taylor & Francis Group, London, ISBN 978-0-367-25803-0

Bubbl.us: A digital mind-mapping tool to promote a mobile-based technology approach in writing courses

Rafidah Abd Karim, Mohd Haniff Mohd Tahir & Airil Haimi Mohd Adnan
Academy of Language Studies, Universiti Teknologi Mara Perak Branch, Tapah Campus, Perak, Malaysia

Noorzaina Idris
Faculty of Education, Universiti Teknologi MARA, Puncak Alam Branch, Selangor, Malaysia

Izwah Ismail
Electrical Engineering Department, Politeknik Ungku Omar, Ipoh, Perak, Malaysia

Abdul Ghani Abu
Faculty of Languages and Communication, Universiti Pendidikan Sultan Idris, Tg. Malim Perak, Malaysia

ABSTRACT: The transformation of technology following the development of mobile devices has given rise to significant changes in today's educational system and daily life. To take advantage of these changes, this study proposes the digital mind-mapping tool Bubbl.us for improving undergraduates' writing abilities via a mobile-based technology approach in a public university. The instrument used for the collection of data was a survey questionnaire. The study aims to investigate students' attitudes pertaining to the use of this digital mind-mapping tool and how it helped them to increase their interests in writing via a mobile-based technology approach. The results revealed that study participants had positive attitudes toward the use of the digital mind-mapping tool via mobile devices. It could be concluded that the use of the digital mind mapping tool among ESL undergraduates should be encouraged, as it helps to motivate undergraduates and create a positive attitude toward writing in English.

1 INTRODUCTION

ESL (English as a second language) or EFL (English as a foreign language) students have face numerous difficulties when writing in English and composing essays (Karim, Abdul, and Khaja, 2017). Writing is a rudimentary expertise that should be aced by all students in the Malaysian English Language educational program (Ministry of Education Malaysia, 2000). Thus, the mind-mapping technique is one of the ways that students can brainstorm ideas. There are several mind-mapping application tools for brainstorming such as MindMup, Coggle, Bubbl.us, Freemind, Milanote, and Popplet, accessible by mobile phones and computers. For this study, the researcher selected and used Bubbl.us, which can help students generate and employ mind maps with colors and sub-headings. According to Daley and Torre (2010), the utilization of computerized mapping programming is more effective than traditional pen and paper techniques. This idea mapping is therefore viable as a learning instrument. This study investigates undergraduates' attitudes toward the use of this digital-mind mapping tool via a mobile-based technology approach and how it helped them to increase their interests in writing. Thus, the research question investigated is what the attitudes of ESL undergraduates are toward using digital mind mapping tool via mobile devices in writing courses.

2 REVIEW OF RELATED LITERATURE

Most past investigations looked at the utilization of computerized mind mapping in training and in defining objectives. The mind maps help to organize and memorize information and prepare to create questions for essays. This technique has the potential to exploit our brain power (Buzan and Buzan, 1996). Hence, it is more useful and effective compared to the traditional approach among students and teachers.

Adam and Mowers (2007) found that visual learners had a 40% higher memory rate compared to verbal learners in writing tasks. This strategy provides several benefits for students and teachers in writing. The impact of technological advancement in education is enormous, which advances the knowledge and the development of teaching and learning environments. Accordingly, Dominic (2014) suggested that the students can generate a mind map using the software, which enables them to manipulate and use colors in creating mind maps. Most students use mobile devices in their learning environment, as they find the devices areeasy to access and very flexible. In another study, Kulska-Hulme (2013) examines Mobile-Assisted Language Learning (MALL), or having learners use mobile devices in language learning.

3 METHODOLOGY

For this study, the quantitative data of respondents were collected through a questionnaire. The questionnaires were disseminated investigate undergraduates' attitudes toward use of digital mind-mapping tools via mobile-based technology and how it helped them to increase their interest in writing.

3.1 *Respondents of the study*

The selected respondents of the small-scale study were 44 ESL undergraduates enrolled in a writing course entitled "Integrated Language Skills III: Writing." The students used mind maps for 2 hours per week for 10 sessions in the last semester. Their ages ranged from 18 to 23, and the group included male and female students.

3.2 *The instrument of the study*

The reviewed questionnaire, derived from from Lee's (2010) questionnaire, was composed of 20 items relating to the use of digital mind mapping via mobile-based devices in undergraduates' writing activities. The items aimed to investigate attitudes toward digital mind-mapping activity, perceptions of the usefulness of digital mind-mapping tools, experience with digital mind-mapping tools, and the experience of digital mind-mapping activities using mobile devices. A rating system was used to determine positive and negative attitudes of respondents. A U response depicts a neutral stand. The questionnaire is based on a five-point scale ranging from strongly agree (5) to strongly disagree (1).

3.3 *Research procedure and data analysis*

The researchers introduced digital mind mapping using Bubbl.us to the respondents. After the writing program, the respondents completed the questionnaire. After the information from the questionnaires were collected, validity was determined by completion of every one of the survey's questions. Toward the end, the results were introduced as a table. The study was piloted during weekly two-hour writing class lessons. For the four-week writing program, the researchers selected several topics on expository essays for undergraduates to write in the class. The topics, including "The importance of healthy diet," "How to reduce stress in daily life," "The impact of social media on youths," and "Similarities and differences between school and university," were selected based on the undergraduates' consistent familiarities.

The majority of the respondents were approached to compose their essay, as indicated by their mind maps, in 200 words and 2 hours' time.

4 RESULTS AND DISCUSSIONS

In this study, 20 items investigates undergraduates' attitudes toward the use of the digital mind-mapping tool and whether it increased their interest in writing.

In Table 1, five items investigate undergraduates' attitudes toward digital mind-mapping activity. Most of the respondents showed positive attitudes: more than 70% stated that they were interested in and enjoyed using Bubbl.us. Most of them mentioned that the digital mind map is not a waste of time, but half, in Item 3, expressed that this tool is a time-consuming task. Table 2 illustrates that respondents had positive perceptions of the usefulness of the digital mind-mapping tool. Generally, more than 70% of the respondents stated that this tool motivated them and helped them to in the writing activity. Moreover, this tool provided them with color and pictures (Item 1) to help them memorize the information better.

In Table 3, the five items of the construct showed that the majority of respondents had positive experiences toward this tool. Most of them agreed that they enjoyed using this tool and that they did not feel bored (Item 4) during the sessions. More than 50% of the respondents indicated that the technique had improved their quality of writing ideas (Item 5).

As seen in Table 4, most of the respondents indicated that they had positive experiences creating digital mind maps using their mobile devices. More than 50% of respondents said they enjoyed using this technique as it made learning easier and more interesting (Item 2). They also revealed that using a mobile device to create digital mind maps helped them to enhance their mobile learning skills.

Table 1. Attitude toward digital mind mapping activity.

Statements	Percentage				
	SA	A	N	D	SD
1 I was interested in the digital mind-mapping app, Bubbl.us, as it was discussed in class.	20.5	63.6	13.6	2.3	2.3
2 I enjoyed drawing digital mind-mapping using *Bubbl.us* to help me create ideas.	20.5	54.5	22.7	0	2.3
3 Creating the digital mind map is a time-consuming task.	11.4	38.6	40.9	6.8	2.3
4 Creating digital mind maps is a waste of time.	0	13.6	20.5	52.3	18.2
5 It makes no difference whether I use the digital mind maps or not.	2.3	27.3	25.0	34.1	11.4

Table 2. Perception of the usefulness of the digital mind-mapping tool.

Statements	Percentage				
	SA	A	N	D	SD
1 Using colors and pictures in the mind map helps me remember the information better.	22.7	52.3	20.5	2.3	2.3
2 I think that digital mind-mapping helped me organize my ideas in writing.	13.6	56.8	22.7	4.5	2.3
3 Creating digital mind maps enhances my motivation to learn writing.	4.5	65.9	25.0	4.5	2.3
4 Using digital mind maps helped me to identify the main ideas and the sub-ideas in a more attractive way.	11.4	68.2	15.9	2.3	2.3
5 The digital mind maps I created helped me organize the information.	6.8	72.7	15.9	4.5	2.3

Table 3. Experience toward digital mind mapping tool.

Statements	Percentage				
	SA	A	N	D	SD
1 While creating the digital mind-mapping using Bubbl.us, I felt the time passed very quickly because I enjoyed creating the maps.	14.0	46.5	30.2	7.0	2.3
2 Creating digital mind maps ensures a relaxed and stress-free atmosphere.	13.6	61.4	20.5	2.3	2.3
3 I found creating digital mind maps were boring and difficult.	4.5	15.9	22.7	45.5	11.4
4 It made me feel tired and bored to use colors and images in creating digital mind maps.	2.3	13.6	18.2	50.0	15.9
5 I felt that digital mind mapping improved the quality of my writing ideas.	4.5	47.7	40.9	4.5	2.3

Table 4. Experience toward digital mind-mapping activity using mobile devices.

Statements	Percentage				
	SA	A	N	D	SD
1 Creating digital mind maps using the mobile device improves my writing skills.	2.3	47.7	45.5	2.3	2.3
2 Using colors, nodes, and links while creating the digital mind maps using the mobile device makes learning easier and more interesting.	18.2	59.1	18.2	2.3	2.3
3 I enjoy using a mobile device to create my digital mind map.	9.1	50.0	29.5	11.4	2.3
4 It would help me to become more skillful at using mobile learning.	9.3	58.1	25.6	4.7	2.3
5 Creating digital mind maps using the mobile device was a boring experience.	0	15.9	18.2	47.7	18.2

5 CONCLUSION

It can be concluded that the undergraduates had positive attitudes and perspectives toward digital mapping tools via mobile devices in the writing class. They believe that the use of digital mind-mapping in writing classrooms makes the lesson more interesting, more exciting, easier, and more effective. Therefore, it is recommended that digital mind-mapping via mobile-based technology should be encouraged more to enhance students' writing, especially in ESL classrooms. Alternatively, further research replicating this investigation ought to be completed to more powerfully decide the legitimacy of such a relationship.

REFERENCES

Adam, A., and Mowers, H. (2007) Get inside their heads with mind mapping. *School Library Journal*, 53(3), 24.

Buzan, T. & Buzan, B. (1996) *The Mind Map Book: How to Use Radiant Thinking to Maximize Your Brain's Untapped Potential*. New York: Dutton, pp. 59, 166, 224, 225, 229.

Daley, B. J., Torre, D. M. (2010) Concept maps in medical education: An analytical literature review. Med Education, 44, 440–448. doi:10.1111/j.1365-2923.2010.03628.x.

Dominic, S. (2014) Mind mapping using semantic technologies. Diploma thesis. Available at: http://dip. felk.cvut.cz/browse/pdfcache/salaidom_2014dipl.pdf.

Karim, R. A., Abu, A. G., and Khaja, F. N. M. (2017) Theoretical perspectives and practices of mobile-assisted language learning and mind-mapping in the teaching of writing in ESL classrooms. *Journal of English Teaching*, 2(1),1–12.

Kukulska-Hulme, A. (2013) *Re-skilling Language Learners for a Mobile World*. Monterey, CA: The International Research Foundation for English Language Education. Available at: http://www/tirfon line.org/wpcontent/uploads/2013/11TIRFMALL_Paper_Kulkuska-Hulme.pdf.

Lee, Y. (2010) Concept mapping strategy to facilitate foreign language writing: A Korean application. (Order No. 3429061, Syracuse University) ProQuest Dissertations and Theses, 201.

Ministry of Education Malaysia. (2000) *Sukatan pelajaran Kurikulum Bersepadu Sekolah Menengah Bahasa Inggeris*. Kuala Lumpur: Kementerian Pendidikan Malaysia.

Emerging Perspectives and Trends in Innovative Technology
for Quality Education 4.0 – Kusmawan et al (eds)
© 2020 Taylor & Francis Group, London, ISBN 978-0-367-25803-0

Effects of constructivist learning on inductive reasoning ability and attitude towards mathematics

Yumiati
Universitas Terbuka, Tangerang Selatan, Indonesia

Saleh Haji
Universitas Bengkulu, Indonesia

ABSTRACT: The purpose of this study was to determine the effects of constructivist learning based on inductive reasoning ability and students' attitude towards Mathematics. Quasi-experimental methods were used with a pretest-posttest control group design. It involved 71 students from class VII at SMPN1 Parung and SMP IT Jabon Mekar. The techniques of data collection included questionnaires and tests of learning outcomes on inductive reasoning. The data collected was analyzed qualitatively and quantitatively. According to the results, constructivist learning could enhance positive effects on the students' attitude towards Mathematics, as well as their inductive reasoning ability. This was proven by data int-table 1.67 being greater than t-table 1.39 with $\alpha = 0.05$, DB = 70, n = 71 and s = 0.72.

1   INTRODUCTION

According to Hers (1997), Inductive and deductive reasoning are aspects developed in learning mathematics since cognitive and analysis are closely related to arithmetic. However, students' reasoning ability in mathematics learning is still a concern because most of them have a low ability in this regard, and cannot make conclusions from various individual cases (inductive reasoning). Besides, several students cannot implement a particular case from a general statement (deductive reasoning).

From Liew, Grishman, Hayes (2008), inductive and deductive reasoning is still a problem for students. This is evident in the learning outcomes of mathematics at SMPN 1 Parung and SMPIT Jabon Parung in Table 1 below.

Several teachers have made efforts to overcome these problems, especially inductive reasoning, including having extra mathematics lessons. However, these efforts have not been in line with the expectations. Also, researchers have made various efforts to overcome the low level of inductive reasoning ability among students. Brigh & Feeney (2014) developed inductive reasoning ability following explanations to the causes of problems. Moreover, Christou & Papageorgiou (2007) established inductive mathematical reasoning through curriculum and assessment tools, while Schwenzer & Mathiak (2012) connected it to other aspects.

Apart from weak inductive reasoning ability, several students are less interested in mathematics, with most of them claiming not to be good at it (Haji, 2012), (Colomeischi, Colomeischi, 2015). Mathematics is presented in an abstract form, containing various symbols and formulas. Besides, learning is passive, and students only receive the materials from the teacher, obey and follow the directions given.

This study develops a constructivist learning approach to overcome the problems of inductive reasoning ability and low attitude of the students toward mathematics. Paynea, Stephensonb, Morrisb, Tempestb, Milehams, Griffinb (2009) explained that constructivist learning could overcome various challenges, such as weaknesses in inductive reasoning. Generally, the reasoning is

Table 1. Average report scores for mathematics subject.

Average Mathematics Score	SMPN 1 Parung		SMPIT Jabon Mekar Parung	
	Experimental Class (Class 8-2)	Control Class (Class 8-2)	Experimental Class (Class 8-B)	Control Class (Class 8-A)
Semester 1	68.5	68.2	65.0	72.2
Semester 2	69.5	67.4	74.5	75.3
Semester 3	70.3	70.1	61.9	62.4

part of the standard process of learning mathematics and adds to other abilities such as problem-solving, proof, communication, connections, and representation (NCTM, 2000).

Indisputably, constructivist learning is related to reasoning (mathematics). According to Watson & Mason (2005), mathematics is a constructive activity. Akyol & Fer (2010) established that constructivist learning is useful in developing the reasoning ability of students of the fifth grade in elementary school. Also, teachers believe that constructivist learning improves students' performance in mathematics (Hursen & Soykara, 2012). It helps students develop perceptions, attitudes, and beliefs, which is critical in education (Ozkal, Tekkaya, Cakiroglu, Sungur, 2008).

Constructivist learning includes a new approach to erudition. It motivates students to utilize their ability to understand and solve a problem using their approaches. Gursesa, Dogarb, Gunesa (2015) stated that teaching based on the constructivist theory is a new approach to learning.

The purpose of this study is to determine the effects of constructivist learning on inductive reasoning ability and the attitude of students towards mathematics at SMP Negeri 1 and SMP Islam Terpadu Jabon Mekar Parung Bogor?

2 LITERATURE REVIEW

2.1 *Inductive reasoning ability and attitude of the students towards mathematics*

According to Suriasumantri (2007), the reasoning is a thought process involving knowledge. Drawing conclusions from a reasoning process might be carried out through in two ways, including 1. Deductions from individual cases, which is referred to as inductive reasoning, and 2. Making conclusions from daily events leading unique things, which is called deductive reasoning. Mathematics knowledge is compiled deductively (Suriasumantri, 2007). For teaching purposes, the materials used might be understood through inductive reasoning. The students easily understand a mathematical concept from special cases to the general things since their mental development is in concrete operation. According to Ruseffendi (1991), at the stage of concrete operations, the students provide inductive reasons, which is very important for mathematics learning. According to Suriasumantri (2007), induction is a way of thinking in which a general conclusion is drawn from several individual cases. Likewise, Bakhtiar (2009) stated that inductive reasoning begins by expressing statements with a specific and limited scope to compile arguments that end with general statements. The conclusions obtained are certain in case the premises used are true and might be fully trusted (Suriasumantri, 2007). Kisworo & Sofana (2017) provided examples of inductive reasoning as follows.

There are some special facts, such as 1. Plants die 2. Animals die, and 3. Humans die. Based on these facts, a general conclusion is obtained, and that is all living things will die.

Inductive reasoning in mathematics learning in the explanation of the topic "Integer Count Operations", is as follows

a. $4 + 5 = 5 + 4$ (Premise 1)
   $5 + 6 = 6 + 5$ (Premise 2)
   $6 + 7 = 7 + 6$ (Premise 3)
   4, 5, 6, and 7 integers
Conclusion: $x + y = y + x$, x and y are integers.

b. $\dfrac{-1}{1} = -1$

$\dfrac{0}{1} = 0$

$\dfrac{1}{1} = 1$

-1, 0, 1 integers

Conclusion: If x is an integer, then1

c. $1 + 3 = 4$
$3 + 5 = 8$
$5 + 7 = 12$
1, 3, 5, 7 odd numbers
4, 8, 12 even numbers
Conclusion: The number of two odd numbers is an even number.

Attitude is the feelings towards an object, situation, concept, or other people as a result of the process of learning or experience (Lestari & Yudhanegara, 2015). The tendency to support something is called a positive attitude, while a dislike is referred to as a negative attitude. A positive attitude towards mathematics is the tendency to be interested in it. In contrast, a negative attitude is the disinterested one has.

2.2 *Constructivist learning*

Constructivist learning is a process that explains how knowledge is compiled in humans based on constructivism philosophy (Hamzah & Muhlisrarini, 2016). According to Suyono & Hariyanto (2011), constructivism is a learning philosophy based on the premise that by reflecting on the experience, humans build knowledge and understanding of the world in which they live. Therefore, students actively construct new knowledge based on their old understanding. Consequently, there is a connection between old and new knowledge, which leads to an understanding of a more meaningful mathematical concept.

Merril (1991) stated that the assumptions which build constructivist learning include 1. Knowledge is built through experience, 2. Learning is an active process in which meaning is developed based on experience, and 3. Learning is a personal interpretation of the real world. Based on these assumptions, learning is defined as an active process of compiling knowledge through interactions with the environment and build a connection between the conceptions possessed and the phenomena under study (Sutrisno, 1994).

Based on understanding the principles of learning, the steps of constructivist learning include 1. Understanding the objective and system, 2. Apperception, 3. Submission of contextual problems, 4. Student activities, 5. Conclusion and summary, 6. Evaluation, and 7. Closing. At the beginning of learning, the teacher explains the objectives to be achieved and the rules to be followed. The ability and the initial experience of the students are determined by providing questions orally and in writing. The teacher conveys contextual problems to be solved, which include the mathematical concepts to be covered, and the students solve the contextual problems on their own. Other activities conducted include reflection, discussions, and asking questions. In the end, they conclude and summarize all the materials before the teacher provides evaluation questions to determine the level of their understanding. Afterward, the teacher closes the lesson by emphasizing the materials and provides direction on the next materials.

3 METHODOLOGY RESEARCH

This was a quasi-experimental study since there was no random determination of the experimental and control groups and used a Pretest-Posttest Control-Group Design (Borg & Gall, 1983). In the experimental group, students were taught using constructivist learning, while common conventional approaches were used in the control groups. The topics were 'Triangle' and 'Prism' at class VII at SMPN 1 Parung and SMPIT Jabon Mekar.

The subjects of this study included 48 and 23 students of SMPN 1 Parung and SMPIT Jabon Mekar Parung Bogor. The instrument of the study was the test on 'Triangle' and 'Prism.'

To determine the effects of the constructivist learning model on changes in the students' attitudes, qualitative descriptive analysis was used based on the questionnaire of 10 items and the observation guidelines. Also, to establish the effects of constructivist learning on the improvement of inductive reasoning, the test on 16 items was used with t-test statistics in case the data were normally distributed (Sugiyono, 2008). However, suppose the data were not normally distributed, a non-parametric statistical technique was used.

4 RESULT AND DISCUSSION

Conventional learning has failed to build a positive attitude of the students towards mathematics. Several students do not like mathematics since it is perceived as a teaching material containing numbers and formulas less related to real life. Besides, learning is monotonous and one-way from teacher to student, with learners not allowed to convey mathematical ideas. Contrastingly, constructivist learning successfully creates a positive attitude towards mathematics. In this regard, students like mathematics and are happy to determine ways of solving problems. Furthermore, they also like discussing with friend's mathematical problems as they try to solve them. Figure 1 presents data on the students' preferences in mathematics through constructivist learning.

Constructivist learning changes the students' preferences in mathematics. For instance, 80% of students like mathematics since it is related to daily life. It guides the students' reasoning and helps them think systematically, logically, and accurately. Haji (2007) established that constructivist learning makes mathematics students happy since they are given the widest opportunity to convey arithmetic ideas.

Apart from developing positive attitudes towards mathematics, constructivist learning also develops the students' inductive reasoning ability in a better way than conventional learning. The average score of the students' inductive reasoning ability in the experimental group was 18, which is greater than the control group, with 4.4 for SMPIT Jabon Mekar. Similarly, in SMPN 1 Parung, the average score of inductive reasoning ability in the experimental group was 41.5 and 24.4 in the control group (Table 2). The results of t-test obtained indicated that t-count (1.67) was greater than t-table (1.39) with $\alpha = 0.05$, DB = 70, n = 71 and s = 0.72. This

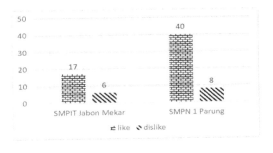

Figure 1. The students' preferences for mathematics.

Table 2. The students' inductive reasoning ability at SMPIT Jabon Mekar.

Score Difference of Pretest-Posttest in Reasoning Ability	SMPIT Jabon Mekar		SMPN 1 Parung	
	Experimental Class	Control Class	Experimental Class	Control Class
Inductive	18.0	4.4	41.5	24.4

showed that there were differences in the students' reasoning abilities in constructivist and conventional learning.

The students taught through constructivist learning could determine the premises of a statement and make correct conclusions. For instance, in solving questions about the weight line of a triangle, they effectively solved related questions. After being provided 3 ABC triangles, they were asked to determine the weight line of each one of them. Using inductive reasoning, they could define and make the weight line of a triangle as well as calculate its length, as shown in Figure 2 below.

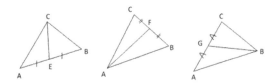

Figure 2. The weight line of a triangle.

Figure 3 presents activities involved in exploring inductive reasoning abilities through constructivist learning. Evidently, 87% and 98% of the students at SMPIT Jabon Mekar and SMPN 1 Parung were active in learning mathematics.

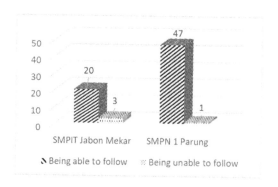

Figure 3. The students' activities in learning on constructivist learning.

5 CONCLUSIONS

This study concluded that constructivist learning had positive effects on the students' inductive reasoning ability and attitudes towards mathematics.

REFERENCES

Akyol, S. & Fer, S. (2010). Effects of social constructivist learning environment design on 5th grade learners learning. Procedia Social and Behavioral Sciences, 9, 948–953.

Bakhtiar, A. (2009). Filsafat Ilmu. Jakarta: PT RajaGrafindo Persada.

Borg, W.R. & Gall, M.D. (1983). Educational Research An Introduction. New York: Longman.

Brigh, A.K., Feeney, A. (2014). Causal knowledge and the development of inductive reasoning. Journal of Experimental Child Psychology, 122, 48–61.

Christou, C. & Papageorgio, E. (2007). A framework of mathematics inductive reasoning. Learning and Instruction, 17, 55–66.

Colomeischi, A.A., Colomeischi, T. (2015). The students' emotional life and their attitude toward mathematics learning. Social and Behavioral Sciences, 150, 744–750.

Gursesa, A., Dogarb, C., Gunesa, K. (2015). A New Approach for Learning: Interactive Direct Teaching Based Constructivist Learning (IDTBCL). Social and Behavioral Sciences, 197, 2384–2389.

Haji, S. (2012). Developing student character through realistic mathematics learning. Proceeding 3th International Seminar 2012, Universitas Pendidikan Indonesia.

Hamzah, A. & Muhlisrarini (2016). Perencanaan dan Strategi Pembelajaran Matematika. Jakarta: Rajawali Pers.

Hersh, R. (1997). What is mathematics, really? New York: Oxford University Press.

Hursen, C. & Soykara, A. (2012). Evaluation of teacher's beliefs towards constructivist learning.practices. Procedia - Social and Behavioral Sciences, 46, 92–100.

Kisworo, M.W. & Sofana, I. (2017). Menulis karya ilmiah. Bandung: Informatika.

Lestari, K.E. & Yudhanegara, M.R. (2015). Penelitian Pendidikan Matematika. Bandung: PT Refika Aditama.

Liew, J., Grisham, J.R., Hayes, B.K. (2008). Inductive and deductive reasoning in obsessive-compulsive disorder. Journal of Behavior Therapy and Experimental Psychiatry, Volume 59, June 2018, Pages79–86.

Merril, M.D. (1991). Constructivism and instructional design. Educational Technology, May-1991.

National Council of Teachers of Mathematics (2000). Principles and Standards for School Mathematics. USA: The National Council of Teachers of Mathematics, Inc.

Ozkal, K., Tekkaya, C., Cakiroglu, J., Sungur, S. (2008). A conceptual model of relationships among constructivist learning environment perceptions, epistemological beliefs, and learning approaches. Learning and Individual Differences. 1041-6080/$ – see front matter © 2008 Elsevier Inc. All rights reserved. doi:10.1016/j.lindif.2008.05.005

Paynea, A.M., Stephensonb,J.E., Morrisb, W.B., Tempestb, H.G., Milehams, A., Griffinb (2009). The use of an e-learning constructivist solution in workplace learning. International Journal of Industrial Ergonomics, 39, 548–553.

Ruseffendi, E.T. (1991). Pengantar Kepada Membantu Guru Mengembangkan Kompetensinya dalam Pengajaran Matematika Untuk Meingkatkan CBSA. Bandung: Tarsito.

Schwenzer, M. & Mathiak, K. (2012). The Correlation of inductive reasoning with multi-dimensional perception. Personality and Individual Differences, 52, 903–907.

Suyono & Hariyanto (2011). Belajar & Pembelajaran. Bandung: PT Remaja Rosdakarya.

Suriasumantri, J.S. (2007). Filsafat ilmu sebuah pengantar popular. Jakarta: Pustaka Sinar Harapan.

Sugiyono. (2008). Statistika untuk Penelitian. Bandung: Alfabeta.

Sutrisno, L. (1994). Pengajaran dengan pendekatan tradisi konstruktivisme. Makalah. Surabaya: Universitas Negeri Surabaya.

Watson, A. & Mason, J. (2005). Mathematics as a constructive activity. London: Lawrence Erlbaum Associates, Publishers.

Engagement and convergence: Leadership for a world-class education in mathematics and technology

P.S. Moyer-Packenham
Utah State University

ABSTRACT: This paper is based on the transcript of a talk given by Patricia Moyer-Packenham, Utah State University, United States of America, at the International Conference on Innovation in Education and Pedagogy: ICIEP 2019, in Tangerang, Indonesia on 5 October 2019. The joint organizers of the conference were Universitas Terbuka Indonesia and Research Synergy Foundation. The focus of the talk was on providing leadership for a World-Class Education in Mathematics and Technology by focusing on two important elements of quality education – the ideas of engagement and convergence.

1 INTRODUCTION

This paper is based on the transcript of a talk given by Patricia Moyer-Packenham, Utah State University, United States of America, at the International Conference on Innovation in Education and Pedagogy: ICIEP 2019, in Tangerang, Indonesia on 5 October 2019. The joint organizers of the conference were Universitas Terbuka Indonesia and Research Synergy Foundation. The focus of the talk was on providing leadership for a World-Class Education in Mathematics and Technology by focusing on two important elements of quality education – the ideas of engagement and convergence.

Thank you to our hosts for the invitation to speak today at this ICIEP conference on quality education. My talk today is titled: Engagement and Convergence: Leadership for a World-Class Education in Mathematics and Technology. To achieve high quality in education, it is important to strive for excellence and provide opportunities in the educational programs for our teachers and students. It is important for our citizens to be prepared to interact locally, nationally, and globally by using technology effectively in mathematics teaching and learning. This means that we need to strive to develop the talent and creativity of our citizens.

2 PURPOSE

The purpose of my talk today is to discuss how we can provide leadership for a World-Class Education in Mathematics and Technology by focusing on two important elements of quality education – the ideas of engagement and convergence. These two key ideas can provide a foundation for promoting excellence in mathematics and technology in the era of IR4.0 (industrial revolution).

3 WORLD CLASS LEARNERS AND EDUCATION

There are many books written about the topic of world-class learners and world-class education. For example, Yong Zhao's (2012) book titled: *World-Class Learners: Educating Creative and Entrepreneurial Students*, focuses on the idea of engaging learners. Some examples from this book on what it means to be a "world-class learner" are to engage learners by providing choice in their learning, by designing personalized learning experiences, and by providing authentic real-world experiences. In the first half of this presentation, I will focus on the idea of engagement and provide some examples of engagement for a high quality education in mathematics and technology.

Another example is a book titled: *A World-Class Education: Learning from International Models of Excellence and Innovation* by Vivien Stewart (2012). Some examples from this book on "world-class education" and the meaning of convergence include looking to examples of educational excellence from around the world, and working together with other researchers through research collaborations. In the second half of this presentation I will provide some examples of convergence for a high quality education in mathematics and technology. Throughout this talk, I will share examples from my own research and from research experts around the world.

4 ENGAGEMENT

First I would like to begin with some recent findings from research on the brain. The first important idea that we know from brain research is that today's student's brains are wired differently than the brains of students in previous generations (Crist, 2017) (see https://www.semel.ucla.edu/longevity/news/mind-what-science-says-about-digital-natives).

Two key things that we know about engaging students, based on brain research, is that today's student's brains have shorter attention spans, and today's students have a better ability to multi-task. This means that for students to be engaged in mathematics and technology learning, they need choice in their learning experiences. These findings from research help us to understand how to design learning experiences when using technology for teaching and learning mathematics.

4.1 Student choice

One of the first important elements of engagement in learning about mathematics and technology is student choice (Anderson, 2016; see http://www.ascd.org/publications/books/116015/chapters/The-Key-Benefits-of-Choice.aspx). Students are more engaged in their learning when they have choices about learning (Dean, 2019; see https://ww2.kqed.org/education/2019/01/10/how-digital-portfolios-increase-student-choice/). Today's students are accustomed to having more choice in how they learn, when they learn, and where they learn. Technology has provided more opportunities for choice and more access to different choices. One element of student choice is simply the environment and the modality of learning. For example, students might choose how they learn by using a computer, a personal touch-screen device or a phone. Students might choose where and when they learn by being at home, at school, or somewhere on the go (like a bicycle or a car or a train). This type of choice has changed the way that learning takes place and it has also increased the opportunities that individuals have for learning.

Another element of student choice is in the different technology apps and games available for learning mathematics. In Figure 1, across the top row, we see examples of three virtual manipulatives for teaching concepts in mathematics such as multiplication of fractions, addition of whole numbers, and fraction equivalence. Across the bottom row, we see examples of three digital mathematics games for teaching decimal numbers, balancing equations and adding integers. Teachers must consider, what is the right choice for selecting the most effective technology environment for teaching mathematics concepts and procedures to students. While educational research can give guidance on making choices about what technologies to use, sometimes the technology selection can become overwhelming because there are so many different choices of technology tools to use.

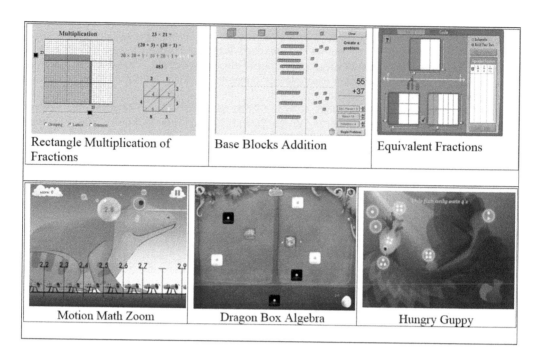

| Rectangle Multiplication of Fractions | Base Blocks Addition | Equivalent Fractions |
| Motion Math Zoom | Dragon Box Algebra | Hungry Guppy |

Figure 1. Examples of virtual manipulatives and digital math games (Moyer, Bolyard & Spikell, 2002; Moyer-Packenham & Bolyard, 2016).

4.2 Personalization

Another important element of engaging the learner is personalization (Hogheim & Reber, 2015; Patall, 2013; Walkington, 2013; Walkington, Sherman, & Howell, 2014). Technology can help educators to rethink how we teach and learn mathematics. We cannot simply use technology in our educational systems to teach mathematics in ways that we have always done in the past with paper and books. Technology can help us to teach mathematics in different ways. Technology can help us to rethink how we teach and learn mathematics. There are two ways that technology can be a benefit to personalize education: (1) One way technology can personalize is that students can focus on a specific area of interest. Students can pursue and learn more about the topics that interest them. (2) Another way technology can personalize is that students can focus on a specific area of weaknesses. Students can select an area of weakness where they want to become more proficient and want to fill in gaps in their knowledge.

Personalization in the use of technology can also be important for teachers too. For example, personalization in learning mathematics through the use of technology can help teachers to focus on mathematics topics where they want to learn more and where they want to focus on using a new technology for their classroom. Personalization can also provide learning opportunities by reaching out to teachers in rural areas. At my own university in the United States, we have many teachers in rural areas. We have created a network of broadcast technology and online technology to reach teachers in rural areas. Through the technology, teachers can earn a degree or an endorsement to add to their teaching license. This provides teachers with personalized learning that meets their needs, and is another way for teachers to be engaged with professional learning in mathematics.

4.3 Authentic experiences

The third important element of engaging students is to provide them with authentic learning experiences (Hill, 2017). There are many examples of providing authentic experiences with mathematics and technology (Picha, 2018; see https://www.edutopia.org/article/effective-tech nology-use-math-class). For example, the use of virtual math manipulatives, digital math games, and computer coding are all ways to provide authentic interactions in mathematics for students.

There are a multitude of animations available to demonstrate difficult to grasp concepts that are best visualized. These animations can be found for any level of mathematics courses and are often available online for no cost. For example, tangent lines are a concept often reserved for calculus courses. However, the concept of increasing and decreasing can be introduced much earlier than the calculus course by using an animation. Seeing the visual example can help to give meaning to the ideas of increasing and decreasing tangent lines. (See http:// upload.wikimedia.org/wikipedia/commons/2/2d/Tangent_function_animation.gif for a gif of this animation.)

Another way that students can engage in authentic experiences is the example of the use of the Texas Instruments (TI) Rover calculator. (See https://www.youtube.com/watch?v=gq9m-gvqrQ4 for information on using the TI Rover.) When students use the TI Rover calculator, they see that learning mathematics is more than just solving equations. Students learn that mathematics is not only an equation to solve on paper, but that mathematics can be brought to life through motion using the TI Rover to explore authentic applications that take the student beyond the worksheet and the book. Showing how equations bring the TI Rover to life helps students to see that mathematics has authentic applications in the real world.

Authentic experiences in mathematics can also come in the form of solving problems collaboratively through mathematical discussions. In the work environment, individuals need to collaborate with others using technology. In a world-class education system, students should not sit passively, waiting for the content to be delivered to them. By providing students with experiences where they collaborate in mathematics to solve authentic problems together, we are preparing world-class learners to work with each other.

As this first part of my presentation shows, the engaged learner will thrive in an environment where they are actively involved in making choices, personalization, and authentic learning experiences.

5 CONVERGENCE

In the second half of my presentation, I would like to focus on the second important element of a high quality, world-class education – the idea of convergence. In the United States we have an organization that funds educational research called the National Science Foundation. In 2017, the National Science Foundation proposed 10 big ideas for the future of education. One of those big ideas was Growing Convergence Research. In this part of the presentation, I will share some examples from my own work showing how educators can contribute to growing convergence research in our own professions.

5.1 *Global perspective*

One way to grow convergence research is to approach our work from a global perspective. This means focusing on research driven by specific and compelling problems that address pressing societal needs. When we converge in our research, we focus on problems that we all share as a global community of educators, teachers, and learners.

One way to grow convergence from a global perspective, and a way that you can contribute to your field, is to define key terminology so that, across different disciplines, every researcher is using the same terminology. In 2002, my colleagues and I defined a virtual manipulative as "an interactive, Web-based visual representation of a dynamic object that presents opportunities for constructing mathematical knowledge" (Moyer, Bolyard, & Spikell, 2002, p. 373). In 2016, we revised and updated the definition of a virtual manipulative to provide more clarity and to take into consideration new touch-screen technologies. The updated and revised definition of a virtual manipulative is "an interactive, technology-enabled visual representation of a dynamic mathematical object, including all of the programmable features that allow it to be manipulated, that presents opportunities for constructing mathematical knowledge" (Moyer-Packenham & Bolyard, 2016, p. 13). By defining a virtual manipulative, and publishing this definition for dissemination, this helped to ensure that all researchers were using the same definition of a virtual manipulative and that there was no confusion and misunderstandings about the construct when it was being investigated in research. As you ponder how to grow convergence research in your own work, here are two key questions for you to consider: What are the key terms that need to be defined in your field? Are there ways that you could contribute to your field through defining a specific construct?

Another way to grow convergence in your research discipline is to conduct a meta-analysis that helps to bring together all of the research on one particular topic to summarize what is known about that topic. This type of research meta-analysis can help your field to move forward by revealing key information about a topic and where there are gaps in the research. For example, in 2013, my colleagues and I conducted a meta-analysis on virtual manipulatives using 66 studies (Moyer-Packenham & Westenskow, 2013). In this meta-analysis, we examined virtual manipulatives as an instructional media compared with other instructional media, and overall, found that the 66 studies showed mostly moderate effects in favor of the use of virtual manipulatives for mathematics instruction. We also examined different mathematical domains, and found that most of the research on virtual manipulatives had been conducted in the topic of geometry. Next we examined different grade levels, and found that most of the research on virtual manipulatives had been conducted in the elementary school grades. We also examined the length of the treatment in each research study, and found that most of the research on virtual manipulatives had been conducted in long-term projects that lasted more than 10 days. Based on what we found in this meta-analysis, we were able to make recommendations to others in the field about where additional research was needed. This meta-analysis helped other researchers in our discipline understand the big picture on the research conducted on virtual manipulatives and where there were gaps in the research that needed to be pursued. Other examples of important convergent research in the areas of mathematics and technology that have made an important contribution to the field include: (1) a meta-analysis by Clark, Tanner-Smith, and Killingsworth (2016) that concluded that digital games significantly improve learning outcomes when they are compared with non-game conditions; and (2) a meta-analysis by Carbonneau, Marley, and Selig (2012) that reported on the efficacy of teaching mathematics with concrete manipulatives. As you consider these convergent research projects and how they have made important contributions, here are two key questions for you to consider: What are the topics that need to be synthesized to contribute to the research in your disciplinary area? How can you help to grow convergence around key research to help your field move forward? Perhaps a meta-analysis is needed to synthesize this research.

Another way that you can grow convergence research, from a global perspective, and a way that you can contribute to your field, is to work collaboratively with other researchers in your peer group who study the same topic that you do. In 2015, during my sabbatical, I contacted researchers from all over the world and asked them to contribute research on virtual manipulatives from each of their countries to a book project titled: *International Perspectives on Teaching and Learning Mathematics with Virtual Manipulatives* (Moyer-Packenham, 2016). The chapters represent the research on virtual manipulatives that is being conducted in different parts of the world by some of the leading researchers on this topic. This book project, with researchers converging from different parts of the world, helped many researchers interested in virtual manipulatives learn about the work being undertaken on this topic in different countries, thereby helping readers to have a global perspective on this research. As you consider growing convergence research on a global scale, here are two key questions for you to consider: Is there are project where you can converge with your colleagues around a specific research topic? Who are the researchers in your peer group who would be excellent collaborators for a convergence project?

5.2 Research collaboration

Another way to grow convergence research is to conduct research collaboratively, which requires integration across disciplines. When we conduct convergence research, we come together as experts from different disciplines and intermingle and integrate our work to pursue common challenges and learn from each other. In the final examples from my own research, I would like to share three large-scale projects where I worked with teams of collaborators across disciplines. Some of my collaborators have included colleagues in mathematics, psychology, technology and special education. I will call these three big projects the Representations Project (N=350, age 8-10), the Captivated Project (N=100, age 3-6), and the Design Features Project (N=193, age 6-12). To conduct a large-scale project, it takes a team of researchers who are willing to converge around a specific problem and set of research questions. The benefits of converging on a project as a team are that you can learn so much from other members on the team, each member of the team helps you to look at the data and the research from a different perspective, and you can conduct much bigger projects than you might be able to conduct if you were just conducting research on your own.

The first large-scale project I would like to share with you was the Representations Project (Moyer-Packenham, Baker, Westenskow, Anderson, Shumway, Rodzon, & Jordan, 2013). In this project, we worked with 350 students, ages 8-10. We had several unique elements that made this project an important contribution. First, there were large numbers of student participants across 17 elementary school classrooms (N=350); Prior research studies comparing virtual manipulatives with other instructional treatments had only included a limited number of students. Next, we used within-class random assignment to experimental groups; Many previous studies on virtual manipulatives did not include random assignment. Another element was that we measured retention effects by using a delayed post-test; Few studies prior to this study had examined student retention of mathematics concepts. We documented instructional fidelity through the use of classroom observers; Few studies prior to this research had reported evidence of what teachers taught during instructional sessions. Finally, we included rigorous instrument development; Few studies prior to ours had reported psychometric properties of the test instruments used in the studies. One of the key findings from this study was that there were no significant differences between the regular Classroom Instruction group and the Virtual Manipulative group in terms of pretest, posttest, and delayed posttest changes. We found that when the amount and frequency of mathematical representations were similar in both the regular Classroom Instruction groups and the Virtual Manipulative groups (e.g., pictorial images, manipulatives, drawings), students who experienced the same representations did just as well on the measures. This showed the importance of representations for teaching and learning mathematics, irrespective of the modality of delivery (i.e., virtual manipulatives, concrete manipulatives, or teacher drawings).

The second large-scale project was the Captivated Project (Moyer-Packenham, Bullock, Shumway, Tucker, Watts, Westenskow, Anderson-Pence, Maahs-Fladung, Boyer-Thurgood, Gulkilik, & Jordan, 2016). In this project we worked with 100 young children, ages 3-6. We conducted individual interviews with children as they used digital mathematics apps on an iPad device. The unique contribution of this study was that we tested some new methods for capturing data. For example, we used a wall-mounted video camera in the interview room to video record the interview, and we used a GoPro camera (worn by each child) to capture the up-close screen perspective of the child's hands manipulating the dynamic mathematical objects (see Figure 2). An observer also recorded field notes based on observations of the interactions. The findings of this study showed that the preschool children became more efficient with the technology only during their interactions with the mathematics apps; While the kindergarten and second-grade children were able to become both more efficient with the technology and to learn the mathematics in the apps.

The third large-scale project was the Design Features Project (Moyer-Packenham, Lommatsch, Litster, Ashby, Bullock, Roxburgh, Shumway, Speed, Covington, Hartmann, Clarke-Midura, Skaria, Westenskow, MacDonald, Symanzik, & Jordan, 2019). In this project we worked with 193 students, ages 6-12. The purpose of this project was to examine how the

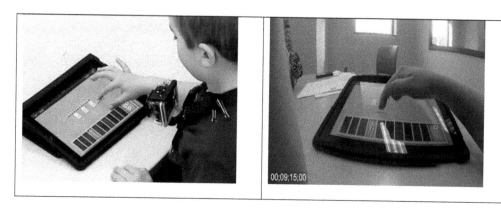

Figure 2.   Examples of a student playing a digital math game while wearing a GoPro camera (left). The view from the GoPro camera of the student's hands on the iPad device (right).

design features in 12 digital math games influenced mathematics learning outcomes. Children played 12 digital math games in a randomized order. We used two data sources – pretests/ posttest based on the mathematics in the apps, and video/audio of the children playing the digital math games. The unique contribution of this study was that it revealed eight prominent categories of design features in the digital games that influenced children's learning. In addition, children were aware of the design features that supported their learning. A key finding in this study was that the two Mathematics Connections categories of design features were only observed in the digital math games that produced significant gains between the pretest and posttest. This means that, while the presence of design features may not guarantee that learning will occur, the absence of design features that focus on mathematics can have negative impacts on mathematics learning. The mathematics design features (those that helped the children to recognize the mathematics) were vital to children's learning when they played the 12 digital math games in the study.

As this part of my presentation highlights, there are many ways to grow convergence research as part of a research collaboration group. Working in collaboration, especially across different disciplines, can support creative interdisciplinary thinking. As you consider possible research collaborations, here is a key question for you to ponder: Who are the colleagues that you can work with on large-scale projects that can help you to look at your research questions and problems in new ways from the perspective of multiple disciplines? Working with others can help you to learn to look at your research topic from a new perspective, to converge important ideas, and make contributions that advance your discipline.

6   CONCLUSION

The focus of my talk today was on two big ideas – the ideas of engagement and convergence for a world-class education and world-class learners in the era of IR4.0. (industrial revolution). As educators think about ways to engage in mathematics and technology, it is important to keep in mind the focus on providing students with a choice in their learning, designing personalized learning experiences based on students' needs, and providing authentic real-world experiences. As we consider ways to grow convergence research, it is useful to think about learning from excellent models and innovation from across different disciplines and from educational successes all over the world. By working together to collaborate on projects and research strategies, we can advance our own research agenda while contributing to the research in our field. Together these key ideas can help us to promote a world-class education in mathematics and technology for our teachers and students.

REFERENCES

Anderson, M. (2016). Learning to choose, choosing to learn. (http://www.ascd.org/publications/books/116015/chapters/The-Key-Benefits-of-Choice.aspx)

Carbonneau, K. J., Marley, S. C., & Selig, J. P. (2013). A meta-analysis of the efficacy of teaching mathematics with concrete manipulatives. Journal of Educational Psychology, 105(2),380–400.

Clark, D. B., Tanner-Smith, E. E., & Killingsworth, S. S. (2016). Digital games, design, and learning: A systematic review and meta-analysis. Review of Educational Research, 86(1), 79–122.

Crist, C. (2017). On the mind: What science says about digital natives. (https://www.semel.ucla.edu/longevity/news/mind-what-science-says-about-digital-natives)

Dean, J. (2019). How digital portfolios increase student choice. (https://ww2.kqed.org/education/2019/01/10/how-digital-portfolios-increase-student-choice/)

Hill, A. M. (2017). Authentic learning and technology education. In M. J. de Vries (Ed.), Handbook of technology education (pp. 473-487) New York: Springer.

Hogheim, S., & Reber, R. (2015). Supporting interest of middle school students in mathematics through context personalization and example choice. Contemporary Educational Psychology, 42, 17–25.

Moyer, P. S., Bolyard, J., J., & Spikell, M. A. (2002). What are virtual manipulatives? Teaching Children Mathematics, 8(6), 372–377.

Moyer-Packenham, P. S. (2016). International perspectives on teaching and learning mathematics with virtual manipulatives. Switzerland: Springer.

Moyer-Packenham, P., Baker, J., Westenskow, A., Anderson, K., Shumway, J., Rodzon, K., & Jordan, K., The Virtual Manipulatives Research Group at Utah State University. (2013). A study comparing virtual manipulatives with other instructional treatments in third- and fourth-grade classrooms. Journal of Education, 193 (2), 25–39.

Moyer-Packenham, P. S., & Bolyard, J. J. (2016). Revisiting the definition of a virtual manipulative. In P. S. Moyer-Packenham (Ed.), International perspectives on teaching and learning mathematics with virtual manipulatives (pp. 3–23). New York: Springer.

Moyer-Packenham, P. S., Bullock, E. P., Shumway, J. F., Tucker, S. I., Watts, C., Westenskow, A., Anderson-Pence, K. L., Maahs-Fladung, C., Boyer-Thurgood, J., Gulkilik, H., & Jordan, K. (2016). The role of affordances in children's learning performance and efficiency when using virtual manipulative mathematics touch-screen apps. Mathematics Education Research Journal, 28(1), 79–105.

Moyer-Packenham, P. S., Lommatsch, C. W., Litster, K., Ashby, J., Bullock, E. K., Roxburgh, A. L., Shumway, J. F., Speed, E., Covington, B., Hartmann, C., Clarke-Midura, J., Skaria, J., Westenskow, A., MacDonald, B., Symanzik, J., & Jordan, K. (2019). How design features in digital math games support learning and mathematics connections. Computers in Human Behavior, 91, 316–332.

Moyer-Packenham, P. S., & Westenskow, A. (2013). Effects of virtual manipulatives on student achievement and mathematics learning. International Journal of Virtual and Personal Learning Environments, 4(3), 35–50.

Patall, E. A. (2013). Constructing motivation through choice, interest, and interestingness. Journal of Educational Psychology, 105(2), 522–534.

Picha, G. (2018). Effective technology use in math class. (https://www.edutopia.org/article/effective-technology-use-math-class)

Stewart, V. (2012). A world-class education: Learning from international models of excellence and innovation. Alexandria, VA: ASCD.

Walkington, C. (2013). Using adaptive learning technologies to personalize instruction to student interests: The impact of relevant contexts on performance and learning outcomes. Journal of Educational Psychology, 105(4), 932–945.

Walkington, C., Sherman, M., & Howell, E. (2014). Personalized learning in algebra. The Mathematics Teacher, 108(4), 272–279.

Zhao, Y. (2012). World class learners: Educating creative and entrepreneurial students. Thousand Oaks, CA: Corwin.

Emerging Perspectives and Trends in Innovative Technology
for Quality Education 4.0 – Kusmawan et al (eds)
© 2020 Taylor & Francis Group, London, ISBN 978-0-367-25803-0

Education 4.0 and intercultural understanding: Perspectives from Australia and Indonesia

Ruth Reynolds

School of Education, The University of Newcastle, Australia

ABSTRACT: This paper is a brief overview of the ideas presented in a presentation at the International Conference on Innovation in Education and Pedagogy: ICIEP 2019, in Tangerang, Indonesia on 5 October 2019. The joint organizers of the conference were Universitas Terbuka Indonesia and research Synergy Foundation. Much is made of the value of STEM disciplines in regard to Industry 4.0 and Education 4.0 but attention must also be paid to how humans interact and dialogue in these disciplines. That too is changing as a result of Industry 4.0 and the increased globalisation of work. This paper examines some aspects of the altered human interactions, and some distinct "cultures" that emerge from a digitally engaged world. It focuses on how global interactions influence changing perspectives in intercultural understanding.

1 DEFINITIONS - INTERCULTURAL UNDERSTANDING

Cultures education is teaching about cultural norms and differences with a view to learning to interact and engage with many different groups. The definition provided by the UNESCO Universal Declaration on Cultural Diversity (2001) is 'culture should be regarded as the set of distinctive spiritual, material, intellectual and emotional features of society or a social group, and that it encompasses, in addition to art and literature, lifestyles, ways of living together, value systems, traditions and beliefs' (p.12). Banks (2001) argued that the overall goal of what is variously called multicultural education and intercultural education is to help students to function in their home communities as well as in the mainstream world - accepting differences. In the Australian Curriculum the General Capability, Intercultural Understanding;

encourages students to make connections between their own worlds and the worlds of others, to build on shared interests and commonalities, and to negotiate or mediate difference. It develops students' abilities to communicate and empathise with others and to analyse intercultural experiences critically. It offers opportunities for them to consider their own beliefs and attitudes in a new light, and so gain insight into themselves and others (ACARA 2013, p. 133).

In Indonesia the ASEAN Master Plan on Connectivity (2025) and Blueprint 25 points for the need to improve people mobility and skills mobility and to provide equitable opportunity to access quality education with an emphasis on ethnic minority groups and elderly/older persons. The 2013 national curriculum in Indonesia has a strong emphasis on the 4Cs (creativity, critical thinking, communication and collaboration) (Directorate General of Primary and Secondary Education, 2017). It is important to link these soft skills, global education and 21st century teaching communities through dialogue and interaction (Crichton & Scarino 2007; Kusmawan, 2015). The Australia–Asia BRIDGE School Partnerships Program (Asia Education Foundation, 2017) has operated between Australia and Indonesian schools for over 10 years with aims of deeper intercultural understanding and real life digital capability. Indonesia is a our largest partner in this project with 164 schools involved.

2 INDUSTRY 4.0 AND EDUCATION 4.0

The concept of Industry 4.0, using transformative technologies to connect the physical world with the digital world, (Australian Government, Department of Industry, Innovation and Science) has led to Education 4.0 - how to address education in this so called 4th industrial revolution. Ideas associated with 21st Century Education (World Economic Forum [WEF](2015), the 'soft skills' such as the 4 Cs as in the Indonesian curriculum (Budiarti, 2019), skills, knowledge and attitudes for global competence (OECD/PISA, 2018), and the Sustainable Development goals [SDGs] (OECD, 2016) are all linked to Education 4.0 and Industry 4.0 because as well as having discipline knowledge in the STEM subjects, future work and future citizens will need to be able to adapt and link these different forms of knowledge and Artificial Intelligence applications to solve problems that have not necessarily needed to be solved before (Reynolds 2018, 2013). As Avis (2018) points out "it is important to recognise that technology, digitalisation and artificial intelligence are entwined with social relations" (p. 356). This paper concentrates on Intercultural Understanding as a key 'soft skill' because it addresses negotiation of cultural differences in the broad sense as per the UNESCO (2001) definition but there is a new cultural imperative associated with negotiating the technological barriers between people who may in many senses be similar in culture but the need to dialogue and interact in a new and constantly changing technological environment requires new intercultural strategies. This latter focus on the culture associated with interaction with technology is of great interest if we are to successfully link industry and education by encouraging proficiencies such as creativity, originality, initiative, resilience, problem solving, leadership, negotiation (WEF, 2018).

3 TECHNOLOGY AS A NEW FRONTIER IN INTERCULTURAL UNDERSTANDING

Kusmawan (2015) emphasised the role of technology in developing global education in Indonesia, especially in its unique position of having a huge diversity of ethnic and religious groups as part of its national character. Technology influences key aspects of global education – it assists building national responses to curriculum to build a curriculum which enables its citizens to be able to interact with the world businesses and political processes and it also assists in developing the "softer" skills of collaboration, intercultural competency, and allows dialoguing about values. He argued that there is a need to stress direct people-to-people connection as the main approach to develop and implement innovative and future-orientated policies. A critical educational focus of debate in Indonesia at the moment concerns distance and online modes of education and the extent to which these education delivery modes can promote and support policy realisation to alleviate inequality between urban and rural districts. As he pointed out employing ICT in education and utilising it to engage students with others across different cultures can be problematic. It is not simply a matter of connecting people technologically. It is what people will do when they do connect and how they will best join with others.

For a start with common communication software such as Skype, Zoom, Collaborate there is a need to learn online etiquette – to be very clear being aware of the audiences involved, strategies to acknowledge changes in the purpose or the conversation (introduction, conclusion, listening, speaking, including) and technical issues that may affect communication. Comas-Quinn, de los Arcos, B. & Mardomingo (2012) point to the different roles teachers and tutors take in these online situations and how new hierarchies and relationships are constantly being established. Using social media of any type requires an understanding of the structures, practices, capabilities and technologies of the partner (Oakley, Pegrum, Xiong, Lim, & Yan, 2018; Chen, 2017; Angelova & Zhao, 2014). In the Virtual Learning environment the individual traits of the user influence the experience and there is a need for the teacher to have both pedagogical and technological knowledge to apply appropriate teaching strategies (Janssen, Tummel, Richert, & Isenhardt, 2016). In addition particular forms of hardware and resultant Apps require special attention and offer different opportunities. In the case of iPads Reynolds (2015) found the mobility of iPads allowed for active approaches to civics and citizenship where student could interact with a wider variety of people and build rapport. On the other hand when iPads are a new innovation in a particular area of study they can be a challenge to teaching. Alshammari (2019) argued that because these tools may be seen as means of amusement and distraction rather than education some parents and teachers in Saudi Arabia did not see them as real teaching tools. Additionally as a whole class exercise the technical issues became quite overwhelming and they were not as useful in some aspects of language teaching. Teachers needed to find the strengths of the various Apps.

4 LEARNING BY DIALOGUE AND ENGAGEMENT TO BUILD COMMUNITY MOBILITY

Intercultural understanding is not simply about technology. Industry 4.0 and Education 4.0 is also about human to human interaction to develop mobility - social mobility and geographical mobility and the skills required in a global community, some of which could be technological skills. Kusmawan, O'Toole, Reynolds & Bourke (2009) found that active involvement in their community helped senior Science students in Indonesia to practice environmental citizenship in their own community which rapidly improved their science knowledge and reduced the distance between beliefs, attitudes and intentions. The connections helped their motivation to learn more and contribute. Preservice teachers were influenced in what they taught in schools by what they saw happening in their schools they attended in Australia (Ferguson-Patrick, Macqueen & Reynolds, 2014). In classrooms where their cooperating teacher felt they needed to teach for standardised tests reduced the preservice teachers' abilities to link actual classroom teaching to global imperatives. When not linked to authentic teaching opportunities their teaching abilities seemed to be reduced. Not understanding how to use curriculum outcomes to focus on 'real' situations, or people or places reduced the impact of their teaching (Reynolds, Macqueen, & Ferguson-Patrick, 2019). Reynolds and Vinterek (2016) compared world views of Swedish and Australian children at various times in the primary school progression. When children were young they knew only about places near to the local context. However as they grew older they were very influenced by social media and technology tools such as Google and so learned more about the world. Places they actually experienced or people they actually met from those places influenced them the most. Young children's views of their world are very influenced by what we call the 'nearby'. The 'nearby' however can be places they can reach easily by plane or by media. Both Australian and Swedish children know about the USA and they both have views on the USA. With their expanding knowledge of the world as they get older comes attitudes and perspectives, influenced strongly by things like climate, friends and family, media, sport and family holidays.

5 CONCLUSION

Intercultural education is required for all aspects of global education or global industry whether it is to be more proficient when handling international work experiences virtually, or actual face to face geographically close interactive experiences in our schools or shopping centres. When we talk of culture we think about beliefs, and attitudes and practices associated with specific groups but in a technologically engaged world those practices appear to be subsumed in a new digital culture. However the digital culture also brings unique ways of thinking that must be addressed in a manner similar to longstanding intercultural understandings are developed. Human to human interactions and human to machine interactions require some fundamental dialoguing skills and practices as well as the need to develop respect and empathy and take responsibility to be a mediator.

REFERENCES

Alshammari, J. (2019). Change is not necessarily always good: Challenges in introducing tablet PCs in English classes at primary schools in Saudi Arabia. Presentation at SSTAR conference, The University of Newcastle, September 30.

Angelova, M. & Zhao, Y. (2014). Using an online collaborative project between American and Chinese students to develop ESL teaching skills, cross-cultural awareness and language skills, Computer Assisted Language Learning, 29(1), 167–185.

ASEAN Secretariat, Jakarta, 2016. ASEAN Socio-Cultural Blueprint 2025. Available at: https://asean.org/asean-socio-cultural/

Asia Education Foundation (AEF) (2017). Australia-Asia BRIDGE School Partnerships. Available at: asiaeducation.edu.au/BRIDGE.

Australian Curriculum and Reporting Authority (ACARA) (2013). *General Capabilities in the Australian Curriculum*. Sydney: ACARA. Available at: www.australiancurriculum.edu.au/generalcapabilities/general%20capabilities.pdf

Australian Government, Department of Industry, Innovation and Science (2019). Industry 4.0. Available at https://www.industry.gov.au/funding-and-incentives/manufacturing/industry-40)

Avis, J. (2018). Socio-technical imaginary of the fourth industrial revolution and its implications for vocational education and training: a literature review. Journal of Vocational Education and Training, 70(3), 337–363.

Banks, J. (2001). Cultural Diversity and Education. Sydney: Allyn and Bacon.

Budiarti, M. (2019). Methodological Challenges while Researching with Teachers and Students in Indonesia. Presentation at SSTAR conference, The University of Newcastle, September 30.

Chen, H. (2017). Intercultural communication in online social networking discourse. Language and Intercultural Communication, 17(2), 166–189.

Comas-Quin, A. de los Arcos, B. & Mardomingo, R. (2012). Virtual learning environments (VLEs) for distance language learning: shifting tutor roles in a contested space for interaction. Virtual learning environments (VLEs) for distance language learning: shifting tutor roles in a contested space for interaction. Computer Assisted Language Learning, 25(2), 129–143.

Crichton, J. & Scarino, A. (2007). How are we to understand the intercultural dimension? The Australian Review of Applied Linguistics, 30(1), 1–21.

Ferguson-Patrick, K., Macqueen, S. & Reynolds, R. (2014). Pre-service teacher perspectives on the importance of global education: world and classroom views. Teachers and Teaching: Theory and Practice. 20(4), 470–482.

Janssen, D., Tummel, C., Richert, A. & Isenhardt, I. (2016). Virtual Environments in Higher Education – Immersion as a Key Construct for Learning 4.0. International Journal of Advanced Corporate Learning (iJAC), 9(2), 20–26.

Kusmawan, U. (2015). Educating diverse teachers in a diverse country. In Reynolds, R., Bradbery, D., Brown, J., Carroll, K., Donnelly, D., Ferguson-Patrick, K., & Macqueen, S. (Eds). Contesting and constructing international perspectives in Global Education. Amsterdam: SENSE Publications.

Kusmawan, U.; O'Toole, M.; Reynolds, R. and Burke, S. (2009). Ecological affinity in Secondary Science students in selected Indonesian schools using different teaching approaches: the link between beliefs and attitudes towards the environment, and student intention to make changes to the environment. International Research in Geographic and Environmental Education, 18(3), 157–169.

Oakley, G.,Pegrum, M., Xiong,X., Lim, C. & Yan, H. (2018). An online Chinese-Australian language and cultural exchange through digital storytelling. Language, Culture and Curriculum, 31(2), 128–149.

Organisation for Economic Cooperation and Development [OECD]/PISA. (2018). https://www.oecd.org/pisa/pisa-2018-global-competence.htm

Organisation for Economic Cooperation and Development [OECD], (2015). OECD and the Sustainable Development Goals. Available at: http://www.oecd.org/greengrowth/sustainable-development-goals.htm

Reynolds, R. (2018). Teaching humanities and social sciences in the primary school (4th Ed.). South Melbourne: Oxford University Press.

Reynolds, R. (2015). Technology for teaching Civics and Citizenship: insights from teacher education. The Social Educator, 33(1), 26–38.

Reynolds, R. (2013). Globalization and international social studies classroom practice. Journal of International Social Studies, 3(1), 1–3.

Reynolds, R., Macqueen, S. & Ferguson-Patrick, K. (2019). Active citizenship in a global world: opportunities in the Australian Curriculum. Curriculum Perspectives, online view article at https://rdcu.be/bQZ3V

Reynolds, R. & Vinterek, M. (2016). Geographical locational knowledge as an indicator of children's views of the world: Research from Sweden and Australia. International Research in Geographical and Environmental, 25(1), 68–83.

UNESCO Universal Declaration on Cultural Diversity (2001). Available at: http://portal.unesco.org/culture/en/ev.php-URL_ID=2977&URL_DO=DO_TOPIC&URL_SECTION=201.html

World Economic Forum (WEF), (2018). The Future of Jobs. Available at: http://www3.weforum.org/docs/WEF_Future_of_Jobs_2018.pdf

World Economic Forum (WEF) (2015). New Vision for Education, Available at: http://www3.weforum.org/docs/WEFUSA_NewVisionforEducation_Report2015.pdf

Emerging Perspectives and Trends in Innovative Technology
for Quality Education 4.0 – Kusmawan et al (eds)
© 2020 Taylor & Francis Group, London, ISBN 978-0-367-25803-0

Culture and local wisdom in the global era: The importance of meaningful learning

Suminto A. Sayuti
Universitas Negeri Yogyakarta, Indonesia

ABSTRACT: Space, opportunity and possibility for change are necessary for the process of cultural scouting and contact to happen. This proposition is by on the reason that something will last if it is open to change and renewal. In this connection, thinking and acting strategically becomes important and urgent. As a result, positioning learning that is strategically meaningful becomes important. Multicultural approach - which avoids one-way, cognitive and exclusive nature; also avoiding superiority, primordialism, and exclusivism of certain values - is one of the ways that can be taken. Through this, an understanding of shared values and collaborative efforts to overcome shared problems is sought, the potential value of which is trans-endangered.

As a nation, we have a culture that is so abundant. Whatever its shape and form, the nation's culture is and becomes the capital and identity, the fortress, and at the same time the "main passport," even more so in today's global relationships and interactions. Through culture, we are known by and introduce ourselves to other nations. Culture is our capital and identity in relating and interacting with "others," those who are not us. Recognition by other nations of the high cultural values that we have, for example, is a "passport" that legitimizes our culture and lends it an equal position with theirs. Into the process of relating and interacting with "others" also enters the inevitability and entry of diverse values, which in some cases often conflict with values that have long been internalized and believed. In this context, the values inherent in the nation's culture function as a fortress. The problem is that materialistic and hedonic tendencies are increasingly becoming more prominent in the midst of community life, and at the same time our national values and character are fading, even though we have strong capital and identity, passports, and cultural fortifications. Is there something wrong in managing cultural systems and mechanisms in the context of our nationality? A series of other cultural questions arise, including those on learning positions that are meaningful as "cultural processes"": have they performed ideological, educative, and cultural imperatives as the main functions in their praxis?

There are at least four reasons that can be put forward in terms of why the meaningful learning position as a "civilizing process" is questionable. First, in its entirety and wholeness, culture is the groundwork and habitat for the seedbed of character, where identity and personality grow and develop. Second, culture requires care, development, and empowerment through meaningful learning. Third, the noble values of the nation's culture will become something foreign to society if the meaningful learning praxis is not regulated by a cultural orientation. Fourth, the function of culture as a source of value may eventually disappear if it is not supported by a conscious and educated community. These four points show the reciprocal and dialectical relationship between meaningful learning on the one hand and culture on the other: a relationship that is important for the exploration and development of local wisdom in the context of nation and character building in the digital age.

The entry of various values originating from "outside" (i.e., the "center") through a variety of modern devices is an inevitable result of the global process, which has given its own color and style to the joints of the nation's cultural life. (If we look closely, the flow of global information is controlled by the West, which has advanced first in various fields. The West has become a kind of "center" that flows global information to "peripheral" regions, i.e., countries that are still in a developmental stage. The frequency and intensity of cross-cultural relations between nations have also increased. Not a single nation on this earth can circumvent this kind of cultural process. The swift global flow from the center to the "periphery," among others, resulted in the emergence of a situation of "cultural underdevelopment." Cultural underdevelopment could have struck most people when they tried to reach the modernist strata along with a flood of information. New cultural symbols are often interpreted incorrectly. Technology that is developing so rapidly and becoming increasingly sophisticated also causes people's communication patterns to change rapidly. Human knowledge and experience are shaped by an array of information that can be stored and transmitted at powerful speed and can reach a vast area. Oral language is replaced by visual images. Utilitarian, materialist, and hedonic attitudes have come to the fore along with the emergence of continuous shifts. As a result, the world view of society is broken, torn, and misplaced (dislocated). This can all be seen as a challenge as well as a threat to our national values, character, and identity.

The process of cultural scouting and contact that is and will continue to happen will truly be dangerous if in cultural systems and mechanisms in the national context no space, opportunity, or possibility for change is provided. Why is this? Something will last if it is open to change and renewal. In this connection, thinking and acting strategically become important and urgent. The design and implementation of various efforts that ultimately lead to the creation of resilience of the nation's identity in dealing with and entering into these processes must be carried out immediately so that cultural values can be cared for, developed, and empowered, while those that are beginning to fade can be reenacted.

When the awareness arises that the local (including the values of existing local wisdom) are always victims of marginalization, the community (ethnic) also needs to redefine itself and its culture. Entering the "enclosure" of local culture, on the one hand, can be taken into account as a basis for efforts to create a national cultural awareness.. This action can, on the other hand, also bring up the paradox that when it is interpreted linearly we will live in the future, not in the past. In fact, when this process becomes exclusive, it becomes a challenge because what is created is no longer shared awareness in the context of the nation-state but the spirit of ethnonationalism.

To anticipate this, the orientation chosen should be directed to the authenticity of human nature as a conscious agent to act to overcome the world and reality that (may be) hostile to and oppressing it, which is entirely in the frame of being together with "the other." The implication, systems, and mechanisms of local and translocal culture must still be treated, developed, and empowered together. A dialectical cross between "the other" and a drive to create and re-create an independent local identity in a process of continuous transformation become imperative. The goal is to prepare a habitat so that the figures involved are able to live up to local values, and at the same time be able to open a space for interactions with "others": to be both local and translocal and global.

In this context, meaningful learning, which demands the presence of a critical, dialogical, and participatory nature, will feel its relevance and significance. However, dilemmas and problems also arise in a number of ways when local (and translocal) values are implemented in and become a meaningful learning base: whether the values are processed creatively (in the sense of being affirmed by the values of "the others") through balanced reconciliation, or just used for the homogenization of values and at the same time domination of other values through subordination reconciliation (Seung, 1982).

Whatever efforts are chosen and carried out should not be trapped into elimination efforts through elimination reconciliation. The effort must always be based on cultural insight, which places education/learning as an effort of civilization (not "pembuaya"). Education has also become a kind of nursery for the seeds of polyphonic and multicultural situations. Through this, a healthy citizenship cultural habitat is prepared, that is, a habitat that requires the availability of space and opportunities for full participation and interaction that is open to all diverse elements of education/learning. This is important because those who continue to live up to their local cultural values are feared to be marginalized and lacking, or not even raised,

256

in the constellation of translocal and global information, and are often less materially disadvantaged. Therefore, efforts to build awareness of local wisdom as a cultural reality, which also functions in positioning identity, ultimately must become a spirit that must not be ignored in the context of maintaining national values so as not to fade, and so that values remain internalized in any situation.

As a diverse nation, Indonesia has two kinds of cultural systems, namely the national cultural system and the local ethnic culture system. Both must be cared for, developed, and empowered together. The national cultural system is a system that applies generally to the entire Indonesian nation, which until now is still in process and is outside the boundaries of any local ethnic culture (Sedyawati, 2007). Cultural values that are formed in the national cultural system welcome the future. These values are essentially "sliced fibers" that are formed when two or more local ethnic cultures emerge, intersect, and enrich each other on the basis of the similarities between them, for example, the value of *"desa kala patra"* (Bali) meets *"empan papan, duga prayoga, angon mangsa"* (Jawa) and *"di mana bumi dipijak, di situ langit dijunjung"* (Minang). Thus, certain local cultural values become translocal/national images because they are combined with other values that are actually derived from old cultural values contained in various local ethnic cultural systems. Therefore, the values of local wisdom are basically the basis for the formation of a national identity that is translocal (national). Local wisdoms that make the culture of the nation have roots.

Local culture often functions as a source or reference for new inventions, for example in language, art, social order, and technology, which are then displayed in cross-cultural fibers. Therefore, the effort to explore local wisdom is basically to find, and finally to establish the nation's identity, which may be lost because of the dialectical crossing process, or because of acculturation and transformation that have been, are, and will continue to happen as inevitable. The effort to find a new national identity based on local wisdom is important for the unification of the nation's culture on the basis of the identity of the archipelago. So, the final end of the culture-conscious situation to be achieved is not a necrophilic situation, that is, the feeling of love for all material things that are not life-minded, but a biophile situation, that is, the feeling of love for everything that is meaningful, which has a life spirit. In this way, all citizens of the nation are expected to have adequate cultural resilience in facing global challenges.

Western culture that has developed economically and technologically inevitably has struck us so strongly that we feel lost (in part) by the traditional identity of the nation. Rebuilding the nation's identity is, in essence, one of the important efforts to select, and not fight, "other" cultural influences. The nativism movement can be seen as naive, but it is a logical reaction if placed in a rapidly changing cultural perspective, such as that which is happening in this era.

By always taking into account local wisdom through and in meaningful learning as a civilizing process, let alone positioning it as a basis, the inevitability of the community trapped in a situation of being a society that is alienated from its reality, which "comes into being" in the sense of "becoming like others and not themselves," can be avoided. So, positioning the results of efforts to explore local wisdom in the process of civilization must always be interpreted in the context of efforts to prepare the community to have cultural awareness, and not in terms of sociocultural tractability.

The value of local wisdom necessitates a strategic function for the formation of character and identity, which in turn will lead to an independent and creative cultural attitude full of initiative. Care, development, and empowerment of relevant and contextual local wisdom have an important meaning for the development of a community, especially when viewed from the point of cultural awareness, as well as having an important meaning for the identity of the region concerned. Cultural works that place local values as a source of creative inspiration will encourage the emergence of a proud attitude among the host community. Creative works can be displayed in the face of translocal discourse so that they have a major contribution to the creation of a new identity for the nation as a whole.

Local wisdom, which also necessitates past cultural content, also serves to build a longing for the lives of the ancestors, who are the cornerstone of present life. The notion that what is relevant to life is only "the here and present" can also be avoided. Local wisdom can be used as a bridge that connects the past and present, generations of ancestors and present generations, in order to prepare the future and future generations. In turn, it can also be used as a kind of glue for intergenerational culture and help prevent an ahistorical situation.

In this context, positioning learning that is strategically meaningful becomes important. A multicultural approach devoid of one-way cognition, as well as avoiding superiority, primordialism, and exclusivity of certain values, is one of the ways that can be taken. Through this, an understanding of shared values and collaborative efforts to overcome shared problems is sought, the potential value of which is transengendered. The value of mutual tolerance is used as the main basis, in addition to the diversity of beliefs, traditions, customs, and other cultural fibers placed appropriately through friendly interactions. All of this can be done if the materials in the education/learning process are counted as the "home" of our human experience. In and through such a process, we "lay off" our experiences that were never singular.

If we are able to implement these points in the framework of nation and character building, undoubtedly social hysteria and situations that are vulnerable to cultural invasion or cultural schizophrenia can be avoided. The formation of character through meaningful learning undoubtedly avoids the formation of fierce-faced humans, whose character and behavior are violent, brutal, and aggressive, one of whose great wills is to antagonize the other, wanting to control and oppress the other. In turn, human characters who are able to respect their dignity and rights will be formed, and not human beings who are only liars for their own conscience.

REFERENCES

Sedyawati, Edi (2007) *Budaya Indonesia. Kajian Arkeologi, Seni, dan Sejarah.* Jakarta: PT RajaGrafindo Perkasa.
Seung, T. K. (1982) *Thematics and Semiotics in Hermeneutics.* New Haven, CT: Yale University Press.

Emerging Perspectives and Trends in Innovative Technology
for Quality Education 4.0 – Kusmawan et al (eds)
© 2020 Taylor & Francis Group, London, ISBN 978-0-367-25803-0

Author Index

Milton Keynes UK
Ingram Content Group UK Ltd.
UKHW051853071024
449327UK00025B/1938

9 780367 545826